Carbonic Anhydrases and Metabolism

Carbonic Anhydrases and Metabolism

Special Issue Editor

Claudiu T. Supuran

MDPI • Basel • Beijing • Wuhan • Barcelona • Belgrade

MDPI

Special Issue Editor
Claudiu T. Supuran
Università degli Studi di Firenze
Italy

Editorial Office
MDPI
St. Alban-Anlage 66
4052 Basel, Switzerland

This is a reprint of articles from the Special Issue published online in the open access journal *Metabolites* (ISSN 2218-1989) from 2017 to 2019 (available at: https://www.mdpi.com/journal/metabolites/special_issues/Carbonic_Anhydrases_Metabolism)

For citation purposes, cite each article independently as indicated on the article page online and as indicated below:

LastName, A.A.; LastName, B.B.; LastName, C.C. Article Title. *Journal Name* **Year**, *Article Number*, Page Range.

ISBN 978-3-03897-800-8 (Pbk)
ISBN 978-3-03897-801-5 (PDF)

Contents

About the Special Issue Editor

Claudiu T. Supuran has been the Professor of Medicinal and Pharmaceutical Chemistry at the University of Florence, Italy, since 1995. He completed his PhD at the University of Bucharest, Romania, and was a visiting scholar at the University of Florida, Gainesville, USA, and at Griffith University, Brisbane, Australia. He was a visiting professor at the University of La Plata, Argentina, and at the University of New South Wales, Sydney, Australia. His main research interest is the medicinal chemistry/biochemistry of carbonic anhydrases, a field in which he has made contributions to the design of many novel classes of enzyme inhibitors and activators, deciphering their mechanism of action at the molecular level; the discovery of new isoforms and their role in disease (cancer, obesity, epilepsy, neuropathic pain, and cognition); and the discovery and characterization of carbonic anhydrases from various organisms (bacteria, fungi, corals, vertebrates other than humans and rodents, etc). Other research interests of his include the X-ray crystallography of metalloenzymes, biologically active organoelement derivatives, QSAR studies, metal-based drugs, cyclooxygenases, serine proteases, matrix metalloproteinases, bacterial proteases, amino acid derivatives, heterocyclic chemistry, and the chemistry of sulfonamides, sulfamates, and sulfamides, among others. He has published more than 1500 papers in these fields, and his Hirsch index is 134. One of the compounds discovered in his laboratory (SLC-0111) is in Phase II clinical trials for the treatment of advanced metastatic solid tumors.

metabolites

MDPI

Editorial

Carbonic Anhydrases and Metabolism

Claudiu T. Supuran [ORCID]

Dipartimento Neurofarba, Sezione di Scienze Farmaceutiche, Laboratorio di Chimica Bioinorganica, Università degli Studi di Firenze, Polo Scientifico, Via U. Schiff 6, Sesto Fiorentino, 50019 Florence, Italy; claudiu.supuran@unifi.it

Received: 18 March 2018; Accepted: 20 March 2018; Published: 21 March 2018

Abstract: Although the role of carbonic anhydrases (CAs, EC 4.2.1.1) in metabolism is well-established, pharmacological applications of this phenomenon started to be considered only recently. In organisms all over the phylogenetic tree, the seven CA genetic families known to date are involved in biosynthetic processes and pH modulation, which may influence metabolism in multiple ways, with both processes being amenable to pharmacologic intervention. CA inhibitors possess antiobesity action directly by inhibiting lipogenesis, whereas the hypoxic tumor metabolism is highly controlled by the transmembrane isoforms CA IX and XII, which contribute to the acidic extracellular environment of tumors and supply bicarbonate for their high proliferation rates. Many of the articles from this special issue deal with the role of cancer CAs in tumor metabolism and how these phenomena can be used for designing innovative antitumor therapies/imaging agents. The metabolic roles of CAs in bacteria and algae are also discussed.

Keywords: carbonic anhydrase; hypoxic tumor; metabolism; carboxylation; bicarbonate; pH regulation; antitumor agent; sulfonamide; bacterial enzymes

Carbonic anhydrases (CAs, EC 4.2.1.1) are a superfamily of metalloenzymes present in all life kingdoms, as they equilibrate the reaction between three simple but essential chemical species: CO_2, bicarbonate, and protons [1–6]. Although discovered 85 years ago, these enzymes are still extensively investigated due to the biomedical application of their inhibitors [7–12] and activators [13] but also because they are an extraordinary example of convergent evolution, with seven genetically distinct CA families that evolved independently in Bacteria, Archaea, and Eukarya, the α-, β-, γ-, δ-, ζ-, η-, and θ-CAs [2,4,5,14–16]. CAs are also among the most efficient enzymes known in nature, probably due to the fact that uncatalyzed CO_2 hydration is a very slow process at neutral pH, and the physiologic demands for its conversion to ionic, soluble species (i.e., bicarbonate and protons) are very high [1–6]. Indeed, CO_2 is generated in most metabolic oxidative processes, and being a gas, it must be converted to soluble products quickly and efficiently. Otherwise, it would tend to accumulate and provoke damage to cells and other organelles in the gaseous state without such an efficient hydration catalyst as the CAs [2,6–8].

Inhibition of the CAs has pharmacologic applications in many fields, such as diuretics [9], antiglaucoma [10], anticonvulsant [7,8,11], antiobesity [11], and anticancer agents/diagnostic tools [1,2,12], but it is also emerging for designing anti-infectives, i.e., antifungal, antibacterial, and antiprotozoan agents with a novel mechanism of action [4,5,8,17,18]. For a long period it has been considered that the pharmacologic effects of CA inhibition or activation are mainly due to effects on pH regulation in cells or tissues where the enzymes are present [1]. Although these phenomena are undoubtedly relevant and take place in most organisms/tissues/cells where these ubiquitous enzymes are found, a lot of recent evidence points to the fact that CAs are true metabolic enzymes at least for two different reasons: (i) due to their direct participation in carboxylating reactions which provide bicarbonate and/or CO_2 to carboxylating enzymes, such as pyruvate carboxylase, acetyl-coenzyme A carboxylase [19,20], phosphoenolpyruvate carboxylase [21], and ribulose-1, 5-bisphosphate carboxylase oxygenase (RUBISCO) [22,23]; and (ii) due to the role that pH itself has on many metabolic reactions, with pH differences as low as 0.1 unit leading to the complete blockade of crucial reactions and thus of entire metabolic pathways [1–3]. For these reasons, the CAs may be considered as important checkpoint enzymes for relevant physiologic processes connected to a host of metabolic pathways, in all types of organisms, from bacteria and archaea [24,25] to algae, plants [26], and other eukaryotes (starting with the simple ones, yeasts and protozoa, and ending with the complex ones, including vertebrates) [1–7].

The metabolic reactions with which CA activity interference has been mostly studied include de novo glucogenesis, urea biosynthesis, and lipogenesis in animals [1–7,19,20] as well as the initial steps of the photosynthetic process in some bacteria, algae, and plants, due to the role that CAs have in providing bicarbonate (through a carbon-concentrating mechanism) to RUBISCO [22,23,26]. In tumors, these metabolic processes are even more complex, as it has been shown that not only do the protons produced by CO_2 hydration contribute to extracellular acidification, typical of most cancers [1,2,12,27], but the bicarbonate is thereafter used as a C_1 carbon source for biosynthetic reactions that convert it into organic compounds (the so-called "organication"), which supplies cancer cells with intermediates useful for sustaining their high proliferation rates [28]. Inhibition of the various CA isoforms/CA enzyme classes involved in these phenomena, mainly with sulfonamides, the most widely used class of CA inhibitors [29–36], has important physiological consequences which motivate their use as pharmacological agents as mentioned earlier. Whereas the use of carbonic anhydrases inhibitors (CAIs) as diuretics and antiglaucoma agents has been well-established for decades [1,7–10], their applications as antiepileptics and antiobesity drugs is more recent [1,7,11,13,20], and only the last decade has seen important advances which have validated CAs as antitumor drug targets [3,12,27,35]. Thus, it is not unexpected that five papers of the special issue deal with the connections between tumor-associated CAs and tumors and the development of new anticancer agents based on them. The first such paper [27] reviews the role that hypoxia has in triggering a diverse metabolism to cancer cells, all of which are orchestrated by the transcription factor hypoxia-inducible factor 1α (HIF-1α), which in the end leads to the overexpression of at least two CA isoforms, which are scarcely present in normal tissues, CA IX and XII. Inhibition of these two enzymes with sulfonamides or coumarins was shown to impair the growth of the primary tumors and metastases and to reduce the population of cancer stem cells, leading thus to a complex and beneficial anticancer action for this class of enzyme inhibitors [27,37–40]. The paper of McDonald et al. [37] discusses CA-mediated regulation of pH together with the recent proteome-wide analyses that have revealed the presence of a complex CA IX interactome in cancer cells, which has multiple roles in metabolite transport and tumor cell migration and invasion. In both these papers [27,37], the various aspects of the development of the first antitumor agent from this class that reached clinical development, SLC-0111, are discussed, considering the fact that these two groups are the discoverers of this new drug.

Mboge et al. [38] consider not only CA IX and XII, the most investigated proteins of this family connected to cancers, but all CA isoforms from mammals and their possible role in tumors as potential targets for cancer therapy. This interesting paper proposes thus that in addition to CA IX/XII, other isoforms, such as the mitochondrial ones CA VA/VB, or some of the cytosolic CAs (isoforms I, II, VII, and XIII) might become anticancer drug targets sooner or later [38].

Ward et al. [39] discuss the various classes and types of CA IX inhibitors which were investigated in detail in various models and systems, together with the fact that this class of pharmacological agents may enhance the effects of anti-angiogenic drugs or chemotherapy agents by different mechanisms that are poorly understood at this moment. Work from their laboratories also showed that CA IX interacts with several of the signaling pathways involved in the cellular response to radiation, suggesting that pH-independent mechanisms may also be important for the role that CA IX inhibitors in combination with radiations have in slowing down tumor progression [39,41].

Iessi et al. [40] discuss the possibility of combining CA IX/XII inhibitors with inhibitors of other proton exchangers and transporters present in tumor cells, such as V-ATPase, Na^+/H^+ exchangers (NHE), and monocarboxylate transporters (MCTs). In fact, recent work suggests that a strong synergistic effect is observed when combining CAIs with proton pump inhibitors of the lansoprazole/omeprazole type [40,42]. Furthermore, the drug delivery of anticancer agents by means of exosomes (natural extracellular nanovesicles), which exploit tumor acidity as a molecular engine, has been proposed by the same group [43].

With the exception of CAs and the cancer metabolism connection, with various aspects discussed in the papers mentioned above [27,37–40], Parkkila's group [26] thoroughly reviewed the roles that different CAs have in the algal model organism *Chlamydomonas reinhardtii*. It is in fact well-known that photosynthetic organisms contain six evolutionarily different classes of CAs, the α-CAs, β-CAs, γ-CAs, δ-CAs, ζ-CAs, and θ-CAs, and many of them possess more than one isoform in the same organism [26]. *Chlamydomonas reinhardtii* contains 15 CAs belonging to three gene families: three α-CAs, nine β-CAs, and three γ-CAs, with quite a different subcellular localization. The review presents the known metabolic roles that some of these enzymes have in the carbon-concentrating mechanism which provides bicarbonate to RUBISCO for photosynthesis, but also predicts functions for some of these CAs for which precise metabolic roles are yet to be discovered [26].

Bacteria also possess CAs belonging to three diverse genetic families, the α-CAs, β-CAs, and γ-CAs [4,5,25]. Supuran and Capasso [25] reviewed the roles that these enzymes have in these organisms, predominantly considering pathogenic bacteria, such as *Escherichia coli*, *Vibrio cholerae*, *Brucella suis*, *Helicobacter pylori*, *Porphyromonas gingivalis*, *Mycobacterium tuberculosis*, and *Burkholderia pseudomallei*. For many of these bacteria, one or more CAs belonging to the three classes were cloned, characterized, and investigated for their inhibition with the main classes of CAIs in the search for antibacterial agents with a new mechanism of action that is free of the drug-resistance problems of currently used antibiotics [25]. Although this field is still in its infancy, substantial progress has been achieved ultimately in understanding the roles that these enzymes have in the life cycle and virulence of many pathogens provoking serious diseases.

Overall, this interesting special issue of *Metabolites* affords a series of interesting reviews which show the multitude of aspects connecting simple enzymes, such as the CAs to metabolic processes, in all types of organisms. They may afford both a better understanding of fundamental processes, such as carbon capture in photosynthesis, tumorigenesis, and the role of pH in metabolism, but also lead to the development of novel therapeutic strategies in areas such as oncology and anti-infective agents.

Acknowledgments: Funding from the author's laboratory was from several EU projects (Euroxy, Metoxia, DeZnIt and Dynano).

Conflicts of Interest: The author declares no conflict of interest.

References

1. Supuran, C.T. Carbonic anhydrases: Novel therapeutic applications for inhibitors and activators. *Nat. Rev. Drug Discov.* **2008**, *7*, 168–181. [CrossRef] [PubMed]
2. Supuran, C.T. Structure and function of carbonic anhydrases. *Biochem. J.* **2016**, *473*, 2023–2032. [CrossRef] [PubMed]
3. Neri, D.; Supuran, C.T. Interfering with pH regulation in tumours as a therapeutic strategy. *Nat. Rev. Drug Discov.* **2011**, *10*, 767–777. [CrossRef] [PubMed]
4. Supuran, C.T.; Capasso, C. New light on bacterial carbonic anhydrases phylogeny based on the analysis of signal peptide sequences. *J. Enzym. Inhib. Med. Chem.* **2016**, *31*, 1254–1260. [CrossRef] [PubMed]
5. Capasso, C.; Supuran, C.T. An overview of the alpha-, beta- and gamma-carbonic anhydrases from Bacteria: Can bacterial carbonic anhydrases shed new light on evolution of bacteria? *J. Enzym. Inhib. Med. Chem.* **2015**, *30*, 325–332. [CrossRef] [PubMed]
6. Supuran, C.T. How many carbonic anhydrase inhibition mechanisms exist? *J. Enzym. Inhib. Med. Chem.* **2016**, *31*, 345–360. [CrossRef] [PubMed]

7. Supuran, C.T. Structure-based drug discovery of carbonic anhydrase inhibitors. *J. Enzym. Inhib. Med. Chem.* **2012**, *27*, 759–772. [CrossRef] [PubMed]
8. Supuran, C.T. Advances in structure-based drug discovery of carbonic anhydrase inhibitors. *Expert Opin. Drug Discov.* **2017**, *12*, 61–88. [CrossRef] [PubMed]
9. Carta, F.; Supuran, C.T. Diuretics with carbonic anhydrase inhibitory action: A patent and literature review (2005–2013). *Expert Opin. Ther. Pat.* **2013**, *23*, 681–691. [CrossRef] [PubMed]
10. Masini, E.; Carta, F.; Scozzafava, A.; Supuran, C.T. Antiglaucoma carbonic anhydrase inhibitors: A patent review. *Expert Opin. Ther. Pat.* **2013**, *23*, 705–716. [CrossRef] [PubMed]
11. Scozzafava, A.; Supuran, C.T.; Carta, F. Antiobesity carbonic anhydrase inhibitors: A literature and patent review. *Expert Opin. Ther. Pat.* **2013**, *23*, 725–735. [CrossRef] [PubMed]
12. Monti, S.M.; Supuran, C.T.; De Simone, G. Anticancer carbonic anhydrase inhibitors: A patent review (2008–2013). *Expert Opin. Ther. Pat.* **2013**, *23*, 737–749. [CrossRef] [PubMed]
13. Supuran, C.T. Carbonic anhydrase activators. *Future Med. Chem.* **2018**, *10*, 561–573. [CrossRef] [PubMed]
14. Supuran, C.T.; Capasso, C. Carbonic Anhydrase from *Porphyromonas Gingivalis* as a Drug Target. *Pathogens* **2017**, *6*, 30. [CrossRef] [PubMed]
15. Del Prete, S.; De Luca, V.; De Simone, G.; Supuran, C.T.; Capasso, C. Cloning, expression and purification of the complete domain of the η-carbonic anhydrase from *Plasmodium falciparum*. *J. Enzym. Inhib. Med. Chem.* **2016**, *31*, 54–59. [CrossRef] [PubMed]
16. Del Prete, S.; De Luca, V.; Vullo, D.; Osman, S.M.; AlOthman, Z.; Carginale, V.; Supuran, C.T.; Capasso, C. A new procedure for the cloning, expression and purification of the β-carbonic anhydrase from the pathogenic yeast *Malassezia globosa*, an anti-dandruff drug target. *J. Enzym. Inhib. Med. Chem.* **2016**, *31*, 1156–1161. [CrossRef] [PubMed]
17. Capasso, C.; Supuran, C.T. Inhibition of Bacterial Carbonic Anhydrases as a Novel Approach to Escape Drug Resistance. *Curr. Top. Med. Chem.* **2017**, *17*, 1237–1248. [CrossRef] [PubMed]
18. De Menezes Dda, R.; Calvet, C.M.; Rodrigues, G.C.; de Souza Pereira, M.C.; Almeida, I.R.; de Aguiar, A.P.; Supuran, C.T.; Vermelho, A.B. Hydroxamic acid derivatives: A promising scaffold for rational compound optimization in Chagas disease. *J. Enzym. Inhib. Med. Chem.* **2016**, *31*, 964–973. [CrossRef] [PubMed]
19. Supuran, C.T. Carbonic anhydrase inhibitors in the treatment and prophylaxis of obesity. *Expert Opin. Ther. Pat.* **2003**, *13*, 1545–1550. [CrossRef]
20. Arechederra, R.L.; Waheed, A.; Sly, W.S.; Supuran, C.T.; Minteer, S.D. Effect of Sulfonamides as Selective Carbonic Anhydrase Va and Vb Inhibitors on Mitochondrial Metabolic Energy Conversion. *Bioorg. Med. Chem.* **2013**, *21*, 1544–1548. [CrossRef] [PubMed]
21. Del Prete, S.; De Luca, V.; Capasso, C.; Supuran, C.T.; Carginale, V. Recombinant thermoactive phosphoenolpyruvate carboxylase (PEPC) from *Thermosynechococcus elongatus* and its coupling with mesophilic/thermophilic bacterial carbonic anhydrases (CAs) for the conversion of CO_2 to oxaloacetate. *Bioorg. Med. Chem.* **2016**, *24*, 220–225. [CrossRef] [PubMed]
22. Tomar, V.; Sidhu, G.K.; Nogia, P.; Mehrotra, R.; Mehrotra, S. Regulatory components of carbon concentrating mechanisms in aquatic unicellular photosynthetic organisms. *Plant Cell Rep.* **2017**, *36*, 1671–1688. [CrossRef] [PubMed]
23. Larkum, A.W.D.; Davey, P.A.; Kuo, J.; Ralph, P.J.; Raven, J.A. Carbon-concentrating mechanisms in seagrasses. *J. Exp. Bot.* **2017**, *68*, 3773–3784. [CrossRef] [PubMed]
24. Zimmerman, S.A.; Ferry, J.G.; Supuran, C.T. Inhibition of the archaeal β-class (Cab) and γ-class (Cam) carbonic anhydrases. *Curr. Top. Med. Chem.* **2007**, *7*, 901–908. [CrossRef] [PubMed]
25. Supuran, C.T.; Capasso, C. An Overview of the Bacterial Carbonic Anhydrases. *Metabolites* **2017**, *7*, 56. [CrossRef] [PubMed]
26. Aspatwar, A.; Haapanen, S.; Parkkila, S. An Update on the Metabolic Roles of Carbonic Anhydrases in the Model Alga *Chlamydomonas reinhardtii*. *Metabolites* **2018**, *8*, 22. [CrossRef] [PubMed]
27. Supuran, C.T. Carbonic Anhydrase Inhibition and the Management of Hypoxic Tumors. *Metabolites* **2017**, *7*, 48. [CrossRef] [PubMed]
28. Santi, A.; Caselli, A.; Paoli, P.; Corti, D.; Camici, G.; Pieraccini, G.; Taddei, M.L.; Serni, S.; Chiarugi, P.; Cirri, P. The effects of CA IX catalysis products within tumor microenvironment. *Cell Commun. Signal.* **2013**, *11*, 81. [CrossRef] [PubMed]

29. Carta, F.; Scozzafava, A.; Supuran, C.T. Sulfonamides: A patent review (2008–2012). *Expert Opin. Ther. Pat.* **2012**, *22*, 747–758. [CrossRef] [PubMed]
30. Scozzafava, A.; Carta, F.; Supuran, C.T. Secondary and tertiary sulfonamides: A patent review (2008–2012). *Expert Opin. Ther. Pat.* **2013**, *23*, 203–213. [CrossRef] [PubMed]
31. Garaj, V.; Puccetti, L.; Fasolis, G.; Winum, J.Y.; Montero, J.L.; Scozzafava, A.; Vullo, D.; Innocenti, A.; Supuran, C.T. Carbonic anhydrase inhibitors: Synthesis and inhibition of cytosolic/tumor-associated carbonic anhydrase isozymes I, II and IX with sulfonamides incorporating 1,2,4-triazine moieties. *Bioorg. Med. Chem. Lett.* **2004**, *14*, 5427–5433. [CrossRef] [PubMed]
32. Supuran, C.T.; Nicolae, A.; Popescu, A. Carbonic anhydrase inhibitors. Part 35. Synthesis of Schiff bases derived from sulfanilamide and aromatic aldehydes: The first inhibitors with equally high affinity towards cytosolic and membrane-bound isozymes. *Eur. J. Med. Chem.* **1996**, *31*, 431–438. [CrossRef]
33. Scozzafava, A.; Menabuoni, L.; Mincione, F.; Supuran, C.T. Carbonic Anhydrase Inhibitors. A General Approach for the Preparation of Water-Soluble Sulfonamides Incorporating Polyamino—Polycarboxylate Tails and of Their Metal Complexes Possessing Long-Lasting, Topical Intraocular Pressure-Lowering Properties. *J. Med. Chem.* **2002**, *45*, 1466–1476. [CrossRef] [PubMed]
34. Scozzafava, A.; Menabuoni, L.; Mincione, F.; Briganti, F.; Mincione, G.; Supuran, C.T. Carbonic anhydrase inhibitors: Perfluoroalkyl/aryl-substituted derivatives of aromatic/heterocyclic sulfonamides as topical intraocular pressure-lowering agents with prolonged duration of action. *J. Med. Chem.* **2000**, *43*, 4542–4551. [CrossRef] [PubMed]
35. Pacchiano, F.; Aggarwal, M.; Avvaru, B.S.; Robbins, A.H.; Scozzafava, A.; McKenna, R.; Supuran, C.T. Selective hydrophobic pocket binding observed within the carbonic anhydrase II active site accommodate different 4-substituted-ureido-benzenesulfonamides and correlate to inhibitor potency. *Chem. Commun.* **2010**, *46*, 8371–8373. [CrossRef] [PubMed]
36. Capasso, C.; Supuran, C.T. Sulfa and trimethoprim-like drugs-antimetabolites acting as carbonic anhydrase, dihydropteroate synthase and dihydrofolate reductase inhibitors. *J. Enzym. Inhib. Med. Chem.* **2014**, *29*, 379–387. [CrossRef] [PubMed]
37. McDonald, P.C.; Swayampakula, M.; Dedhar, S. Coordinated Regulation of Metabolic Transporters and Migration/Invasion by Carbonic Anhydrase IX. *Metabolites* **2018**, *8*, 20. [CrossRef] [PubMed]
38. Mboge, M.Y.; Mahon, B.P.; McKenna, R.; Frost, S.C. Carbonic Anhydrases: Role in pH Control and Cancer. *Metabolites* **2018**, *8*, 19. [CrossRef] [PubMed]
39. Ward, C.; Meehan, J.; Gray, M.; Kunkler, I.H.; Langdon, S.P.; Argyle, D.J. Carbonic Anhydrase IX (CAIX), Cancer, and Radiation Responsiveness. *Metabolites* **2018**, *8*, 13. [CrossRef] [PubMed]
40. Iessi, E.; Logozzi, M.; Mizzoni, D.; Di Raimo, R.; Supuran, C.T.; Fais, S. Rethinking the Combination of Proton Exchanger Inhibitors in Cancer Therapy. *Metabolites* **2018**, *8*, 2. [CrossRef] [PubMed]
41. Ward, C.; Langdon, S.P.; Mullen, P.; Harris, A.L.; Harrison, D.J.; Supuran, C.T.; Kunkler, I.H. New strategies for targeting the hypoxic tumour microenvironment in breast cancer. *Cancer Treat. Rev.* **2013**, *39*, 171–179. [CrossRef] [PubMed]
42. Federici, C.; Lugini, L.; Marino, M.L.; Carta, F.; Iessi, E.; Azzarito, T.; Supuran, C.T.; Fais, S. Lansoprazole and carbonic anhydrase IX inhibitors sinergize against human melanoma cells. *J. Enzym. Inhib. Med. Chem.* **2016**, *31*, 119–125. [CrossRef] [PubMed]
43. Kusuzaki, K.; Matsubara, T.; Murata, H.; Logozzi, M.; Iessi, E.; Di Raimo, R.; Carta, F.; Supuran, C.T.; Fais, S. Natural extracellular nanovesicles and photodynamic molecules: Is there a future for drug delivery? *J. Enzym. Inhib. Med. Chem.* **2017**, *32*, 908–916. [CrossRef] [PubMed]

metabolites

MDPI

Review

Carbonic Anhydrase Inhibition and the Management of Hypoxic Tumors

Claudiu T. Supuran

Università degli Studi di Firenze, Dipartimento Neurofarba, Sezione di Scienze Farmaceutiche e Nutraceutiche, Via U. Schiff 6, 50019 Sesto Fiorentino, Florence, Italy; claudiu.supuran@unifi.it; Tel.: +39-055-457-3729; Fax: +39-055-457-3385

Received: 31 August 2017; Accepted: 15 September 2017; Published: 16 September 2017

Abstract: Hypoxia and acidosis are salient features of many tumors, leading to a completely different metabolism compared to normal cells. Two of the simplest metabolic products, protons and bicarbonate, are generated by the catalytic activity of the metalloenzyme carbonic anhydrase (CA, EC 4.2.1.1), with at least two of its isoforms, CA IX and XII, mainly present in hypoxic tumors. Inhibition of tumor-associated CAs leads to an impaired growth of the primary tumors, metastases and reduces the population of cancer stem cells, leading thus to a complex and beneficial anticancer action for this class of enzyme inhibitors. In this review, I will present the state of the art on the development of CA inhibitors (CAIs) targeting the tumor-associated CA isoforms, which may have applications for the treatment and imaging of cancers expressing them. Small molecule inhibitors, one of which (SLC-0111) completed Phase I clinical trials, and antibodies (girentuximab, discontinued in Phase III clinical trials) will be discussed, together with the various approaches used to design anticancer agents with a new mechanism of action based on interference with these crucial metabolites, protons and bicarbonate.

Keywords: tumor; metabolism; carbonic anhydrase; isoforms IX and XII; inhibitor; sulfonamide; antibody

---

1. Introduction

A salient feature of many tumors is the fact that they are hypoxic and acidic compared to normal tissues of the same type. This has been known for many decades as the Warburg effect [1,2] but has been explained at the molecular level only recently, after the discovery of a transcription factor regulating these phenomena, the hypoxia inducible factor 1α, HIF-1α [3–5].

As seen from Figure 1, in normoxic conditions HIF-1α is unstable, being degraded rapidly by a well understood biochemical process: under the action of prolyl hydroxylases (PHD), a proline residue from the transcription factor is hydroxylated, being then recognized by a protein possessing ubiquitin ligase E3 activity, more precisely the von Hippel Lindau protein (VHL), which targets it to ubiquitylation and degradation within the proteosomes (Figure 1) [5–8].

However, in hypoxia, which as mentioned above is frequent in many tumor cells [1–3], an accumulation of HIF-1α occurs, followed by its translocation from the cytosol to the nucleus, where it forms a dimer with a constitutive subunit, HIF-1β, leading to an active transcription factor, which, by interaction with a hypoxia responsive element (HRE) found on different genes, leads to overexpression of proteins involved in aerobic glycolysis (such as, for example, the glucose transporters GLUT1-3), angiogenesis (such as, for example, the vascular endothelial growth factor, VEGF), erythropoesis (such as, for example, erythropoetin 1) and pH regulation (such as the tumor-associated enzymes CA IX and XII) [5–11].

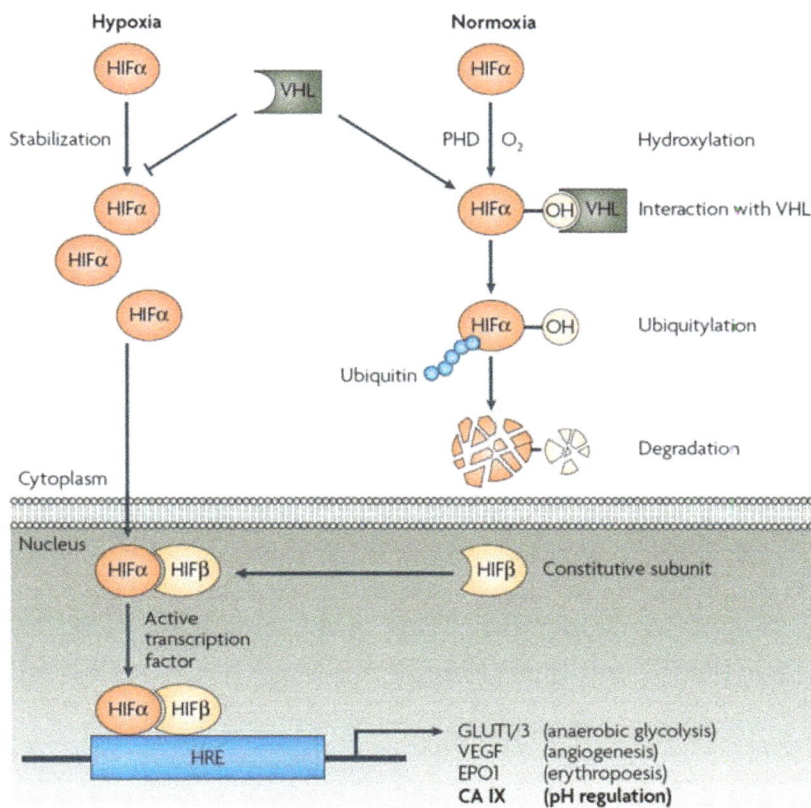

Figure 1. Mechanism by which the transcription factor HIF-1α (abbreviated as HIFα) orchestrates the overexpression of proteins involved in aerobic glycolysis, angiogenesis, erythropoesis and pH regulation in hypoxic tumors. In normoxia HIFα is hydroxylated at a Pro residue and targeted for degradation by the proteasome (PHD, prolyl-hydroxylase; VHL, von Hippel-Landau factor, HRE, hypoxia responsive element). In hypoxia, its accumulation leads to overexpression of the proteins involved in tumorigenesis mentioned above [5–8].

The overexpression of these proteins has profound effects on the metabolism of cancer cells, which on one hand are deprived of oxygen for the normal metabolism involving the oxidative phosphorylation [1,2], and on the other one, have an enhanced uptake of glucose (due to the overexpression of the glucose transporters GLUT1-GLUT3, which import the sugar within the cell), which cannot undergo the oxidative pathways for the generation of ATP [5–8]. Thus, an alternative pathway, the glycolytic one, occurs, with the formation of pyruvic (and lactic acids) from glucose, which generates less ATP (compared to the oxidative pathway), but which seems to be enough for the cancer cells to survive in hypoxic conditions [1–4]. The formed organic acids are extruded from the cells through the monocarboxylate transporters MCT1-MCT4 (some of which are overexpressed in tumors [4]), leading to an acidification of the extracellular milieu, up to pH values as low as 6.5 [4–8]. Additional perturbations of the extra- and intracellular pH equilibrium of the tumor cells are also furnished by other proteins which are involved in this process (Figure 2), among which the sodium-proton exchanger (Na^+–H^+ antiporter) NHE, which may import or export protons in exchange for sodium ions, the plasma membrane proton pump H^+-ATPase (V-ATPase), the various isoforms of the anion exchangers (chloride-bicarbonate exchangers) AE1–AE3, the sodium bicarbonate channels NBCs, which transport sodium and bicarbonate out of the cell or import it within the cell, various other bicarbonate transporters BT, as well as several isoforms of the metalloenzyme CA, such as the cytosolic CA II, and the transmembrane CA IX/XII, which efficiently catalyze CO_2 hydration to bicarbonate and protons [4–11]. By the coupling of all these effects, a slightly alkaline intracellular pH is achieved (of around 7.2) and an acidic extracellular pH of the tumor is formed, with values as low as 6.5 [4–11] (Figure 2). The extracellular acidosis (coupled with the hypoxia) is beneficial for the growth of the tumor cell and impairs the growth of the normal cells, leading thus to a massive proliferation, invasion and subsequently metastasis of the primary tumors [12–15].

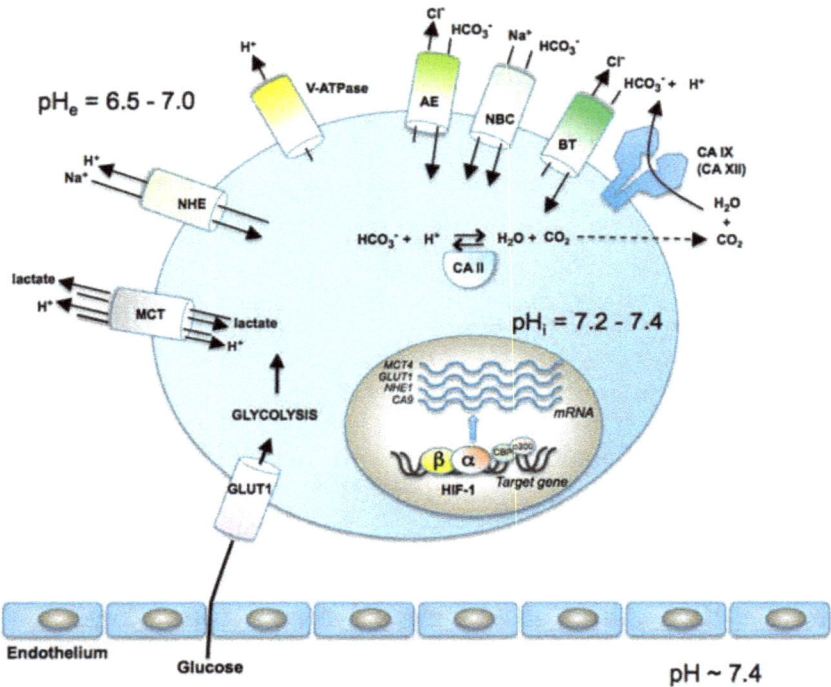

Figure 2. Proteins involved in pH regulation in tumors: GLUT1, the glucose transporter isoform 1; MCT, monocarboxylate transporter, which extrude lactic acid and other monocarboxylates formed by the glycolytic degradation of glucose; NHE, sodium-proton exchanger (Na^+–H^+ antiporter); V-ATPase, plasma membrane proton pump; AE, anion exchanger (chloride-bicarbonate exchanger); NBC, sodium bicarbonate channels; BT, bicarbonate transporter; CA II (cytosolic) and CA IX/XII, which catalyze CO_2 hydration to bicarbonate and protons [4–15].

Data of Figures 1 and 2 show the multitude of proteins involved in these processes, which in the end lead to features of the tumor cells which are quite different from those of the normal ones, and could thus be exploited for designing novel anticancer therapies. Among those who proposed this approach for the first time was Pouysségur et al. [4] who initially considered the NHE inhibitors as the most interesting pharmacological agents for interfering with tumor hypoxia/acidosis [16]. However, the significant toxicity of this class of drugs, or the lack of isoform-selective ones for other proteins involved in these processes (such as the MCTs, AEs, V-ATPase, etc. [17,18]) led to most of the work being concentrated on the metalloenzyme involved in pH regulation, i.e., the carbonic anhydrase (CA, EC 4.2.1.1) [7,8,19,20]. It should be however mentioned that H^+/K^+-inhibitors of the omeprazole type were shown to possess, alone or in combination with CA inhibitors (CAIs) significant antitumor effects [21–23]. Here I shall review the field of the CAIs as theragnostic agents for the management of hypoxic, metastatic tumors, without considering the other valuable approaches found in the literature which target other of the many proteins involved in these processes, and which have been reviewed by other researchers [6,13,16].

2. Validation of CA IX/XII as Antitumor Drug Targets

CA IX was discovered by Pastorek et al. in 1994 [10] and CA XII by Tureci et al. in 1998 [11], and it became immediately obvious that they differ considerably from other members of this family of proteins, which includes 15 isoforms in humans, hCA I-hCA VA, hCA VB, hCA VI-hCA XIV [7,8,20]. The first unusual feature of CA IX and XII was that the two enzymes are transmembrane, multi-domain proteins incorporating a short intra-cytosolic tail, a transmembrane short domain, and an extracellular catalytic domain, rather homologous to the one found in the cytosolic, mitochondrial, secreted or membrane-anchored CA isoforms known at that time [10,11,24–27]. Furthermore, CA IX has an additional domain at its *N*-terminus, termed the proteoglycan (PG) domain, which seems to play important functions connected with the role of CA IX in tumorigenesis being present only in this CA isoform [28,29] (Figure 3). In fact, all domains of this molecule, the intracellular tail [30], the catalytic domain [25,29] and the PG domain play diverse functions in tumorigenesis, making CA IX one of the key proteins involved in such processes in hypoxic tumors [7,8,15,17,24–30]. It is also interesting to note that CA IX seems to be an even more complicated protein: recent proteomic/interactomic studies suggests that at a stage in the cell's life CA IX possibly has a nuclear localization [31] interacting with proteins involved in nuclear/cytoplasmic transport processes, gene transcription, and protein stability, among which cullin-associated NEDD8-dissociated protein 1 (CAND1), which is itself involved in gene transcription and assembly of ubiquitin ligase complexes [32]. The precise role of these interactions of CA IX with this type of proteins is poorly understood at this moment but may lead to significant drug design developments in the future.

Returning to the main function of CA IX/XII, that of catalyzing the hydration of CO_2 to bicarbonate and protons [7,8], the validation of these proteins as drug targets followed the usual steps that most drug targets experience. They are summarized below:

(1) recombinant CA IX and XII were shown to possess a significant catalytic activity (in vitro) for the physiologic reaction (hydration of carbon dioxide to bicarbonate and protons), being among the most effective catalysts known in nature, with the following kinetic parameters: for human (h) CA IX (full length): k_{cat} of 1.1×10^6 s^{-1}, k_{cat}/K_M of 1.5×10^8 M^{-1} s^{-1} [24], whereas for hCA XII (catalytic domain) these parameters are k_{cat} of 4.2×10^5 s^{-1}, k_{cat}/K_M of 3.5×10^7 M^{-1} s^{-1} [33].

(2) potent in vitro CAIs of the sulfonamide type have been identified for both hCA IX [34] and hCA XII [33], followed by a large number of drug design studies of such agents [35] which have been reviewed recently and will be not detailed here [36–38]. As a consequence of such studies a large number of sulfonamide, sulfamate and sulfamides showing effective hCA IX/XII inhibitory potency (in vitro) and sometimes also some selectivity for inhibiting these two isoforms over the cytosolic, off-target and widespread ones hCA I and II, became available for in vivo studies [33–39].

The drug design of CAIs targeting isoform IX were highly favored by the report of the X-ray crystal structure of the protein (its catalytic domain) by De Simone's group in 2009 [25]. This 3D structure allowed the identification of similarities and differences between CA IX and the other members of the family, which led to the identification soon thereafter of highly isoform-selective inhibitors belonging to a variety of chemical classes, such as the sulfonamides, sulfamates, sulfamides, coumarins, polyeamines, etc. [36–38].

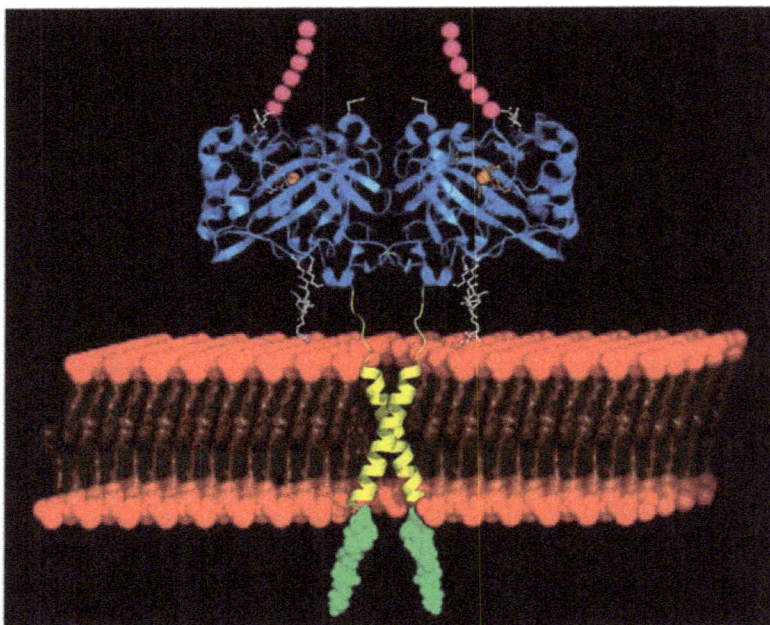

Figure 3. CA IX X-ray crystal structure of the catalytic domain (in blue), the PG domain (cartoon in pink), plasma membrane (in red), the transmembrane domain in yellow (modeled) and the intracytosolic tail (modeled, in green) [29].

(3) Pastorekova's group [29] demonstrated the role of CA IX in extracellular acidification of hypoxic tumors, and the possibility to reverse this effect by inhibiting the enzyme activity with sulfonamides. Furthermore, in the same studies it was observed that a fluorescent potent sulfonamide CA IX/II inhibitor accumulated only in the hypoxic cells, whereas it did not bind in cells expressing CA IX, but in normoxic conditions [29,39]. This effect has been explained as being due to the PG domain of the protein, which in normoxic conditions closes the active site. The opening of the active site is triggered by hypoxia, making it available for inhibitors to bind, but only in hypoxic conditions [39]. This makes CA IX an ideal drug target, as this phenomenon will lead to the inhibition of only the CA IX present in tumors, leading thus to drugs with fewer side effects compared to the classical chemotherapeutic agents [29,39].

(4) Dubois et al. [40,41] then published the proof-of-concept studies showing that in xenograft animal models of hypoxic tumors it is possible to image the hypoxic regions rich in CA IX/XII by using fluorescent sulfonamide CAIs possessing the same structural elements as the compounds used in the study of Pastorekova's group, mentioned above [29].

(5) The first study showing an in vivo antitumor effect due to CA IX inhibition was from Neri's group [42], followed shortly thereafter by similar studies from different laboratories, on diverse models and cancer types, which demonstrated that sulfonamide/sulfamate [42–46] or coumarin [47] CA IX/XII inhibitors have a profound effect in inhibiting the growth of the primary tumors and the metastases expressing CA IX/XII. Probably the most interesting studies are those from Dedhar's group [44,45,47] who rigorously showed the involvement of CA IX/XII in the antitumor/antimetastatic effects of the inhibitors of the sulfonamide or coumarin types. In fact, as it will be shown shortly, one of the compounds described in such studies progressed to clinical trials and completed Phase I trials in 2016 [45].

(6) Dedhar's group [48] also discovered another important phenomenon connected to CA IX/XII inhibition, i.e., the depletion of cancer stem cell population within the hypoxic tumors, which is considered to be a very positive feature of an antitumor agent, considering the fact that most such therapies lead to an increase of this stem cell population, hypothesized to be one of the reasons for the recurrence of some cancers [49]. The same group recently elucidated [50] the mechanism used by the hypoxic tumors for invasion, which reinforces the role played by CA IX in tumor progression and clinical outcome of cancer patients harboring CA IX-positive tumors. This relevant study demonstrated an association between CA IX and matrix metalloproteinase 14 (MMP14), with the first protein furnishing H$^+$ ions used in the proteolytic cleavage of collagen mediated by MMP14, which leads to tissue degradation. This study showed that CA IX is one of the metabolic components of the cellular migration and invasion mechanisms in hypoxic tumors, and provides new mechanistic insights into the role played by this enzyme in tumor cell biology, with the possibility to design dual agents, targeting both these enzymes (CA IX and MMP14) as new antitumor drugs [50].

3. Small Molecule CA IX/XII Inhibitors as Antitumor Agents

Among the huge number of sulfonamide, sulfamate, sulfamide and coumarin CA IX/XII inhibitors reported to date [4,7,34–38], few compounds were investigated in detail in animal tumor models, and only one such derivatives, SLC-0111 (also known as WBI-5111) progressed to clinical trials [45,51].

As seen from Figure 4, SLC-0111 is a simple, ureido-substituted benzenesulfonamide derivative which has significant hCA IX and XII inhibitory properties in vitro (K$_I$s of 45 nM against hCA IX and of 4.5 nM against hCA XII), being much less effective as an inhibitor of hCA I and II, widespread cytosolic CAs in many organs [45]. The CA IX/XII-selective inhibitory properties of this sulfonamide and of some of its congeners were explained at the molecular level by using X-ray crystallography of enzyme-inhibitor adducts [52]. This study allowed to observe that the tail of the inhibitors (in the case of SLC-0111, the tail is a 4-fluorophenyl moiety) adopts very different conformations when the sulfonamide is bound within the enzyme active site cavity, and is orientated towards the exit of the cavity, which is the most variable part of the different CA isoforms present in mammals [52]. As a consequence, this class of sulfonamide CAIs show some of the highest selectivity ratios for inhibiting the tumor-associated over the cytosolic isoforms [52]. In vivo studies showed SLC-0111 to potently inhibit the growth of tumors harboring CA IX/XII, whereas tumors that did not express these enzymes were unaffected [44,45]. The metastases formation was also inhibited in the T4 murine breast cancer model [44], and important antitumor effects were observed also in combination with other anticancer agents used clinically, such as paclitaxel, doxorubicine, etc. [44,45]. As mentioned above, a notable depletion of the cancer stem cell population was also evident after the treatment with this compound. Although the results of the Phase I clinical trial are not yet published, the compound has been scheduled for Phase II trials which will start late in 2016 [51].

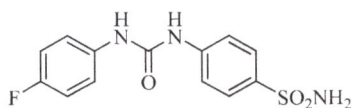

Figure 4. Structure of SLC-0111 (WBI-5111), the sulfonamide CA IX/XII inhibitor in Phase I/II clinical trials.

Although there are many other highly effective in vitro CA IX inhibitors reported so far, only a few of them were investigated in vivo in details. In one such study [53], important inhibition of growth of osteosarcoma was observed after inhibiting CA IX with positively charged pyridinium sulfonamides, suggesting their potential use for this refractory, difficult to treat tumor. In another study, [54], the CA IX and AP endonuclease-1/redox effector factor 1 (APE1/Ref-1) dual targeting was shown to be synergistic in pancreatic ductal adenocarcinomas (PDACs), another difficultly treatable tumor. A different and innovative approach has been used on the other hand by Neri's group [55], who conjugated maytansinoid DM1, a cytotoxic natural product payload, to a sulfonamide, more precisely a derivative of acetazolamide (a clinically used CAI drug for decades [7,8]), as targeting ligand for CA IX recognition. This conjugate molecule exhibited a potent in vivo antitumor effect in SKRC52 renal cell carcinomas [55].

It is probable that many other small molecule CA IX/XII inhibitors may enter soon in clinical trials, but probably, most researchers/companies wait for results of the clinical trials of the first-in-the-class such compound (SLC-0111) to be released.

4. Antibodies Targeting CA IX and XII as Antitumor Agents

4.1. Anti-CA IX Antibodies

The renal cell carcinoma (RCC)-associated protein G250 was recognized by its discoverers to be an anti-CA IX monoclonal antibody (Mab) and proposed as a possible antitumor target for RCC [56]. Indeed, G250, formulated as chimeric IgG1 monoclonal antibody and denominated girentuximab, was the first CA IX inhibitor to enter clinical trials [57], being actually in Phase III, although it seems that its development has been interrupted due to lack of efficacy [58]. Thus, no other details will be discussed about this Mab, but Pastorekova's group [59,60] proposed several interesting approaches based both on antibodies that inhibit the catalytic activity as well as those that target the PG domain of CA IX (and do not inhibit the CO_2 hydrase activity of the enzyme). For example the mouse monoclonal antibody VII/20 was shown to bind to the catalytic domain of CA IX, leading to an efficient receptor-mediated internalization of the antibody-enzyme conjugate, which is the main process that regulates abundance and signaling of cell surface proteins [60]. This internalization has a considerable impact on immunotherapy and in this particular case elicited significant anticancer effects in a mouse xenograft model of colorectal cancer [60]. The same group [59] demonstrated that the monoclonal antibody M75 (targeting the PG domain of CA IX and widely used as a reagent in immune-histochemical studies [10,17]) can be encapsulated into alginate microbeads or microcapsules made of sodium alginate, cellulose sulfate, and poly(methylene-co-guanidine), which afforded a rapid M75 antibody release at pH 6.8 (characteristic of the acidic tumors) compared to pH 7.4 (the physiologic, normal pH) [59].

4.2. Anti-CA XII Antibodies

There are far fewer studies to target CA XII with Mabs compared to CA IX. The most significant one comes from Zeidler's group [61] who discovered 6A10, the first monoclonal antibody that binds to the catalytic domain of CA XII and also acts as an inhibitor of the enzyme. 6A10 was shown to be a low nanomolar CA XII inhibitor and to inhibit the growth of tumor cells in spheroids and in vivo, in a mouse xenograft model of human cancer [61,62].

5. Imaging CA IX/XII Positive Tumors

The initial imaging strategy (after the fluorescent sulfonamides used for the proof-of-concept study mentioned above, which cannot be used to image human cancers [41]) was to incorporate ^{99m}Tc or ^{18}F as positron-emitting isotopes in the molecules of sulfonamide or coumarin CAIs in order to obtain agents useful for positron emission tomography (PET) [63–66]. The initial sulfonamides or coumarins labeled with these isotopes were not highly efficient imaging agents, probably due to pharmacokinetic-related problems. For example the SLC-0111 analog labeled with ^{18}F as well as a coumarin derivative incorporating the same isotope, although highly potent as in vitro CA IX inhibitors in vivo, in HT-29 (colorectal) xenografts in mice did not accumulate in the tumor, but were principally present in the blood, liver and nose of the animals, making them inappropriate as PET agents. However, the next generation inhibitors labeled with ^{18}F (trimeric sulfonamides [67] or positively-charged sulfonamides [68]) or ^{68}Ga-labelled sulfonamides (originally reported by Bénard's group [69] and soon thereafter by Poulsen's group [70]) showed that such ^{68}Ga-polyaminocarboxylate chelator-conjugated sulfonamides do accumulate preferentially within the hypoxic tumor, making them excellent candidates for clinical studies [69]. In HT-29 colorectal xenograft tumors in mice, the gallium-containing sulfonamides showed an excellent and specific tumor accumulation, coupled with a low uptake in blood and clearing intact into the urine, making them of great interest for further development [69,70].

Antibodies were also proposed as imaging agents for CA IX-positive tumors, originally by Neri's group [71]. By using the phage technology, high-affinity Mabs targeting hCA IX were generated (denominated A3 and CC7) which were used for imaging purposes in animal models of colorectal cancer (LS174T cell line). Such imaging studies with the two anti-hCA IX Mabs disclosed by this group closely matched the pimonidazole (an azole agent which accumulates in hypoxic regions of tumors) staining of these tumors, furnishing the proof-of-concept study that, in addition to the small molecule CA IX inhibitors, the antibodies can also be used for non-invasive imaging of hypoxic tumors. There are in fact many other similar imaging studies of the Mab in clinical trials mentioned above, girentuximab, which has been labeled with various isotopes for these purposes. For example, [111]In- [72], [99m]Tc- [73] and [124]I [74]—labeled girentuximab as well as dual-labeled Mab with a radionuclide and a fluorescence tag [75] have been developed and used for hypoxic tumor imaging with various degrees of success. However, antibodies have some problematic pharmacological aspects that must be considered attentively when used, and most probably small molecule CA IX/XII inhibitors may be more useful and easier to develop for a possible theragnostic agent targeting these enzymes.

6. Conclusions

Discovered at the beginning of the 90s, CA IX (and subsequently CA XII) were shown to possess crucial roles in tumorigenesis due to their involvement in the metabolism of hypoxic, acidic tumors. Overexpressed in tumor cells as a consequence of the HIF-1 cascade, these enzymes generate H^+ and bicarbonate ions, the simplest metabolites known, from CO_2 as substrate, being involved in many processes connected to tumorigenesis, from the regulation of the internal/external tumor cell pH, to migration, invasion, metastases formation as well as regulation of the cancer stem cell population. Many of these fascinating phenomena, discovered in the last decade, were shown to be useful for obtaining antitumor therapies/tumor imaging agents with a novel mechanism of action, by targeting these enzymes either with small molecule inhibitors or antibodies. The initial success of the clinical trials started with these agents, which is a continuing story, constitutes an excellent example of how fundamental research discoveries, thought to only explain some intricate biochemical/physiologic processes, may lead to innovative therapeutic strategies for fighting tumors.

Acknowledgments: Research from author's group was financed by several European Union projects (EUROXY, METOXIA, DeZnIT and Dynano, in the period 2004–2014) and by Signal Life Sciences (in the period 2013–2015). No funds were received for covering the costs to publish in open access this paper.

Conflicts of Interest: The author declares conflict of interest, being one of the inventors of SLC-0111.

References

1. Warburg, O. On respiratory impairment in cancer cells. *Science* **1956**, *124*, 269–270. [PubMed]
2. Schwartz, L.; Supuran, C.T.; Alfarouk, K.O. The Warburg effect and the hallmarks of cancer. *Anticancer Agents Med. Chem.* **2017**, *17*, 164–170. [CrossRef] [PubMed]
3. Semenza, G.L. Hypoxia-inducible factor 1: Oxygen homeostasis and disease pathophysiology. *Trends Mol. Med.* **2001**, *7*, 345–350. [CrossRef]
4. Pouysségur, J.; Dayan, F.; Mazure, N.M. Hypoxia signalling in cancer and approaches to enforce tumour regression. *Nature* **2006**, *441*, 437–443. [CrossRef] [PubMed]
5. Hockel, M.; Vaupel, P. Tumor hypoxia: Definitions and current clinical, biologic, and molecular aspects. *J. Natl. Cancer Inst.* **2001**, *93*, 266–276. [CrossRef] [PubMed]

6. Kremer, G.; Pouysségur, J. Tumor cell metabolism: Cancer's Achilles' heel. *Cancer Cell.* **2008**, *13*, 472–482. [CrossRef] [PubMed]

7. Supuran, C.T. Carbonic anhydrases: novel therapeutic applications for inhibitors and activators. *Nat. Rev. Drug Discov.* **2008**, *7*, 168–181. [CrossRef] [PubMed]

8. Neri, D.; Supuran, C.T. Interfering with pH regulation in tumours as a therapeutic strategy. *Nat. Rev. Drug Discov.* **2011**, *10*, 767–777. [CrossRef] [PubMed]

9. Wykoff, C.C.; Beasley, N.J.; Watson, P.H.; Turner, K.J.; Pastorek, J.; Sibtain, A.; Wilson, G.D.; Turley, H.; Talks, K.L.; Maxwell, P.H.; et al. Hypoxia-inducible expression of tumor-associated carbonic anhydrases. *Cancer Res.* **2000**, *60*, 7075–7083. [PubMed]

10. Pastorek, J.; Pastoreková, S.; Callebaut, I.; Mornon, J.P.; Zelník, V.; Opavský, R.; Zat'ovicová, M.; Liao, S.; Portetelle, D.; Stanbridge, E.J. Cloning and characterization of MN, a human tumor-associated protein with a domain homologous to carbonic anhydrase and putative helix-loop-helix DNA binding segment. *Oncogene* **1994**, *9*, 2877–2888. [PubMed]

11. Tureci, O.; Sahin, U.; Vollmar, E.; Siemer, S.; Göttert, E.; Seitz, G.; Parkkila, A.K.; Shah, G.N.; Grubb, J.H.; Pfreundschuh, M.; et al. Human carbonic anhydrase XII: cDNA cloning, expression, and chromosomal localization of a carbonic anhydrase gene that is overexpressed in some renal cell cancers. *Proc. Natl. Acad. Sci. USA* **1998**, *95*, 7608–7613. [CrossRef] [PubMed]

12. Boron, W.F. Regulation of intracellular pH. *Adv. Physiol. Educ.* **2004**, *28*, 160–179. [CrossRef] [PubMed]

13. Gatenby, R.A.; Gillies, R.J. A microenvironmental model of carcinogenesis. *Nat. Rev. Cancer.* **2008**, *8*, 56–61. [CrossRef] [PubMed]

14. Gatenby, R.A.; Gillies, R.J. Why do cancers have high aerobic glycolysis? *Nat. Rev. Cancer.* **2004**, *4*, 891–899. [CrossRef] [PubMed]

15. Fiaschi, T.; Giannoni, E.; Taddei, M.L.; Cirri, P.; Marini, A.; Pintus, G.; Nativi, C.; Richichi, B.; Scozzafava, A.; Carta, F.; et al. Carbonic anhydrase IX from cancer-associated fibroblasts drives epithelial-mesenchymal transition in prostate carcinoma cells. *Cell Cycle* **2013**, *12*, 1791–1801. [CrossRef] [PubMed]

16. Parks, S.K.; Chiche, J.; Pouysségur, J. Disrupting proton dynamics and energy metabolism for cancer therapy. *Nat. Rev. Cancer* **2013**, *13*, 611–623. [CrossRef] [PubMed]

17. Pettersen, E.O.; Ebbesen, P.; Gieling, R.G.; Williams, K.J.; Dubois, L.; Lambin, P.; Ward, C.; Meehan, J.; Kunkler, I.H.; Langdon, S.P.; et al. Targeting tumour hypoxia to prevent cancer metastasis. From biology, biosensing and technology to drug development: the METOXIA consortium. *J. Enzyme Inhib. Med. Chem.* **2015**, *30*, 689–721. [CrossRef] [PubMed]

18. Perez-Sayans, M.; Garcia-Garcia, A.; Scozzafava, A.; Supuran, C.T. Inhibition of V-ATPase and carbonic anhydrases as interference strategy with tumor acidification processes. *Curr. Pharm. Des.* **2012**, *18*, 1407–1413. [CrossRef] [PubMed]

19. Alterio, V.; Di Fiore, A.; D'Ambrosio, K.; Supuran, C.T.; De Simone, G. Multiple binding modes of inhibitors to carbonic anhydrases: How to design specific drugs targeting 15 different isoforms? *Chem. Rev.* **2012**, *112*, 4421–4468. [CrossRef] [PubMed]

20. Supuran, C.T. Structure and function of carbonic anhydrases. *Biochem. J.* **2016**, *473*, 2023–2032. [CrossRef] [PubMed]

21. De Milito, A.; Marino, M.L.; Fais, S. A rationale for the use of proton pump inhibitors as antineoplastic agents. *Curr. Pharm. Des.* **2012**, *18*, 1395–1406. [CrossRef] [PubMed]

22. Lugini, L.; Federici, C.; Borghi, M.; Azzarito, T.; Marino, M.L.; Cesolini, A.; Spugnini, E.P.; Fais, S. Proton pump inhibitors while belonging to the same family of generic drugs show different anti-tumor effect. *J. Enzyme Inhib. Med. Chem.* **2016**, *31*, 538–545. [CrossRef] [PubMed]

23. Federici, C.; Lugini, L.; Marino, M.L.; Carta, F.; Iessi, E.; Azzarito, T.; Supuran, C.T.; Fais, S. Lansoprazole and carbonic anhydrase IX inhibitors sinergize against human melanoma cells. *J. Enzyme Inhib. Med. Chem.* **2016**, *31*, 119–125. [CrossRef] [PubMed]

24. Hilvo, M.; Baranauskiene, L.; Salzano, A.M.; Scaloni, A.; Matulis, D.; Innocenti, A.; Scozzafava, A.; Monti, S.M.; Di Fiore, A.; De Simone, G.; et al. Biochemical characterization of CA IX: One of the most active carbonic anhydrase isozymes. *J. Biol. Chem.* **2008**, *283*, 27799–27809. [CrossRef] [PubMed]

25. Alterio, V.; Hilvo, M.; Di Fiore, A.; Supuran, C.T.; Pan, P.; Parkkila, S.; Scaloni, A.; Pastorek, J.; Pastorekova, S.; Pedone, C.; et al. Crystal structure of the extracellular catalytic domain of the tumor-associated human carbonic anhydrase IX. *Proc. Natl. Acad. Sci. USA* **2009**, *106*, 16233–16238. [CrossRef] [PubMed]

26. Innocenti, A.; Pastorekova, S.; Pastorek, J.; Scozzafava, A.; De Simone, G.; Supuran, C.T. The proteoglycan region of the tumor-associated carbonic anhydrase isoform IX acts as an intrinsic buffer optimizing CO_2 hydration at acidic pH values characteristic of solid tumors. *Bioorg. Med. Chem. Lett.* **2009**, *19*, 5825–5828. [CrossRef] [PubMed]

27. Pastorek, J.; Pastorekova, S. Hypoxia-induced carbonic anhydrase IX as a target for cancer therapy: From biology to clinical use. *Semin. Cancer Biol.* **2015**, *31*, 52–64. [CrossRef] [PubMed]

28. Svastová, E.; Zilka, N.; Zaťovicová, M.; Gibadulinová, A.; Ciampor, F.; Pastorek, J.; Pastoreková, S. Carbonic anhydrase IX reduces E-cadherin-mediated adhesion of MDCK cells via interaction with beta-catenin. *Exp. Cell Res.* **2003**, *290*, 332–345. [CrossRef]

29. Svastová, E.; Hulíková, A.; Rafajová, M.; Zaťovicová, M.; Gibadulinová, A.; Casini, A.; Cecchi, A.; Scozzafava, A.; Supuran, C.T.; Pastorek, J.; et al. Hypoxia activates the capacity of tumor-associated carbonic anhydrase IX to acidify extracellular pH. *FEBS Lett.* **2004**, *577*, 439–445. [CrossRef] [PubMed]

30. Ditte, P.; Dequiedt, F.; Svastova, E.; Hulikova, A.; Ohradanova-Repic, A.; Zatovicova, M.; Csaderova, L.; Kopacek, J.; Supuran, C.T.; Pastorekova, S.; et al. Phosphorylation of carbonic anhydrase IX controls its ability to mediate extracellular acidification in hypoxic tumors. *Cancer Res.* **2011**, *71*, 7558–7567. [CrossRef] [PubMed]

31. Buanne, P.; Renzone, G.; Monteleone, F.; Vitale, M.; Monti, S.M.; Sandomenico, A.; Garbi, C.; Montanaro, D.; Accardo, M.; Troncone, G.; et al. Characterization of carbonic anhydrase IX interactome reveals proteins assisting its nuclear localization in hypoxic cells. *J. Proteome Res.* **2013**, *12*, 282–292. [CrossRef] [PubMed]

32. Buonanno, M.; Langella, E.; Zambrano, N.; Succoio, M.; Sasso, E.; Alterio, V.; Di Fiore, A.; Sandomenico, A.; Supuran, C.T.; Scaloni, A.; et al. Disclosing the interaction of carbonic anhydrase IX with cullin-associated NEDD8-dissociated protein 1 by molecular modeling and integrated binding measurements. *ACS Chem. Biol.* **2017**, *12*, 1460–1465. [CrossRef] [PubMed]

33. Vullo, D.; Innocenti, A.; Nishimori, I.; Pastorek, J.; Scozzafava, A.; Pastoreková, S.; Supuran, C.T. Carbonic anhydrase inhibitors. Inhibition of the transmembrane isozyme XII with sulfonamides-a new target for the design of antitumor and antiglaucoma drugs? *Bioorg. Med. Chem. Lett.* **2005**, *15*, 963–969. [CrossRef] [PubMed]

34. Vullo, D.; Franchi, M.; Gallori, E.; Pastorek, J.; Scozzafava, A.; Pastorekova, S.; Supuran, C.T. Carbonic anhydrase inhibitors: Inhibition of the tumor-associated isozyme IX with aromatic and heterocyclic sulfonamides. *Bioorg. Med. Chem. Lett.* **2003**, *13*, 1005–1009. [CrossRef]

35. Guler, O.O.; De Simone, G.; Supuran, C.T. Drug design studies of the novel antitumor targets carbonic anhydrase IX and XII. *Curr. Med. Chem.* **2010**, *17*, 1516–1526. [CrossRef] [PubMed]

36. Supuran, C.T. How many carbonic anhydrase inhibition mechanisms exist? *J. Enzyme Inhib. Med. Chem.* **2016**, *31*, 345–360. [CrossRef] [PubMed]

37. Supuran, C.T. Advances in structure-based drug discovery of carbonic anhydrase inhibitors. *Expert Opin. Drug Discov.* **2017**, *12*, 61–88. [CrossRef] [PubMed]

38. Supuran, C.T. Structure-based drug discovery of carbonic anhydrase inhibitors. *J. Enzyme Inhib. Med. Chem.* **2012**, *27*, 759–772. [CrossRef] [PubMed]

39. Cecchi, A.; Hulikova, A.; Pastorek, J.; Pastoreková, S.; Scozzafava, A.; Winum, J.Y.; Montero, J.L.; Supuran, C.T. Carbonic anhydrase inhibitors. Design of fluorescent sulfonamides as probes of tumor-associated carbonic anhydrase IX that inhibit isozyme IX-mediated acidification of hypoxic tumors. *J. Med. Chem.* **2005**, *48*, 4834–4841. [CrossRef] [PubMed]

40. Dubois, L.; Douma, K.; Supuran, C.T.; Chiu, R.K.; van Zandvoort, M.A.; Pastoreková, S.; Scozzafava, A.; Wouters, B.G.; Lambin, P. Imaging the hypoxia surrogate marker CA IX requires expression and catalytic activity for binding fluorescent sulfonamide inhibitors. *Radiother. Oncol.* **2007**, *83*, 367–373. [CrossRef] [PubMed]

41. Dubois, L.; Lieuwes, N.G.; Maresca, A.; Thiry, A.; Supuran, C.T.; Scozzafava, A.; Wouters, B.G.; Lambin, P. Imaging of CA IX with fluorescent labelled sulfonamides distinguishes hypoxic and (re)-oxygenated cells in a xenograft tumour model. *Radiother. Oncol.* **2009**, *92*, 423–428. [CrossRef] [PubMed]

42. Ahlskog, J.K.; Dumelin, C.E.; Trüssel, S.; Mårlind, J.; Neri, D. In vivo targeting of tumor-associated carbonic anhydrases using acetazolamide derivatives. *Bioorg. Med. Chem. Lett.* **2009**, *19*, 4851–4856. [CrossRef] [PubMed]

43. Dubois, L.; Peeters, S.; Lieuwes, N.G.; Geusens, N.; Thiry, A.; Wigfield, S.; Carta, F.; McIntyre, A.; Scozzafava, A.; Dogné, J.M.; et al. Specific inhibition of carbonic anhydrase IX activity enhances the in vivo therapeutic effect of tumor irradiation. *Radiother. Oncol.* **2011**, *99*, 424–431. [CrossRef] [PubMed]

44. Lou, Y.; McDonald, P.C.; Oloumi, A.; Chia, S.; Ostlund, C.; Ahmadi, A.; Kyle, A.; Auf dem Keller, U.; Leung, S.; Huntsman, D.; et al. Targeting tumor hypoxia: Suppression of breast tumor growth and metastasis by novel carbonic anhydrase IX inhibitors. *Cancer Res.* **2011**, *71*, 3364–3376. [CrossRef] [PubMed]

45. Pacchiano, F.; Carta, F.; McDonald, P.C.; Lou, Y.; Vullo, D.; Scozzafava, A.; Dedhar, S.; Supuran, C.T. Ureido-substituted benzenesulfonamides potently inhibit carbonic anhydrase IX and show antimetastatic activity in a model of breast cancer metastasis. *J. Med. Chem.* **2011**, *54*, 1896–1902. [CrossRef] [PubMed]

46. Gieling, R.G.; Babur, M.; Mamnani, L.; Burrows, N.; Telfer, B.A.; Carta, F.; Winum, J.Y.; Scozzafava, A.; Supuran, C.T.; Williams, K.J. Antimetastatic effect of sulfamate carbonic anhydrase IX inhibitors in breast carcinoma xenografts. *J. Med. Chem.* **2012**, *55*, 5591–5600. [CrossRef] [PubMed]

47. Touisni, N.; Maresca, A.; McDonald, P.C.; Lou, Y.; Scozzafava, A.; Dedhar, S.; Winum, J.Y.; Supuran, C.T. Glycosyl coumarin carbonic anhydrase IX and XII inhibitors strongly attenuate the growth of primary breast tumors. *J. Med. Chem.* **2011**, *54*, 8271–8277. [CrossRef] [PubMed]

48. Lock, F.E.; McDonald, P.C.; Lou, Y.; Serrano, I.; Chafe, S.C.; Ostlund, C.; Aparicio, S.; Winum, J.Y.; Supuran, C.T.; Dedhar, S. Targeting carbonic anhydrase IX depletes breast cancer stem cells within the hypoxic niche. *Oncogene.* **2013**, *32*, 5210–5219. [CrossRef] [PubMed]

49. McDonald, P.C.; Chafe, S.C.; Dedhar, S. Overcoming hypoxia-mediated tumor progression: combinatorial approaches targeting ph regulation, angiogenesis and immune dysfunction. *Front. Cell Dev. Biol.* **2016**, *4*, 27. [CrossRef] [PubMed]

50. Swayampakula, M.; McDonald, P.C.; Vallejo, M.; Coyaud, E.; Chafe, S.C.; Westerback, A.; Venkateswaran, G.; Shankar, J.; Gao, G.; Laurent, E.M.N.; et al. The interactome of metabolic enzyme carbonic anhydrase IX reveals novel roles in tumor cell migration and invadopodia/MMP14-mediated invasion. *Oncogene.* **2017**, in press. [CrossRef] [PubMed]

51. A phase I, multi-center, open-label, study to investigate the safety, tolerability and pharmacokinetic of SLC-0111 in subjects with advanced solid tumours. 2016. Available online: https://clinicaltrials.gov/ct2/show/NCT02215850 (accessed on 15 September 2017).

52. Pacchiano, F.; Aggarwal, M.; Avvaru, B.S.; Robbins, A.H.; Scozzafava, A.; McKenna, R.; Supuran, C.T. Selective hydrophobic pocket binding observed within the carbonic anhydrase II active site accommodate different 4-substituted-ureido-benzenesulfonamides and correlate to inhibitor potency. *Chem. Commun.* **2010**, *46*, 8371–8373. [CrossRef] [PubMed]

53. Perut, F.; Carta, F.; Bonuccelli, G.; Grisendi, G.; Di Pompo, G.; Avnet, S.; Sbrana, F.V.; Hosogi, S.; Dominici, M.; Kusuzaki, K.; et al. Carbonic anhydrase IX inhibition is an effective strategy for osteosarcoma treatment. *Expert Opin. Ther. Targets.* **2015**, *19*, 1593–1605. [CrossRef] [PubMed]

54. Logsdon, D.P.; Grimard, M.; Luo, M.; Shahda, S.; Jiang, Y.; Tong, Y.; Yu, Z.; Zyromski, N.; Schipani, E.; Carta, F.; et al. Regulation of HIF1α under hypoxia by APE1/Ref-1 impacts CA9 expression: Dual targeting in patient-derived 3D pancreatic cancer models. *Mol. Cancer Ther.* **2016**, *15*, 2722–2732. [CrossRef] [PubMed]

55. Krall, N.; Pretto, F.; Decurtins, W.; Bernardes, G.J.; Supuran, C.T.; Neri, D. A small-molecule drug conjugate for the treatment of carbonic anhydrase IX expressing tumors. *Angew. Chem. Int. Ed. Engl.* **2014**, *53*, 4231–4235. [CrossRef] [PubMed]

56. Grabmaier, K.; Vissers, J.L.; De Weijert, M.C.; Oosterwijk-Wakka, J.C.; Van Bokhoven, A.; Brakenhoff, R.H.; Noessner, E.; Mulders, P.A.; Merkx, G.; Figdor, C.G.; et al. Molecular cloning and immunogenicity of renal cell carcinoma-associated antigen G250. *Int. J. Cancer* **2000**, *85*, 865–870. [CrossRef]

57. Siebels, M.; Rohrmann, K.; Oberneder, R.; Stahler, M.; Haseke, N.; Beck, J.; Hofmann, R.; Kindler, M.; Kloepfer, P.; Stief, C. A clinical phase I/II trial with the monoclonal antibody cG250 (RENCAREX®) and interferon-alpha-2a in metastatic renal cell carcinoma patients. *World J. Urol.* **2011**, *29*, 121–126. [CrossRef] [PubMed]

58. A Randomized, Double Blind Phase III Study to Evaluate Adjuvant cG250 Treatment Versus Placebo in Patients with Clear Cell RCC and High Risk of Recurrence (ARISER). Available online: https://clinicaltrials.gov/ct2/show/NCT00087022?term=girentuximab&rank=6 (accessed on 15 September 2017).

59. Takacova, M.; Hlouskova, G.; Zatovicova, M.; Benej, M.; Sedlakova, O.; Kopacek, J.; Pastorek, J.; Lacik, I.; Pastorekova, S. Encapsulation of anti-carbonic anhydrase IX antibody in hydrogel microspheres for tumor targeting. *J. Enzyme Inhib. Med. Chem.* **2016**, *31*, 110–118. [CrossRef] [PubMed]

60. Zatovicova, M.; Jelenska, L.; Hulikova, A.; Csaderova, L.; Ditte, Z.; Ditte, P.; Goliasova, T.; Pastorek, J.; Pastorekova, S. Carbonic anhydrase IX as an anticancer therapy target: Preclinical evaluation of internalizing monoclonal antibody directed to catalytic domain. *Curr. Pharm. Des.* **2010**, *16*, 3255–3263. [CrossRef] [PubMed]

61. Battke, C.; Kremmer, E.; Mysliwietz, J.; Gondi, G.; Dumitru, C.; Brandau, S.; Lang, S.; Vullo, D.; Supuran, C.; Zeidler, R. Generation and characterization of the first inhibitory antibody targeting tumour-associated carbonic anhydrase XII. *Cancer Immunol. Immunother.* **2011**, *60*, 649–658. [CrossRef] [PubMed]

62. Gondi, G.; Mysliwietz, J.; Hulikova, A.; Jen, J.P.; Swietach, P.; Kremmer, E.; Zeidler, R. Antitumor efficacy of a monoclonal antibody that inhibits the activity of cancer-associated carbonic anhydrase XII. *Cancer Res.* **2013**, *73*, 6494–6503. [CrossRef] [PubMed]

63. Akurathi, V.; Dubois, L.; Lieuwes, N.G.; Chitneni, S.K.; Cleynhens, B.J.; Vullo, D.; Supuran, C.T.; Verbruggen, A.M.; Lambin, P.; Bormans, G.M. Synthesis and biological evaluation of a ^{99m}Tc-labelled sulfonamide conjugate for in vivo visualization of carbonic anhydrase IX expression in tumor hypoxia. *Nucl. Med. Biol.* **2010**, *37*, 557–564. [CrossRef] [PubMed]

64. Akurathi, V.; Dubois, L.; Celen, S.; Lieuwes, N.G.; Chitneni, S.K.; Cleynhens, B.J.; Innocent, A.; Supuran, C.T.; Verbruggen, A.M.; Lambin, P.; et al. Development and biological evaluation of ^{99m}Tc-sulfonamide derivatives for in vivo visualization of CA IX as surrogate tumor hypoxia markers. *Eur. J. Med. Chem.* **2014**, *71*, 374–384. [CrossRef] [PubMed]

65. Peeters, S.G.; Dubois, L.; Lieuwes, N.G.; Laan, D.; Mooijer, M.; Schuit, R.C.; Vullo, D.; Supuran, C.T.; Eriksson, J.; Windhorst, A.D.; et al. [^{18}F]VM$_4$–o$_{37}$ MicroPET imaging and biodistribution of two in vivo CAIX-expressing tumor models. *Mol. Imaging Biol.* **2015**, *17*, 615–619. [CrossRef] [PubMed]

66. Pan, J.; Lau, J.; Mesak, F.; Hundal, N.; Pourghiasian, M.; Liu, Z.; Bénard, F.; Dedhar, S.; Supuran, C.T.; Lin, K.S. Synthesis and evaluation of ^{18}F-labeled carbonic anhydrase IX inhibitors for imaging with positron emission tomography. *J. Enzyme Inhib. Med. Chem.* **2014**, *29*, 249–255. [CrossRef] [PubMed]

67. Lau, J.; Liu, Z.; Lin, K.S.; Pan, J.; Zhang, Z.; Vullo, D.; Supuran, C.T.; Perrin, D.M.; Bénard, F. Trimeric radiofluorinated sulfonamide derivatives to achieve in vivo selectivity for carbonic anhydrase IX-targeted PET imaging. *J. Nucl. Med.* **2015**, *56*, 1434–1440. [CrossRef] [PubMed]

68. Zhang, Z.; Lau, J.; Zhang, C.; Colpo, N.; Nocentini, A.; Supuran, C.T.; Bénard, F.; Lin, K.S. Design, synthesis and evaluation of ^{18}F-labeled cationic carbonic anhydrase IX inhibitors for PET imaging. *J. Enzyme Inhib. Med. Chem.* **2017**, *32*, 722–730. [CrossRef] [PubMed]

69. Lau, J.; Zhang, Z.; Jenni, S.; Kuo, H.T.; Liu, Z.; Vullo, D.; Supuran, C.T.; Lin, K.S.; Bénard, F. PET imaging of carbonic anhydrase IX expression of HT-29 tumor xenograft mice with ^{68}Ga-labeled benzenesulfonamides. *Mol. Pharm.* **2016**, *13*, 1137–1146. [CrossRef] [PubMed]

70. Sneddon, D.; Niemans, R.; Bauwens, M.; Yaromina, A.; van Kuijk, S.J.; Lieuwes, N.G.; Biemans, R.; Pooters, I.; Pellegrini, P.A.; Lengkeek, N.A.; et al. Synthesis and in vivo biological evaluation of ^{68}Ga-labeled carbonic anhydrase IX targeting small molecules for positron emission tomography. *J. Med. Chem.* **2016**, *59*, 6431–6443. [CrossRef] [PubMed]

71. Ahlskog, J.K.; Schliemann, C.; Mårlind, J.; Qureshi, U.; Ammar, A.; Pedley, R.B.; Neri, D. Human monoclonal antibodies targeting carbonic anhydrase IX for the molecular imaging of hypoxic regions in solid tumours. *Br. J. Cancer* **2009**, *101*, 645–657. [CrossRef] [PubMed]

72. Huizing, F.J.; Hoeben, B.A.W.; Franssen, G.; Lok, J.; Heskamp, S.; Oosterwijk, E.; Boerman, O.C.; Bussink, J. Preclinical validation of ^{111}In-girentuximab-F(ab')$_2$ as a tracer to image hypoxia related marker CA IX expression in head and neck cancer xenografts. *Radiother Oncol.* **2017**, in press. [CrossRef]

73. Honarvar, H.; Garousi, J.; Gunneriusson, E.; Höidén-Guthenberg, I.; Altai, M.; Widström, C.; Tolmachev, V.; Frejd, F.Y. Imaging of CAIX-expressing xenografts in vivo using ^{99m}Tc-HEHEHE-ZCAIX 1 affibody molecule. *Int. J. Oncol.* **2015**, *46*, 513–520. [CrossRef] [PubMed]

74. Khandani, A.H.; Rathmell, W.K.; Wallen, E.M.; Ivanovic, M. PET/CT with ^{124}I-cG250: Great potential and some open questions. *AJR Am. J. Roentgenol.* **2014**, *203*, 261–262. [CrossRef] [PubMed]
75. Muselaers, C.H.; Rijpkema, M.; Bos, D.L.; Langenhuijsen, J.F.; Oyen, W.J.; Mulders, P.F.; Oosterwijk, E.; Boerman, O.C. Radionuclide and fluorescence imaging of clear cell renal cell carcinoma using dual labeled anti-carbonic anhydrase IX antibody G250. *J. Urol.* **2015**, *194*, 532–538. [CrossRef] [PubMed]

metabolites

MDPI

Review

Coordinated Regulation of Metabolic Transporters and Migration/Invasion by Carbonic Anhydrase IX

Paul C. McDonald [1], Mridula Swayampakula [1] and Shoukat Dedhar [1,2,*]

[1] Department of Integrative Oncology, BC Cancer Research Centre, Vancouver, BC V5Z 1L3, Canada;
    pmcdonal@bccrc.ca (P.C.M.); mswayampak@bccrc.ca (M.S.)
[2] Department of Biochemistry and Molecular Biology, University of British Columbia,
    Vancouver, BC V6T 1Z3, Canada
* Correspondence: sdedhar@bccrc.ca; Tel.: +1-604-675-8029

Received: 16 February 2018; Accepted: 6 March 2018; Published: 8 March 2018

Abstract: Hypoxia is a prominent feature of the tumor microenvironment (TME) and cancer cells must dynamically adapt their metabolism to survive in these conditions. A major consequence of metabolic rewiring by cancer cells in hypoxia is the accumulation of acidic metabolites, leading to the perturbation of intracellular pH (pHi) homeostasis and increased acidosis in the TME. To mitigate the potentially detrimental consequences of an increasingly hypoxic and acidic TME, cancer cells employ a network of enzymes and transporters to regulate pH, particularly the extracellular facing carbonic anhydrase IX (CAIX) and CAXII. In addition to the role that these CAs play in the regulation of pH, recent proteome-wide analyses have revealed the presence of a complex CAIX interactome in cancer cells with roles in metabolite transport, tumor cell migration and invasion. Here, we explore the potential contributions of these interactions to the metabolic landscape of tumor cells in hypoxia and discuss the role of CAIX as a hub for the coordinated regulation of metabolic, migratory and invasive processes by cancer cells. We also discuss recent work targeting CAIX activity using highly selective small molecule inhibitors and briefly discuss ongoing clinical trials involving SLC-0111, a lead candidate small molecule inhibitor of CAIX/CAXII.

Keywords: hypoxia; carbonic anhydrase IX; cancer metabolism; transporter; integrin; MMP14; migration; invasion; metastasis

1. Introduction

As tumors develop, cancer cells must reprogram their metabolism to meet the demands of energy production and biosynthesis. The availability of nutrients and the configuration of the cellular metabolic network collaborate to determine how cancer cells perform core metabolic functions, including energy production, biomass accumulation and control of the redox state [1]. The expanding knowledge base surrounding metabolism in cancer has resulted in the identification of several hallmarks of cancer metabolism, including, but not limited to, deregulated uptake of glucose and amino acids, use of opportunistic modes of nutrient acquisition and dynamic metabolic interactions with the tumor microenvironment (TME) [2].

A consequence of the proliferation of cancer cells beyond the reach of established blood vessels is the development of intratumoral hypoxia, defined as regions exhibiting low partial pressure of oxygen (O_2) [3]. Hypoxia is a prominent feature of the TME and its presence results in the stabilization by cancer cells of hypoxia-inducible factor 1 alpha (HIF-1α), the master regulator of the hypoxic response, leading to the upregulation of a plethora of gene products geared toward the protection of tumor cells against hypoxic stress [4]. Biological responses of the tumor to hypoxia include the induction of angiogenesis, resulting in the formation of a dysfunctional vasculature that serves to perpetuate poor perfusion and exacerbate hypoxia, and dynamic adaptation of cancer cell metabolism to enable the acquisition and use of nutrients and metabolites from an increasingly nutrient-poor, low-O_2 environment, thereby maintaining viability and enabling continued proliferation [2,5].

Metabolically, hypoxia reduces the amount of O_2 available for oxidative phosphorylation and cancer cells respond to this challenge by shifting toward the use of glycolysis for respiration [1–3]. This shift is coupled with the use of alternative fuel sources, including glutamine and fatty acids, optimization of the efficiency of oxidative phosphorylation and use of the tricarboxylic acid (TCA) cycle to generate metabolic precursors [1,2]. A major consequence of metabolic rewiring by cancer cells in hypoxia is the increased production and accumulation of acidic metabolites, particularly lactate, carbon dioxide (CO_2) and protons (H^+) [5]. The development of acidosis in the hypoxic TME leads to the perturbation of intracellular pH (pHi) homeostasis, a situation which rapidly impinges on cellular viability and drives cancer cells to engage compensatory survival mechanisms.

To mitigate the potentially detrimental consequences of an increasingly hypoxic and acidic TME, cancer cells employ a network of enzymes and transporters that work in concert to provide effective pH regulation [5]. Critical components of this pH regulatory system are carbonic anhydrases (CAs), particularly extracellular-facing carbonic anhydrase IX (CAIX) and CAXII [6–8]. In particular, CAIX is a HIF-1α-induced, cell-surface enzyme that regulates pHi and promotes tumor cell survival [6,8]. In addition to hypoxia, which is a major driver of CAIX expression by cancer cells [6,9], CAIX can also be induced in normoxia by high cell-density-mediated pseudohypoxia [10,11], and by hypoxia-independent mechanisms such as lactate- [12] and redox-mediated [13] stabilization of HIF-1α. CAIX is widely regarded as a prominent biomarker of poor patient prognosis and treatment resistance for many solid cancers [9]. Several studies have now demonstrated the critical role of CAIX in the growth and metastasis of multiple types of cancers [14–17], and recent data have suggested an important role of CAIX in tumor cell migration [18,19] and invasion [17,20–22].

The multifaceted role of CAIX in cancer cell biology, coupled with the relative paucity of available data regarding physiologically-relevant associations between CAIX and putative interacting proteins in cancer cells, has driven the need for studies aimed at elucidating the components comprising the CAIX interactome. As part of this research focus, recent investigations have been undertaken to identify proximal CAIX-interacting proteins using an unbiased proteomic screen centered on proximity-dependent biotin identification (BioID) technology [22]. These studies have uncovered the presence of an intricate CAIX interactome in cancer cells that controls important functional parameters essential to the processes of pH regulation, transport of metabolic intermediates, cell migration and invasion [22]. In particular, the data have revealed the presence of two major classes of membrane proteins that are associated with CAIX, specifically metabolic transport proteins and cell adhesion/migration/invasion proteins. Here, we explore the potential contributions of these interactions to the metabolic landscape of tumor cells in hypoxia [23,24] and discuss the role of CAIX as a key hub for the coordinated regulation of metabolic, migratory and invasive processes by cancer cells. We also discuss recent work targeting CAIX activity in pre-clinical models using small molecule inhibitors of CAIX and briefly describe ongoing clinical trials using a lead compound, SLC-0111.

2. Membrane-Localized Metabolic Transport Proteins

Among the CAIX-associating proteins identified using the BioID platform, associations with membrane-localized metabolic transporters from several functional classes, including bicarbonate (HCO_3^-) transporters and amino acid (AA) transporters were observed [22]. Specifically, CAIX is proximally associated with the sodium-dependent electroneutral bicarbonate transporter n1 (NBCn1), encoded by the gene solute-like carrier (SLC) 4A7 (*SLC4A7*). CAIX also associates with a suite of proteins involved in AA transport, including the L-type AA transporter, LAT1 (*SLC7A5*), the AA transport heavy chain subunit, CD98hc (*SLC3A2*) and the glutamine transporters alanine-serine-cysteine-preferring transporter 2 (ASCT2; *SLC1A5*) and sodium-coupled neutral amino acid transporter 2 (SNAT2; *SLC38A2*). The interaction of CAIX with this diverse array of membrane-localized metabolic transport proteins suggests that it may serve as a central regulator of metabolic processes by cancer cells during hypoxic stress (Figure 1). Each of these interactions is discussed further below.

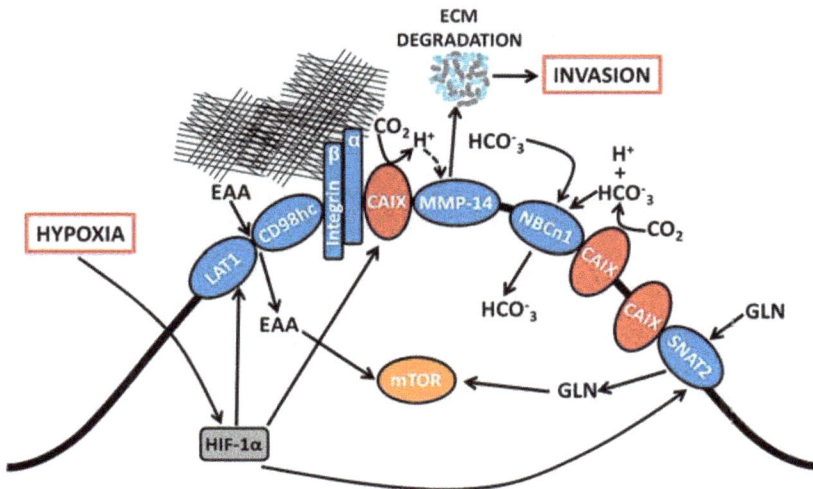

Figure 1. Coordinated regulation by CAIX of amino acid and bicarbonate transporters, and migration/ invasion through interaction with integrins and MMP14. Proteomic analyses revealed associations between CAIX and several membrane-bound transport proteins. CAIX couples with bicarbonate transporters to facilitate influx of HCO_3^-. CAIX associates with amino acid transporters important for the import of both essential amino acids and the conditionally essential amino acid glutamine, which serve as alternative metabolic fuels and biosynthetic precursors for use by cancer cells. CAIX also forms novel associations with collagen- and laminin-binding integrins localized at pseudopodia-like protrusions at the leading edge of migrating cells. Finally, CAIX potentiates MMP14 activity at invadopodia through donation of H^+ released by CAIX-mediated CO_2 hydration. EAA, essential amino acids; LAT1, large neutral amino acid transporter 1; CD98hc, cluster of differentiation 98 heavy chain; CAIX, carbonic anhydrase IX; NBCn1, sodium-dependent electroneutral bicarbonate transporter n1; SNAT2, sodium-coupled neutral amino acid transporter 2; MMP14, matrix metallopeptidase 14; GLN, glutamine; HCO_3^-, bicarbonate; CO_2, carbon dioxide; H^+, proton; ECM, extracellular matrix.

2.1. Bicarbonate Transporters

A key functional parameter of pHi regulation is the efficient, effective capture and import of HCO_3^- produced by CAIX-mediated hydration of CO_2 at the extracellular surface to buffer intracellular acidosis. CAIX has previously been proposed to associate with Na^+/HCO_3^- co-transporters to form a transport "metabolon", defined as a protein complex composed of metabolic enzymes that function to optimize the transfer of metabolic intermediates between active sites, bypassing the need for equilibration with bulk buffer [5]. While functional coordination of enzymes in metabolons has been reported [24], the existence of a proximal or physical association between the enzymes in metabolons generally and between CAIX and HCO_3^- transporters, specifically, has been controversial [5]. Studies have suggested that the expression of various HCO_3^- transporters is upregulated in hypoxia in a complex, cell-type-dependent fashion [25,26]. Furthermore, hypoxia-induced expression of the HCO_3^- transporter *SLC4A9* has been identified as having an essential role in tumor progression [26], while constitutive expression of electrogenic Na^+/HCO_3^- co-transporter (*SLC4A4*) has been reported to play a role in breast and colon cancer cell proliferation, migration and pHi regulation [25]. However, these studies did not directly investigate the presence of an association between CAIX and the HCO_3^- transporters.

Interrogation of the components of the CAIX interactome in triple negative breast cancer cells by employing unbiased proteome-wide strategies such as BioID has now affirmed the presence of a proximal association of CAIX with the electroneutral bicarbonate transporter NBCn1 (*SLC4A7*) (Figure 1) [22]. The NBCn1 transporter is implicated in breast cancer susceptibility [27] and knockout of NBCn1 in a mouse model of breast cancer has been reported to increase the latency of tumor development and impair tumor growth [28], indicating the importance of this transporter in breast cancer and alluding to the pathophysiological relevance of an interactive metabolon involving CAIX and HCO_3^- transporters. The proximal association of CAIX with NBCn1 is congruent with its participation in the setting of a metabolon and indicates that CAIX induced by hypoxia may couple with HCO_3^- transporters already present on the cell surface. While the precise molecular mechanisms remain to be determined, the presence of the CAIX-NBCn1 association demonstrates that CAIX can couple with specific HCO_3^- transporters, potentially facilitating the local production of HCO_3^- for subsequent capture and transfer into the cell to efficiently regulate pHi.

2.2. Essential Amino Acid Transporters

In addition to associating with proteins that rely directly on the catalytic function of CAIX, proteome-wide analyses have uncovered novel associations between CAIX and components of the amino acid (AA) transport system (Figure 1), including the large neutral amino acid transporter 1 (LAT1) and CD98 heavy chain (CD98hc), which themselves form a heterodimeric complex and function as a transporter of essential AAs (EAAs) that cannot be produced de novo by mammalian cells, such as leucine (leu), and the glutamine transporters ASCT2 and SNAT2 (discussed in Section 2.3 below) [22]. Such interactions raise the exciting possibility that cancer cells which have undergone metabolic reprogramming and require augmented capacity for nutrient acquisition to support cell growth in hypoxia may recruit CAIX to assist in the coordinated regulation of AA transport. While the molecular mechanisms and functional contribution of coupling CAIX to AA transporters remains an area for future investigation, it is clear that, in addition to skewing their metabolism toward glucose utilization, cancer cells rely on additional fuels to carry out core metabolic functions, including energy production and biosynthetic processes [1]. The presence of hypoxia further limits nutrient acquisition from the TME [3], making the capacity to acquire and utilize alternative fuels and nutrients particularly advantageous. Furthermore, overexpression of LAT1 is a negative prognostic indicator for many cancers [29] and LAT1 activity was found to be required for tumor growth in conditions of hypoxia and nutrient depletion [30], similar to conditions that induce CAIX expression.

It is notable that LAT1 has been shown to promote the activity of mTORC1, a master regulator of cell growth and metabolism [30], and sustained activation of mTORC1 requires the presence of intracellular leucine (leu), an EAA imported by LAT1 [29]. Recently published studies have shown that pharmacologic inhibition of CAIX activity in vivo in a model of glioblastoma multiforme (GBM)—when used in combination with the standard of cancer chemotherapy, temozolomide—results in an altered flux of AAs, including essential AAs such as leucine [31], potentially linking CAIX to EAA transport through an association with LAT1, although the mechanism remains to be determined. These findings suggest that CAIX may play a role in regulating metabolic functions in cancer cells beyond pH homeostasis and point to the potential involvement of CAIX in coordinating the regulation of EAA transport by cancer cells in hypoxia.

2.3. Glutamine Transporters

While metabolic reprogramming by cancer cells clearly results in a shift toward the use of glucose as fuel source, it is now recognized that tumor cells are capable of using diverse array of nutrients, in particular glutamine, to support metabolic and biosynthetic functions. The increased use of glutamine by cancer cells as an alternative fuel source, combined with nutrient delivery inadequacies brought on by the deregulated tumor vascular supply, results in the selective depletion of glutamine from the TME [32]. Under these conditions, glutamine is considered to be a conditionally essential AA and cancer cells must find ways to augment glutamine acquisition and uptake. For example, cancer cells may upregulate ASCT2, the major transporter for glutamine uptake [32,33]. Furthermore, cancer cells may upregulate other sodium neutral amino acid transporters, including SNAT1 and SNAT2, as a way of supplementing uptake by ASCT2 [33,34].

Metabolically, glutaminolysis contributes to the production of intracellular CO_2 via decarboxylation of metabolic intermediates [5]. Thus, association of CAIX with glutamine transporters such as ASCT2 and SNAT2 potentially couples glutamine import and metabolism with the effective management of CO_2 production (Figure 1). Furthermore, the association of CAIX with the glutamine transporter, SNAT2, an interaction specifically identified by the BioID analysis [22], together with its association with the essential AA transporter LAT1, suggests an active role for CAIX in coordinating the regulation of AA flux in general, especially since the import of EAA such as leu is coupled with glutamine efflux [2]. Thus, it is tempting to speculate that CAIX contributes functionally to cell energetics and biosynthesis in the context of hypoxia.

3. Cell Adhesion/Migration/Invasion Proteins

A growing body of evidence supports a role of CAIX as a key regulator of cancer cell migration, invasion and metastasis. For example, studies have demonstrated that the genetic depletion of CAIX reduces breast cancer invasion and metastasis [15,17] and pharmacologic inhibition of CAIX activity serves to inhibit metastasis in pre-clinical models of cancer [15,17,35]. Furthermore, recent analyses have shown that CAIX is a critical functional mediator of invasion in vitro in the biologically-relevant context of hypoxia [22]. Indeed, recent investigations have shown that CAIX is requisite for the invasion of tumor cells through matrices, including matrigel and type 1 collagen. Furthermore, it is probable that multiple regions of the CAIX protein, including the proteoglycan-(PG)-like domain and intracellular domain, contribute functionally to invasion, and that the catalytic activity of CAIX is necessary for the invasive process, given that invasion is abrogated in the presence of a small molecule inhibitor of CAIX [22].

In support of a role of CAIX in migration and invasion, proteomic and co-immunoprecipitation analyses have revealed novel associations of CAIX with several proteins involved in cell adhesion, matrix remodeling and invasion. In particular, CAIX associates with a compendium of integrin subunits, specifically integrin β1 (ITGB1), integrin α2 (ITGA2), integrin α3 (ITGA3), integrin α5 (ITGA5) and integrin α6 (ITGA6) [22], highlighting a potential role of CAIX in coordinating the regulation of collagen- and laminin-binding integrins to control cancer cell adhesion, a critical process involved in migration and invasion (Figure 1). CAIX also associates with MMP14, a key player in the matrix degradation process required for successful invasion by cancer cells (Figure 1). The association of CAIX with these well-recognized effectors of migration and invasion suggests that these interactions may be functionally relevant in the formation and/or activity of protrusive invasive structures such as pseudopodia [36] and invadopodia [37] in hypoxia. The association of CAIX with integrins and MMP14 within these protrusive structures, together with potential mechanistic consequences of the interaction with MMP14, is discussed further below.

3.1. Association with Integrins and MMP14 in Pseudopodia

Examination of the membrane extensions formed by breast cancer cells cultured on collagen in hypoxia has demonstrated spatial localization of CAIX in association with integrins ITGB1 and ITGA2 in actin- and cofilin-positive, pseudopodia-like protrusions resembling lamellipodia (Figure 1) [22]. The association of CAIX with integrins was very evident at the leading edges of cells with a migratory phenotype, but was distinctly absent from focal adhesions, suggesting that CAIX associates with these proteins specifically in cellular regions involved in migration [22]. In addition to integrins, immunofluorescence analyses have shown that CAIX associates with MMP14, a matrix metalloprotease that counts collagen type I among its substrates, at pseudopodia-like protrusions resembling lamellipodia, suggesting that CAIX may function to regulate MMP14-mediated matrix degradation at these structures (Figure 1) [22]. Further analysis using a proximity ligation assay (PLA) has confirmed these results, demonstrating that CAIX and MMP14 reside in close proximity to one another in breast cancer cells (Figure 2).

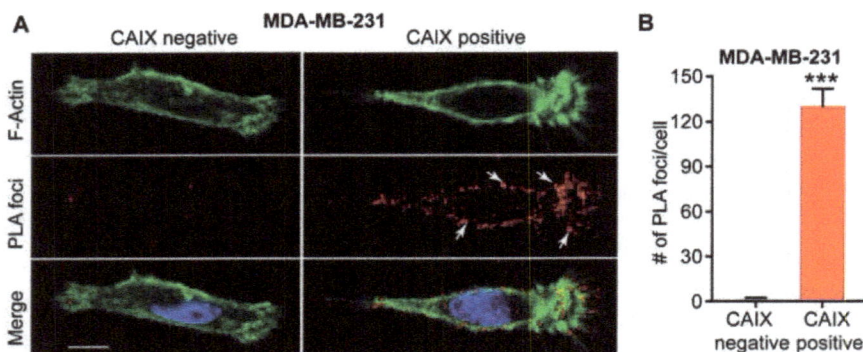

Figure 2. Interaction between CAIX and MMP14 as observed by proximity ligation assay (PLA). (**A**) Immunofluorence images showing the interaction of CAIX and MMP14 by PLA (red foci; arrows) in MDA-MB-231 cells depleted of CAIX using CRISPR-Cas technology (CAIX negative) or similar cells constitutively expressing CAIX (CAIX positive). Actin (green) and nuclei (blue) are shown for purposes of orientation. PLA-positive signals are concentrated in pseudopodia-like protrusions at the leading edge of migrating, CAIX-positive cells. Scale bar = 10 μm; (**B**) Quantification of PLA-positive foci. Data show the mean ± sem of n = 74 cells and are representative of 2 independent experiments. *** $p < 0.001$.

It is interesting that CAIX associates both with integrins and, as discussed above, with CD98hc. While CD98hc is a component of the AA transport system, evidence also suggests that it plays a role in regulating integrin-mediated tissue stiffness [38]. Given these findings, it is possible that CAIX may provide a link between AA transport and integrin-mediated adhesion at membrane protrusions. The potential functional relevance of the interactions between CAIX, integrins and CD98hc remain to be elucidated by future research.

3.2. Functional Role of a CAIX-MMP14 Interaction at Invadopodia

In addition to its presence in pseudopodia-like protrusions, MMP14 is a well-established component of invadopodia, the protrusive, matrix degrading structures on the ventral surface of cells that concentrate and release proteases to enable ECM degradation, thereby facilitating invasion and metastasis by cancer cells [37,39,40]. Congruent with the findings that MMP14 is a component of the CAIX interactome in cancer cells and that CAIX localizes with MMP14 in pseudopodia-like protrusions, immunofluorescence analyses have now shown that, in breast cancer cells, CAIX specifically co-localizes with MMP14 at functional invadopodia, where it functions to regulate collagen degradation [22]. Furthermore, detailed biochemical examination of this interaction has revealed that the intracellular domain of CAIX interacts with MMP14 and that the interaction depends on putative phosphorylation sites positioned within the intracellular domain of CAIX [22]. Mechanistically, CAIX enhances MMP14-mediated collagen degradation by providing a local reservoir of H^+ required for MMP14 catalytic activity [22] (Figure 1). Importantly, this novel mechanism for the regulation of MMP14-mediated invasion by CAIX is highly biologically relevant, given that extracellular acidosis is thought to activate proteases [41] and it has been reported that collagen degradation by MMP14 is increased in acidic pH [42]. Furthermore, the contribution of CAIX to the regulation of MMP14 activity at invadopodia is of particular importance in hypoxia. It is now understood that the pH regulatory protein Na^+/H^+ exchanger 1 (NHE1) is recruited to invadopodia, where it regulates invadopodia function by modulating pHi [43,44] and drives cofilin-dependent actin polymerization and recruitment of MMPs, including MMP14 [45]. As a consequence of its activity, NHE1 extrudes H^+ into the extracellular environment, thereby contributing to extracellular acidosis. In regions of hypoxia, however, NHE1 activity is reduced [46] and NHE1 gene expression is reported to be low in basal type and triple negative breast cancers [47], whereas CAIX is expressed in over 50% of patients with this breast cancer subtype [15], a patient subset that also expresses MMP-14. Thus, the MMP14-CAIX interaction at invadopodia provides a putative mechanism for potentiation of MMP14 degradative activity in situations where the activity of NHE1 may be compromised.

It is also interesting that while the presence of CAIX at protrusive structures at the leading edge of cancer cells allude to a possible functional contribution by CAIX to the process of cancer cell migration; the interaction between CAIX and MMP14 at invadopodia suggests a scenario whereby CAIX actively modulates invasion via mechanisms that are independent of migration and that involve localized stimulation of MMP14 activity to regulate the degradation of collagen [22]. Recent studies have reported the presence and/or upregulation of CAIX at the invasive front of carcinomas in patients [15,48]. Similarly, MMP14 expression is associated with tumor progression, invasion and metastasis, and poor prognosis [49–52]. Taken together, these data suggest that the localization and association of CAIX and MMP14 may lead to MMP14 activation at invadopodia, providing a novel mechanism of invasion that can be incorporated into the arsenal of functional processes used by cancer cells for invasion and metastasis.

4. Targeting CAIX Activity in Hypoxic Solid Tumors

Substantive research efforts in recent years have focused on the development, pre-clinical and clinical evaluation of therapeutic strategies targeting CAIX (and CAXII) in solid malignancies. The HIF-1-mediated, tumor-specific upregulation of CAIX, its localization at the cell surface, its highly restricted expression in normal tissues, the association of CAIX upregulation with poor prognosis and its functional relevance to tumor biology all serve as key properties for its use as a therapeutic target. To date, several studies have provided validation of targeting CAIX in multiple tumor models [14–17] and an array of potential therapeutic strategies to target CAIX have been developed, including the use of small molecule inhibitors of CAIX/CAXII activity [6,7].

Among a large number of CAIX/CAXII inhibitor compounds reported to date, the ureido-substituted benzenesulfonamides has been evaluated extensively for anti-tumor activity [15,17,31,53]. Pre-clinical studies in models of breast cancer have demonstrated the efficacy of the lead candidate CAIX/CAXII inhibitor, SLC-0111, in reducing tumor growth and inhibiting the formation of metastases [15,17]. Furthermore, administration of SLC-0111 in combination with conventional chemotherapy agents such as paclitaxel further inhibited tumor growth, compared to individual treatments [17]. Moving beyond breast cancer, recently published data demonstrate that targeting CAIX/CAXII is a promising therapeutic avenue in glioblastoma multiforme (GBM), when used in combination with standard of care chemotherapy. The Cancer Genome Atlas (TCGA) data from patients with GBM showed that high CAIX correlated with markedly reduced survival [31]. Treatment of a model of recurrent GBM with SLC-0111 in combination with temozolomide significantly delayed tumor growth, while treatment of an orthotopic patient derived xenograft (PDX) model of GBM with the combination resulted in increased survival [31]. Current studies using SLC-0111 in combination with gemcitabine in models of pancreatic cancer are yielding congruent data. Taken together, these findings indicate the potential for the broad use of CAIX/CAXII inhibitors in combination with standard of care chemotherapy to enhance therapeutic response, reduce toxicity, and combat therapeutic resistance across multiple cancer types [9]. In fact, SLC-0111 has now completed a multi-centre Phase I clinical trial (NCT02215850) in cancer patients. While the results of this first human trial are still to be published, SLC-0111 was shown to have a favorable safety profile and is now the subject of a Phase Ib trial targeting patients with CAIX-positive, metastatic pancreatic cancer.

5. Conclusions

In addition to the well-recognized role of CAIX in pH regulation, global proteomic initiatives have uncovered novel interactions of CAIX involved in metabolite transport and tumor cell migration and invasion, and new functional roles in regulating hypoxia-induced cancer cell invasion. In particular, CAIX associates with two protein classes, membrane-bound metabolic transport proteins and cell adhesion/migration/invasion proteins. The association of CAIX with metabolic transporters, especially with those that transport EAAs and glutamine, suggests an increasingly complex role of CAIX in coordinated regulation of cancer cell metabolism in hypoxia. Moreover, associations with integrins and MMP14 indicate novel roles of CAIX in modulating motility and invasion. The multifunctional capacity driven by the complexity of the CAIX interactome suggests that targeting its activity will exact substantive therapeutic benefits by interfering with several aspects of cancer biology, including metabolism, pH regulation, invasion and metastasis.

Acknowledgments: This research was supported by grants to SD from the Canadian Institutes of Health Research (CIHR #FDN-143318) and the Canadian Cancer Society Research Institute (CCSRI #703191). No funds were received for covering the costs to publish this paper in open access format.

Author Contributions: P.C.M. and S.D. wrote the paper; M.S. performed the experiments and analyzed the data; P.C.M., S.M. and S.D. reviewed the final manuscript.

Conflicts of Interest: S.D. and P.C.M. are inventors of SLC-0111 and declare no other conflicts of interest. M.S. declares no conflict of interest. The funding sponsors had no role in the design of the study; in the collection, analyses, or interpretation of data; in the writing of the manuscript, and in the decision to publish the results.

References

1. Vander Heiden, M.G.; DeBerardinis, R.J. Understanding the Intersections between Metabolism and Cancer Biology. *Cell* **2017**, *168*, 657–669. [CrossRef] [PubMed]
2. Pavlova, N.N.; Thompson, C.B. The Emerging Hallmarks of Cancer Metabolism. *Cell Metab.* **2016**, *23*, 27–47. [CrossRef] [PubMed]
3. Nakazawa, M.S.; Keith, B.; Simon, M.C. Oxygen availability and metabolic adaptations. *Nat. Rev. Cancer* **2016**, *16*, 663–673. [CrossRef] [PubMed]
4. Xie, H.; Simon, M.C. Oxygen availability and metabolic reprogramming in cancer. *J. Biol. Chem.* **2017**, *292*, 16825–16832. [CrossRef] [PubMed]
5. Corbet, C.; Feron, O. Tumour acidosis: From the passenger to the driver's seat. *Nat. Rev. Cancer* **2017**, *17*, 577–593. [CrossRef] [PubMed]
6. McDonald, P.C.; Winum, J.Y.; Supuran, C.T.; Dedhar, S. Recent Developments in Targeting Carbonic Anhydrase IX for Cancer Therapeutics. *Oncotarget* **2012**, *3*, 84–97. [CrossRef] [PubMed]
7. Neri, D.; Supuran, C.T. Interfering with pH regulation in tumours as a therapeutic strategy. *Nat. Rev. Drug Discov.* **2011**, *10*, 767–777. [CrossRef] [PubMed]
8. Parks, S.K.; Chiche, J.; Pouyssegur, J. Disrupting proton dynamics and energy metabolism for cancer therapy. *Nat. Rev. Cancer* **2013**, *13*, 611–623. [CrossRef] [PubMed]
9. McDonald, P.C.; Chafe, S.C.; Dedhar, S. Overcoming Hypoxia-Mediated Tumor Progression: Combinatorial Approaches Targeting pH Regulation, Angiogenesis and Immune Dysfunction. *Front. Cell Dev. Biol.* **2016**, *4*, 27. [CrossRef] [PubMed]
10. Kaluzova, M.; Kaluz, S.; Stanbridge, E.J. High cell density induces expression from the carbonic anhydrase 9 promoter. *Biotechniques* **2004**, *36*, 228–235. [PubMed]
11. Chen, C.L.; Chu, J.S.; Su, W.C.; Huang, S.C.; Lee, W.Y. Hypoxia and metabolic phenotypes during breast carcinogenesis: Expression of HIF-1alpha, GLUT1, and CAIX. *Virchows Arch.* **2010**, *457*, 53–61. [CrossRef] [PubMed]
12. Panisova, E.; Kery, M.; Sedlakova, O.; Brisson, L.; Debreova, M.; Sboarina, M.; Sonveaux, P.; Pastorekova, S.; Svastova, E. Lactate stimulates CA IX expression in normoxic cancer cells. *Oncotarget* **2017**, *8*, 77819–77835. [CrossRef] [PubMed]

13. Fiaschi, T.; Giannoni, E.; Taddei, M.L.; Cirri, P.; Marini, A.; Pintus, G.; Nativi, C.; Richichi, B.; Scozzafava, A.; Carta, F.; et al. Carbonic anhydrase IX from cancer-associated fibroblasts drives epithelial-mesenchymal transition in prostate carcinoma cells. *Cell Cycle* **2013**, *12*, 1791–1801. [CrossRef] [PubMed]

14. Chiche, J.; Ilc, K.; Laferriere, J.; Trottier, E.; Dayan, F.; Mazure, N.M.; Brahimi-Horn, M.C.; Pouyssegur, J. Hypoxia-inducible carbonic anhydrase IX and XII promote tumor cell growth by counteracting acidosis through the regulation of the intracellular pH. *Cancer Res.* **2009**, *69*, 358–368. [CrossRef] [PubMed]

15. Lou, Y.; McDonald, P.C.; Oloumi, A.; Chia, S.; Ostlund, C.; Ahmadi, A.; Kyle, A.; Auf dem Keller, U.; Leung, S.; Huntsman, D.; et al. Targeting tumor hypoxia: Suppression of breast tumor growth and metastasis by novel carbonic anhydrase IX inhibitors. *Cancer Res.* **2011**, *71*, 3364–3376. [CrossRef] [PubMed]

16. McIntyre, A.; Patiar, S.; Wigfield, S.; Li, J.L.; Ledaki, I.; Turley, H.; Leek, R.; Snell, C.; Gatter, K.; Sly, W.S.; et al. Carbonic anhydrase IX promotes tumor growth and necrosis in vivo and inhibition enhances anti-VEGF therapy. *Clin. Cancer Res.* **2012**, *18*, 3100–3111. [CrossRef] [PubMed]

17. Lock, F.E.; McDonald, P.C.; Lou, Y.; Serrano, I.; Chafe, S.C.; Ostlund, C.; Aparicio, S.; Winum, J.Y.; Supuran, C.T.; Dedhar, S. Targeting carbonic anhydrase IX depletes breast cancer stem cells within the hypoxic niche. *Oncogene* **2013**, *32*, 5210–5219. [CrossRef] [PubMed]

18. Svastova, E.; Pastorekova, S. Carbonic anhydrase IX: A hypoxia-controlled "catalyst" of cell migration. *Cell Adhes. Migr.* **2013**, *7*, 226–231. [CrossRef] [PubMed]

19. Csaderova, L.; Debreova, M.; Radvak, P.; Stano, M.; Vrestiakova, M.; Kopacek, J.; Pastorekova, S.; Svastova, E. The effect of carbonic anhydrase IX on focal contacts during cell spreading and migration. *Front. Physiol.* **2013**, *4*, 271. [CrossRef] [PubMed]

20. Ward, C.; Meehan, J.; Mullen, P.; Supuran, C.; Dixon, J.M.; Thomas, J.S.; Winum, J.Y.; Lambin, P.; Dubois, L.; Pavathaneni, N.K.; et al. Evaluation of carbonic anhydrase IX as a therapeutic target for inhibition of breast cancer invasion and metastasis using a series of in vitro breast cancer models. *Oncotarget* **2015**, *6*, 24856–24870. [CrossRef] [PubMed]

21. Radvak, P.; Repic, M.; Svastova, E.; Takacova, M.; Csaderova, L.; Strnad, H.; Pastorek. J.; Pastorekova, S.; Kopacek, J. Suppression of carbonic anhydrase IX leads to aberrant focal adhesion and decreased invasion of tumor cells. *Oncol. Rep.* **2013**, *29*, 1147–1153. [CrossRef] [PubMed]

22. Swayampakula, M.; McDonald, P.C.; Vallejo, M.; Coyaud, E.; Chafe, S.C.; Westerback, A.; Venkateswaran, G.; Shankar, J.; Gao, G.; Laurent, E.M.N.; et al. The interactome of metabolic enzyme carbonic anhydrase IX reveals novel roles in tumor cell migration and invadopodia/MMP14-mediated invasion. *Oncogene* **2017**, *36*, 6244–6261. [CrossRef] [PubMed]

23. Jamali, S.; Klier, M.; Ames, S.; Barros, L.F.; McKenna, R.; Deitmer, J.W.; Becker, H.M. Hypoxia-induced carbonic anhydrase IX facilitates lactate flux in human breast cancer cells by non-catalytic function. *Sci. Rep.* **2015**, *5*, 13605. [CrossRef] [PubMed]

24. Deitmer, J.W.; Theparambil, S.M.; Ruminot, I.; Becker, H.M. The role of membrane acid/base transporters and carbonic anhydrases for cellular pH and metabolic processes. *Front. Neurosci.* **2014**, *8*, 430. [CrossRef] [PubMed]

25. Parks, S.K.; Pouyssegur, J. The Na^+/HCO_3^- Co-Transporter SLC4A4 Plays a Role in Growth and Migration of Colon and Breast Cancer Cells. *J. Cell. Physiol.* **2015**, *230*, 1954–1963. [CrossRef] [PubMed]

26. McIntyre, A.; Hulikova, A.; Ledaki, I.; Snell, C.; Singleton, D.; Steers, G.; Seden, P.; Jones, D.; Bridges, E.; Wigfield, S.; et al. Disrupting Hypoxia-Induced Bicarbonate Transport Acidifies Tumor Cells and Suppresses Tumor Growth. *Cancer Res.* **2016**, *76*, 3744–3755. [CrossRef] [PubMed]

27. Ahmed, S.; Thomas, G.; Ghoussaini, M.; Healey, C.S.; Humphreys, M.K.; Platte, R.; Morrison, J.; Maranian, M.; Pooley, K.A.; Luben, R.; et al. Newly discovered breast cancer susceptibility loci on 3p24 and 17q23.2. *Nat. Genet.* **2009**, *41*, 585–590. [CrossRef] [PubMed]

28. Lee, S.; Axelsen, T.V.; Andersen, A.P.; Vahl, P.; Pedersen, S.F.; Boedtkjer, E. Disrupting Na^+, HCO_3^--cotransporter NBCn1 (Slc4a7) delays murine breast cancer development. *Oncogene* **2016**, *35*, 2112–2122. [CrossRef] [PubMed]

29. Parks, S.K.; Cormerais, Y.; Pouyssegur, J. Hypoxia and cellular metabolism in tumour pathophysiology. *J. Physiol.* **2017**, *595*, 2439–2450. [CrossRef] [PubMed]

30. Cormerais, Y.; Giuliano, S.; LeFloch, R.; Front, B.; Durivault, J.; Tambutte, E.; Massard, P.A.; de la Ballina, L.R.; Endou, H.; Wempe, M.F.; et al. Genetic Disruption of the Multifunctional CD98/LAT1 Complex Demonstrates the Key Role of Essential Amino Acid Transport in the Control of mTORC1 and Tumor Growth. *Cancer Res.* **2016**, *76*, 4481–4492. [CrossRef] [PubMed]

31. Boyd, N.H.; Walker, K.; Fried, J.; Hackney, J.R.; McDonald, P.C.; Benavides, G.A.; Spina, R.; Audia, A.; Scott, S.E.; Libby, C.J.; et al. Addition of carbonic anhydrase 9 inhibitor SLC-0111 to temozolomide treatment delays glioblastoma growth in vivo. *JCI Insight* **2017**, *2*. [CrossRef] [PubMed]

32. Zhang, J.; Pavlova, N.N.; Thompson, C.B. Cancer cell metabolism: The essential role of the nonessential amino acid, glutamine. *EMBO J.* **2017**, *36*, 1302–1315. [CrossRef] [PubMed]

33. Van Geldermalsen, M.; Wang, Q.; Nagarajah, R.; Marshall, A.D.; Thoeng, A.; Gao, D.; Ritchie, W.; Feng, Y.; Bailey, C.G.; Deng, N.; et al. ASCT2/SLC1A5 controls glutamine uptake and tumour growth in triple-negative basal-like breast cancer. *Oncogene* **2016**, *35*, 3201–3208. [CrossRef] [PubMed]

34. Broer, A.; Rahimi, F.; Broer, S. Deletion of Amino Acid Transporter ASCT2 (SLC1A5) Reveals an Essential Role for Transporters SNAT1 (SLC38A1) and SNAT2 (SLC38A2) to Sustain Glutaminolysis in Cancer Cells. *J. Biol. Chem.* **2016**, *291*, 13194–13205. [CrossRef] [PubMed]

35. Gieling, R.G.; Babur, M.; Mamnani, L.; Burrows, N.; Telfer, B.A.; Carta, F.; Winum, J.Y.; Scozzafava, A.; Supuran, C.T.; Williams, K.J. Antimetastatic effect of sulfamate carbonic anhydrase IX inhibitors in breast carcinoma xenografts. *J. Med. Chem.* **2012**, *55*, 5591–5600. [CrossRef] [PubMed]

36. Yamaguchi, H.; Pixley, F.; Condeelis, J. Pseudopodia and adhesive structures. In *Adhesive Interactions in Normal and Transformed Cells*; Rovensky, Y.A., Ed.; Springer: Berlin, Germany, 2006; pp. 37–56.

37. Yamaguchi, H. Pathological roles of invadopodia in cancer invasion and metastasis. *Eur. J. Cell Biol.* **2012**, *91*, 902–907. [CrossRef] [PubMed]

38. Estrach, S.; Lee, S.A.; Boulter, E.; Pisano, S.; Errante, A.; Tissot, F.S.; Cailleteau, L.; Pons, C.; Ginsberg, M.H.; Feral, C.C. CD98hc (SLC3A2) loss protects against ras-driven tumorigenesis by modulating integrin-mediated mechanotransduction. *Cancer Res.* **2014**, *74*, 6878–6889. [CrossRef] [PubMed]

39. Gould, C.M.; Courtneidge, S.A. Regulation of invadopodia by the tumor microenvironment. *Cell Adhes. Migr.* **2014**, *8*, 226–235. [CrossRef]

40. Leong, H.S.; Robertson, A.E.; Stoletov, K.; Leith, S.J.; Chin, C.A.; Chien, A.E.; Hague, M.N.; Ablack, A.; Carmine-Simmen, K.; McPherson, V.A.; et al. Invadopodia are required for cancer cell extravasation and are a therapeutic target for metastasis. *Cell Rep.* **2014**, *8*, 1558–1570. [CrossRef] [PubMed]

41. Estrella, V.; Chen, T.; Lloyd, M.; Wojtkowiak, J.; Cornnell, H.H.; Ibrahim-Hashim, A.; Bailey, K.; Balagurunathan, Y.; Rothberg, J.M.; Sloane, B.F.; et al. Acidity generated by the tumor microenvironment drives local invasion. *Cancer Res.* **2013**, *73*, 1524–1535. [CrossRef] [PubMed]

42. Gioia, M.; Fasciglione, G.F.; Monaco, S.; Iundusi, R.; Sbardella, D.; Marini, S.; Tarantino, U.; Coletta, M. pH dependence of the enzymatic processing of collagen I by MMP-1 (fibroblast collagenase), MMP-2 (gelatinase A), and MMP-14 ectodomain. *J. Biol. Inorg. Chem.* **2010**, *15*, 1219–1232. [CrossRef] [PubMed]

43. Busco, G.; Cardone, R.A.; Greco, M.R.; Bellizzi, A.; Colella, M.; Antelmi, E.; Mancini, M.T.; Dell'Aquila, M.E.; Casavola, V.; Paradiso, A.; et al. NHE1 promotes invadopodial ECM proteolysis through acidification of the peri-invadopodial space. *FASEB J.* **2010**, *24*, 3903–3915. [CrossRef] [PubMed]

44. Magalhaes, M.A.; Larson, D.R.; Mader, C.C.; Bravo-Cordero, J.J.; Gil-Henn, H.; Oser, M.; Chen, X.; Koleske, A.J.; Condeelis, J. Cortactin phosphorylation regulates cell invasion through a pH-dependent pathway. *J. Cell Biol.* **2011**, *195*, 903–920. [CrossRef] [PubMed]

45. Beaty, B.T.; Wang, Y.; Bravo-Cordero, J.J.; Sharma, V.P.; Miskolci, V.; Hodgson, L.; Condeelis, J. Talin regulates moesin-NHE-1 recruitment to invadopodia and promotes mammary tumor metastasis. *J. Cell Biol.* **2014**, *205*, 737–751. [CrossRef] [PubMed]

46. Amith, S.R.; Fong, S.; Baksh, S.; Fliegel, L. Na$^+$/H$^+$ exchange in the tumour microenvironment: Does NHE1 drive breast cancer carcinogenesis? *Int. J. Dev. Biol.* **2015**, *59*, 367–377. [CrossRef] [PubMed]

47. Amith, S.R.; Vincent, K.M.; Wilkinson, J.M.; Postovit, L.M.; Fliegel, L. Defining the Na$^+$/H$^+$ exchanger NHE1 interactome in triple-negative breast cancer cells. *Cell Signal.* **2017**, *29*, 69–77. [CrossRef] [PubMed]

48. Lloyd, M.C.; Cunningham, J.J.; Bui, M.M.; Gillies, R.J.; Brown, J.S.; Gatenby, R.A. Darwinian dynamics of intratumoral heterogeneity: Not solely random mutations but also variable environmental selection forces. *Cancer Res.* **2016**, *76*, 3136–3144. [CrossRef] [PubMed]

49. Hauff, S.J.; Raju, S.C.; Orosco, R.K.; Gross, A.M.; Diaz-Perez, J.A.; Savariar, E.; Nashi, N.; Hasselman, J.; Whitney, M.; Myers, J.N.; et al. Matrix-metalloproteinases in head and neck carcinoma-cancer genome atlas analysis and fluorescence imaging in mice. *Otolaryngol. Head Neck Surg.* **2014**, *151*, 612–618. [CrossRef] [PubMed]
50. Rosse, C.; Lodillinsky, C.; Fuhrmann, L.; Nourieh, M.; Monteiro, P.; Irondelle, M.; Lagoutte, E.; Vacher, S.; Waharte, F.; Paul-Gilloteaux, P.; et al. Control of MT1-MMP transport by atypical PKC during breast-cancer progression. *Proc. Natl. Acad. Sci. USA* **2014**, *111*, E1872–E1879. [CrossRef] [PubMed]
51. Macpherson, I.R.; Rainero, E.; Mitchell, L.E.; van den Berghe, P.V.; Speirs, C.; Dozynkiewicz, M.A.; Chaudhary, S.; Kalna, G.; Edwards, J.; Timpson, P.; et al. CLIC3 controls recycling of late endosomal MT1-MMP and dictates invasion and metastasis in breast cancer. *J. Cell Sci.* **2014**, *127*, 3893–3901. [CrossRef] [PubMed]
52. Lodillinsky, C.; Infante, E.; Guichard, A.; Chaligne, R.; Fuhrmann, L.; Cyrta, J.; Irondelle, M.; Lagoutte, E.; Vacher, S.; Bonsang-Kitzis, H.; et al. p63/MT1-MMP axis is required for in situ to invasive transition in basal-like breast cancer. *Oncogene* **2015**, *35*, 344. [CrossRef] [PubMed]
53. Pacchiano, F.; Carta, F.; McDonald, P.C.; Lou, Y.; Vullo, D.; Scozzafava, A.; Dedhar, S.; Supuran, C.T. Ureido-Substituted Benzenesulfonamides Potently Inhibit Carbonic Anhydrase IX and Show Antimetastatic Activity in a Model of Breast Cancer Metastasis. *J. Med. Chem.* **2011**, *54*, 1896–1902. [CrossRef] [PubMed]

metabolites

MDPI

Review

Carbonic Anhydrases: Role in pH Control and Cancer

Mam Y. Mboge *, Brian P. Mahon, Robert McKenna and Susan C. Frost *

University of Florida, College of Medicine, Department of Biochemistry and Molecular Biology, P.O. Box 100245, Gainesville, FL 32610, USA; brian.mahon@nih.gov (B.P.M.); rmckenna@ufl.edu (R.M.)
* Correspondence: mammboge@ufl.edu (M.Y.M.); sfrost@ufl.edu (S.C.F.)

Received: 30 November 2017; Accepted: 22 February 2018; Published: 28 February 2018

Abstract: The pH of the tumor microenvironment drives the metastatic phenotype and chemotherapeutic resistance of tumors. Understanding the mechanisms underlying this pH-dependent phenomenon will lead to improved drug delivery and allow the identification of new therapeutic targets. This includes an understanding of the role pH plays in primary tumor cells, and the regulatory factors that permit cancer cells to thrive. Over the last decade, carbonic anhydrases (CAs) have been shown to be important mediators of tumor cell pH by modulating the bicarbonate and proton concentrations for cell survival and proliferation. This has prompted an effort to inhibit specific CA isoforms, as an anti-cancer therapeutic strategy. Of the 12 active CA isoforms, two, CA IX and XII, have been considered anti-cancer targets. However, other CA isoforms also show similar activity and tissue distribution in cancers and have not been considered as therapeutic targets for cancer treatment. In this review, we consider all the CA isoforms and their possible role in tumors and their potential as targets for cancer therapy.

Keywords: tumors; pH; carbonic anhydrases; metalloenzymes; carbonic anhydrase IX; carbonic anhydrase XII; cancer therapeutics; metabolism; tumor microenvironment; drug discovery

1. Introduction

Carbonic anhydrases (CAs) have been studied for over 90 years. Since their first discovery in 1933, CAs have been at the forefront of scientific discovery; from basic enzymology, to the application of structural biology and in silico approaches to study protein dynamics. In addition, CAs have been shown to be important in drug discovery and clinical medicine. CAs comprise a family of metalloenzymes that catalyze the reversible hydration of CO_2, in the presence of water, to HCO_3^- with the release of a proton a mechanism that was first inferred in 1933 [1]. CAs are grouped into six evolutionary distinct classes (α, β, γ, δ, ζ, and η), which implies convergent evolution of a biochemical reaction essential for life processes, encompassing both prokaryotic and eukaryotic organisms and viruses. These ubiquitous enzymes play important roles in ion transport, acid–base regulation, gas exchange, photosynthesis and CO_2 fixation [2,3]. The α-class, which is the best characterized of the six classes, is found primarily in vertebrates. The β-class is present in higher plants and some prokaryotes, γ is expressed in higher plants and prokaryotes, and the δ and ζ have only been observed in diatoms [2,4]. The most recently identified CA family is the η-class, which was discovered in the malaria pathogen, *Plasmodium falciparum* [5].

The human CAs (α-class) share the same 3D tertiary structure, but differ in sequence (Table 1). Furthermore, CAs are expressed in specific tissues and cellular compartments that differ in pH and metabolic rate, properties that drive the contributions of catalyzed CO_2 reactions in many physiological processes. In humans, 15 CA isoforms are encoded. Among these, only 12 coordinate a zinc in the active site making them catalytically active (CAs I–IV, CAs VA–VB, CAs VI–VII, CA IX, and CAs XII–XIV). Isoforms CA VIII, X, and XI are termed CA-related proteins (CA-RPs) as they lack the required metal ion within the active site [6,7]. While the α-CAs were initially purified from bovine erythrocytes, studies in human erythrocytes revealed three CA isoforms designated as A and B, which we now know are identical, and C (now named CA I and CA II, respectively) [8,9]. Amino acid sequences and X-ray crystallography studies for both isoforms were reported during the 1970s [6,7,10,11]. During the same decade, CA III, a sulfonamide-resistant CA isoform, was discovered and purified from rabbit skeletal muscle [12]. In 1979, the secreted CA VI isoform was isolated from the saliva of sheep and later characterized in humans [13,14]. CA IV, a membrane-associated isoform, was purified during the 1980s [15–18], while cytosolic CA VII isoform [19], the membrane-bound isoforms CA IX [20,21], CA XII [22], and CA XIV [23,24], and the mitochondrial isoforms CA VA and CA VB [25–27] were all discovered in the 1990s. During the early 2000s, CA XIII was purified and characterized [28], whereas the CA-RPs were shown to be cytosolic proteins [29].

The α-CAs are involved in many physiological processes including respiration, pH regulation, Na^+ retention, calcification, bone resorption, signal transduction, electrolyte secretion, gluconeogenesis, ureagenesis, and lipogenesis. Due to their broad roles in metabolism, CAs have served as therapeutic targets for several diseases, including glaucoma and epileptic seizures, altitude sickness, and more recently in the treatment of obesity and pain [30–32]. In addition, two of the membrane-bound CAs have been shown to be important for tumorigenesis [3,33–41]. CAs have also gained industrial interest as biocatalysts for carbon sequestration of fuel-gas and CO_2 gas exchange in artificial lungs [42–47]. All of these applications are possible because of the favorable properties of CAs such as "fast" enzyme kinetics, easy expression, high solubility, and intermediate heat resistance [4,48]. The kinetic properties of the CAs are listed in Table 2.

Of the 15 CA isoforms expressed in humans, only CA IX and CA XII have been implicated in cancer. These enzymes are transmembrane proteins in which their extracellular domain contains the catalytic activity, positioning them in the regulation of the tumor microenvironment. CA IX is of particular interest because of its high expression in solid tumors while exhibiting low expression in normal tissues [3,48–50]. Yet, reducing activity of either CA IX/XII activity appears to affect the pH of the tumor microenvironment reducing tumor cell survival and proliferation [33,51]. Taken together, these characteristics make CA IX/XII attractive as anti-cancer targets. Other isoforms may have targeting potential with respect to cancer, but little is known about their specific function even though there is evidence of their expression and upregulation in tumors.

Because of the established roles that both CA IX and CA XII play in the process of tumorigenesis, cancer cell signaling, tumor progression, acidification, and metastasis, many classes of CA IX/CA XII-targeted inhibitors and biologics have been studied in the preclinical setting. These studies have yielded promising results showing that inhibition of CA IX/CA XII catalytic activity decreased the growth, proliferation, and metastatic potential of several aggressive cancers both in vitro and in vivo [36,52–55]. The most successful to date include the use of sulfonamide-based compounds and monoclonal antibodies for the treatment of cancers that overexpress CA IX or CA XII. Please see the extensive reviews published by us and others [32,50,56,57]. Two of the sulfonamide-based inhibitors (SLC-0111 and E7070/Indisulam) are currently in clinical trials. Clinical trials involving immunotherapy using monoclonal antibodies (G250 and its chimeric derivative cG250) alone or combined with other therapeutic techniques are also currently under development [58–61]. In addition, immuno-detection strategies have also been adopted to target CA IX for molecular imaging of tumor hypoxia [62–64]. The goal of detecting hypoxia in a non-invasive manner, could predict patient outcome to drug therapy and serve as a tool to inform treatment decisions.

In this review, the current understanding of CAs role in tumor physiology and pH regulation is discussed. The expression and function of each CA isoform in both normal and tumor cells/tissues, as well as some of the commonly found mutations in tumor specimens, will be discussed. The goal of this review is to provide a comprehensive overview of the CA family and their combined role in cancer and current anti-cancer therapies.

Table 1. Primary sequence identity (%) (italicized, bottom left) and number of conserved residues (right, top) between the catalytically active CA isoforms.

	I	II	III	IV	VA	VB	VI	VII	IX	XII	XIII	XIV
I	-	154	141	78	126	128	82	132	83	91	154	85
II	*60*	-	152	88	133	138	90	147	85	89	157	96
III	*54*	*58*	-	82	120	117	87	130	80	86	151	90
IV	*30*	*33*	*32*	-	89	93	97	90	84	91	84	62
VA	*48*	*51*	*45*	*24*	-	184	93	131	83	84	124	88
VB	*47*	*52*	*43*	*23*	*59*	-	82	134	89	79	131	88
VI	*32*	*33*	*32*	*27*	*28*	*24*	-	93	107	104	90	106
VII	*51*	*56*	*50*	*32*	*48*	*49*	*35*	-	95	103	139	97
IX	*33*	*34*	*31*	*27*	*32*	*33*	*39*	*37*	-	101	90	113
XII	*36*	*34*	*32*	*28*	*32*	*30*	*38*	*33*	*39*	-	91	123
XIII	*59*	*60*	*58*	*28*	*46*	*48*	*33*	*53*	*35*	*35*	-	98
XIV	*34*	*36*	*34*	*29*	*32*	*29*	*36*	*36*	*44*	*46*	*37*	-

Information adapted from Pinard et al. [48].

Table 2. Catalytic efficiency of the CA isoforms.

Isoform	K_{cat} (s^{-1})	K_M (mM)	K_{cat}/K_M (M^{-1} s^{-1})
I	2.0×10^5	4.0	5.0×10^7
II	1.4×10^6	9.3	1.5×10^8
III	1.3×10^4	33.3	4.0×10^5
IV	1.1×10^6	21.5	5.1×10^7
VA	2.9×10^5	10.0	2.9×10^7
VB	9.5×10^5	9.7	9.8×10^7
VI	3.4×10^5	6.9	4.9×10^7
VII	9.5×10^6	11.4	8.3×10^7
IX	1.1×10^6	6.9	1.6×10^8
XII	4.2×10^5	12.0	3.5×10^7
XIII	1.5×10^5	13.8	1.1×10^7
XIV	3.1×10^5	7.9	3.9×10^7

2. pH—The Role of CAs in Tumors

2.1. Differential pH Creates the Ideal Conditions for Tumor Cell Proliferation and Survival

Primary tumors are often described as heterogeneous, in that cells of different types, metabolic states, and stages within the cell cycle can exist at any given time-point [65,66]. This diversity complicates anti-cancer treatment targeting the primary tumors, and becomes even more complicated as tumors reach later stages and/or become highly aggressive. One of the most problematic situations occurs in the context of hypoxia. Tumor hypoxia has been well characterized and is initiated through environmental or genetic factors causing a metabolic shift towards rapid aerobic glycolysis [67,68]. This process is commonly known as the "Warburg Effect" and is defined by increased glucose consumption via glycolysis diverting glucose carbons to lactic acid, even in the presence of oxygen [69,70]. This limits ATP production via oxidative phosphorylation but increases ATP production by glycolysis [71]. There is evidence that oxidative phosphorylation is supported by anepleurotic reactions particularly through glutamate dehydrogenase, which converts glutamate (derived from glutamine) to α-ketoglutarate another commonly observed metabolic alteration in cancer [72–74]. Additionally, glycolysis supports the synthesis of phospholipids by providing the glycerol phosphate backbone via the glycolytic intermediate, dihydroxycetone phosphate. This is important for membrane biogenesis, which underlies the success of cancer cell replication [75]. When tumor cells transition to a hypoxic, or the aerobic glycolytic state, there is a measurable pH difference between the extracellular and intracellular pH (pH$_e$ and pH$_i$, respectively). In part, this is postulated to be due to the over production of lactate because of high glycolytic rates and inhibition of pyruvate decarboxylation in the mitochondria. Export of metabolic acids ultimately lowers pH_e [76].

In most normal cells, a pH differential is maintained between pH_i and pH_e such that the extracellular space maintains a slightly more basic environment ($pH_e \geq 7.3$) relative to the intracellular environment ($pH_i = 7.2$) [77–79]. This gradient permits the function of normal metabolic, transport, and regulatory processes. In hypoxic tumor cells, however, pH_e drops to values ranging from 6.5–7.1 with only a marginal decrease in pH_i (≥ 7.2) [77,78,80]. This activates a cascade of events that provide an advantage for tumor cell survival and proliferation. Specifically, the acidic pH_e becomes favorable for extracellular matrix (ECM) remodeling, limits HCO_3^- dependent dynamic buffering, and induces acid activation and expression of proteases, resulting in the facilitation of tumor cell dissemination and invasion. Additionally, the slight decrease in pH_e favors tumor cell proliferation, metabolic adaptation, migration pathways, and results in evasion of apoptosis. This ultimately sets up conditions that benefit tumor cell survival and proliferation, resulting in an unfavorable prognosis for cancer patients [67,76,78,81]. The increase in pH_i, when compared to the more acidic pH_e, favors flux through glycolysis and inhibition of gluconeogenesis (mostly in the liver and pancreas) [82,83]. Specifically, $pH_i \geq 7.2$ stimulates lactate dehydrogenase (LDH) activity, which has an in vitro pH optimum of ~pH 7.5. This enzyme mediates the conversion of pyruvate to lactate and regenerates NAD^+, which is required for continued glycolytic activity [84,85]. Furthermore, the increased pH_i increases expression of several glycolytic enzymes, thus contributing to the high rate of observed glycolytic activity within the tumor cell. Alternatively, a lower pH_i (<7.2) will reverse these conditions and decrease expression of glycolytic enzymes such as LDH, and transporters like GLUT1 [86,87].

The decrease in pH_e is caused by the rapid extrusion of lactic acid and free protons from tumor cells, resulting from the upregulation of glycolysis in the cytosolic compartment. It has been postulated that glycolytic enzymes cluster at the inner surface of cell membranes and interact with ion and proton transporters at the cell surface. This allows for a rapid transport of protons both in and out of the cell depending on the shift in metabolic equilibrium achieved within the cellular microenvironment [88,89]. In addition, an acidic pH_e establishes a favorable environment for cell metastasis and invasion [90]. Specifically, acidic pH_e enhances expression and activities of ECM reorganizational proteases, such as matrix metalloproteinases (MMPs) and cathepsin B [91,92]. In combination with this, an increased $pH_i \geq 7.2$ creates an environment that favors de novo actin filament formation through the expression and activation of actin-binding proteins such as cofilin, villin, profilin, twinfilin, and talin [93–98]. This, in turn, promotes metastatic and invasive tumor cell behavior [93–98].

This unique pH profile in tumor cells also permits cancer cell proliferation through bypassing cell cycle check points and evasion of apoptotic pathways [99,100]. When $pH_i \geq 7.2$, there is an increase in the activity of CDKs, specifically CDK1, which increases the efficiency of MAPK pathways [101]. This stimulates the rate of progression through the G2/M phase and into the S phase, where tumor cells become more adapted to proliferation and less sensitive to chemo- and radiation therapies [101]. In addition, the increased pH_i suppresses DNA damage checkpoints that would typically slow the progression of a cell through the G2/M phase and restrict proliferation [78]. In combination with this, a $pH_i \geq 7.2$ favors a suppression of apoptotic pathways [100,102]. In normal cells (where $pH_i = 7.2$), a reduction in pH_i to < 7.2 would result in a conformational change in the pro-apoptotic factor, BAX, causing its activation and interaction with the mitochondrial membrane [103]. This interaction causes the release of cytochrome c from the inner mitochondrial membrane and activation of other pro-apoptotic factors such as the caspases [100]. Caspase activity achieves optimal efficiency near pH 6.8 in vitro. When the pH_i becomes slightly more alkaline, these pathways are suppressed, allowing the tumor cells to resist apoptosis [100]. In addition, with a $pH_i \geq 7.2$, which is the case in most tumor cells, there is a high probability that this anti-apoptotic pH level will be maintained, even in cases where there may exist a small influx of protons. Taken together with pH-induced ECM and metabolic transitions, it is clear that the unique pH differential across membranes drives tumor cell proliferation and survival.

2.2. CAs' Role in Creating a Tumor Cell pH Differential

If we can understand the complex factors that establish the unique pH environment of tumor cells, it may be possible to develop therapeutics strategies that can reduce or inhibit these factors to re-normalize pH. In recent years, the carbonic anhydrase family has been shown to create and maintain the pH differential in tumor cells [104]. Evidence over the last two decades has determined CAs play a pivotal role in tumor cell metabolism, migration and invasion, and also in cell survival [3,50,105]. This has prompted efforts by several groups to determine targeting strategies against CAs to inhibit cancer progression. Despite the progress that has been made, there remains many unanswered questions about the role CAs play in tumor cell biology, and the mechanistic details that correlate CA inhibition with the observable therapeutic effects on cancer.

It is widely accepted that CAs are the driving force in pH regulation in primary tumor cells as they are for normal tissue. These enzymes function, through their hydratase/CA activity, to regulate the production of bicarbonate, the universal physiological buffer. Along with this, CAs produce or sequester protons. In addition to their hydratase activity, CAs also have a slower esterase activity, which is mediated by the same catalytic pocket with a mechanism similar to that of the hydratase/CA activity [106]. Thus, many investigators use esterase activity as an indicator of the hydratase activity [106,107]. The role of this esterase reaction in cancer, however, is currently unknown. The two-step CA reaction mechanism is given below (Equations (1) and (2)), and reviewed in detail by Lindskog and Coleman [108], and again by Frost and McKenna [3]. The process, common to reversible biochemical pathways, can be thought of as a two-step equilibrium, whereby substrate/product concentrations determine the reaction direction. In most normal tissues, it is predicted that this results in a constant supply of buffering agents for the maintenance of physiological pH. How does this process contribute to distinct regulation in the hypoxic tumor cell? Two hypotheses have been proposed to describe how CA activity can regulate and foster the unique pH differential within a tumor cell.

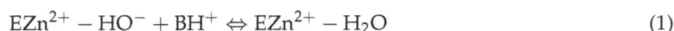

$$EZn^{2+} - HO^- + BH^+ \Leftrightarrow EZn^{2+} - H_2O \tag{1}$$

$$EZn^{2+} - H_2O + HCO_3^+ \xoverset{H_2O}{\Longleftrightarrow} EZn^{2+} - HCO_3^- \Leftrightarrow EZn^{2+} - OH^- + CO_2 \tag{2}$$

Hypothesis 1 (Figure 1A). *This hypothesis suggests cooperative activity between cytosolic (such as CAs I and II) and extracellular CAs (IX, XII, and perhaps XIV) that 'cycle' substrates for pH regulation between the extracellular and intracellular tumor environment. Specifically, extracellular CAs take advantage of the large quantity of CO_2 present in the extracellular space, and convert this to HCO_3^- and protons. Adjacent ion transporters, such as anion exchangers (AE1–AE3), can then traffic HCO_3^- [109] into the cytosol while the free protons remain outside the cell thus lowering the pH_e. Inside the cell, the newly imported HCO_3^- can be converted back to CO_2, which can be used in metabolic pathways or diffuse back outside of the cell carrying that sequester proton. This restores pH_i to more normal levels (Figure 1A) [30,57,110]. This pathway relies on several assumptions. First, extracellular CA activity must remain active at both an alkaline pH of 7.4 and as the pH_e decreases (<6.5). Originally, it was proposed that CA IX, one of the most prevalent extracellular CAs in hypoxic tumors, retained its activity in acidic and basic environments through the use of an 'internal buffer', which exists as an extended N-terminal domain, called the PG domain [111]. The PG domain of CA IX has a large quantity of aspartate residues [111]. It is thought that these residues could act as titratable proton sinks that prevent protons from interacting with the active site of the enzyme and therefore allow it to retain activity in acidic environments [111,112]. Recent evidence has determined that the PG domain is not necessary for the enzyme to retain activity in acidic conditions, as the catalytic domain alone of CA IX can retain activity in pH as low as 5.0 [51].*

The above hypothesis also relies on the ability of CA IX (or other membrane-associated CA) to cluster close enough to bicarbonate transporters, such that newly synthesized HCO_3^- can be readily transported into the cell rather than diffuse away. This is supported by recent evidence showing that glycolytic enzymes, in association with CA, cluster near transporters at the cellular membrane [89,113–115]. However, for this hypothesis to be valid in hypoxic tumors, it would also have to account for the protons produced by extracellular CAs, and a rationale for how these free protons are not rapidly transported into the cytosol given there is an established intra/extracellular proton concentration gradient and the reversible nature of some transporters. It is possible that there is a significant difference in proton concentration between the intra- and extracellular space, where both transporters and glycolytic enzymes are clustered, and as a result, proton expulsion is favored rather than uptake into the cell. This would account for the observed pH differential within the tumor microenvironment and further reinforce the feasibility of this hypothesis depicting the role CA plays in hypoxic tumor cells.

Another factor that must be taken into account is extracellular lactate concentration and its effect on CA IX activity. In 2001, Innocenti et al. showed that CA IX has low sensitivity to inhibition by lactate ($K_i > 150$ mM) representing an evolutionary adaption of the enzyme to harsh conditions such as high lactate levels, increased acidity and hypoxia [116]. It is also interesting to note that pyruvate, the oxidized analog of L-lactate, is a CA IX inhibitor (K_i of 1.12 mM) more potent than HCO_3^- (K_i of 13 mM) where as its reduced form, under hypoxic conditions, is not [116]. These data suggest an evolutionary adaptation of CA IX to assure an efficient CO_2 hydration activity in tumors exposed to more harsh conditions than in the physiological setting.

Hypothesis 2 (Figure 1B). *This hypothesis is similar to Hypothesis 1, in that both extracellular transmembrane CAs and cytosolic CAs play a role in tumor cell pH regulation. However, in this alternative hypothesis, it is proposed that extracellular CAs do not necessarily lower pH_e and produce substrate (HCO_3^-) for cytosolic CAs, but instead buffer pH_e to levels (between 6.5–7.1) that promote tumor cell survival and avoid tumor necrosis. In this case, the tumor cell has begun to transition to a hypoxic state and its metabolic shift toward glycolysis has taken place. This causes a rapid extrusion of lactic acid and a drastic reduction in pH_e, which has been recorded at levels as low as 5.5. At this pH, the cell would soon undergo necrosis resulting in its death. But we know that cancer cells continue to survive and proliferate within the acidic microenvironment. It is postulated that the extracellular CAs act to sequester protons in the form of CO_2 to raise and maintain the pH_e to a level that favors tumor cell growth, proliferation, and survival (Figure 1B). This has been postulated by work from Li et al. [33] and Mahon et al. [51] These studies showed that extracellular CA activity, especially CA IX, from both purified enzyme and in membranes isolated from MDA-MB-231 cells (which display hypoxia-dependent CA IX upregulation), favors HCO_3^- dehydration at lower pH. Thus, CA IX under proton load will sequester protons through its catalytic activity in the form of CO_2 in the extracellular space, which increases pH_e. Together, these studies have shown that CA IX is able to both retain activity at low pH levels (pH < 5.0) and preserve the core, folded structure (as low as pH = 2.0). Both of these attributes are necessary for the enzyme to stabilize pH_e in an acidic environment. Sweitach et al. has also suggested that the net effect of CA IX on pH_e will depend on the emission of CO_2 relative to lactic acid [117]. In their studies, CA IX was shown to be important in the formation of extracellular and intracellular pH gradients in multicellular spheroid growths of cancer cells [117]. When CA IX expression was reduced, spheroids developed very low pH_i (~6.3) and pH_e was measured at pH ~6.9 at the core [117]. However, with CA IX expression, an increase in pH_i (~6.6) was observed and extracellular acidity was increased (pH_e ~6.6) [117]. While these data appear to support Hypothesis 1, the authors conclude that the activity of CA IX can reduce pH_e but the direction of the reaction is ultimately dependent on substrate availability, i.e., lactate levels and the pH of the tumor microenvironment.*

2.3. Further Considerations

Recent evidence found by Jamali et al. [68] has suggested that extracellular CA acts to sequester protons via its catalytic histidine residue (200 using CA IX numbering, commonly referred to as His 64 using CA II numbering). Specifically, the study finds that a single monomer of an extracellular CA can sequester a proton via residue (Figure 2). This result favors Hypothesis 2, whereby extracellular CAs act more favorably to raise pH_e in tumors. Furthermore, this result suggests that the presence of the PG domain of CA IX has limited involvement in pH regulatory processes, in contrast to previous thought. If it is true that extracellular CA IX acts to regulate pH through sequestering a single proton via a His residue then several questions arise: (1) what is the maximum pH_e regulating potential of all extracellular CAs on the tumor cell surface? And (2) is this significant enough to contribute to the pH differential observed in tumor cells, which would validate this hypothesis? To answer these questions, the pH regulating potential, or change in pH_e (ΔpH_e), induced by an extracellular CA acting as a proton-sequestering agent can be estimated. Consider that the concentrations of CA IX and/or XII dimer (common extracellular CAs in tumor cells) is ~19 nM for a single hypoxic tumor cell with a diameter = 20 μm and an initial pH_e of 6.5. In addition, assume that a monomer of extracellular CA can sequester protons in a ~1:1 ratio via a single active site histidine. With these assumptions, the following expression can be used to establish the relationship between pH and proton concentration in a solution:

$$\Delta pH_e = -log([H^+]_{tot} - n[CA]) - pH_e. \tag{3}$$

Here, $[H^+]_e$ represents the concentration of total protons in the extracellular environment, $n[CA]$ represents the CA concentration multiplied by an integer that is equivalent to the CA oligomeric state (considering CA IX and XII are dimers, this would be $n = 2$), and pH_e represents the initial pH of the extracellular environment (in this case is 6.5). Considering only these parameters, the total quantity of extracellular CA in a single tumor cell contributes an overall ΔpH_e of 0.05 units over time. Considering that a change in pH of ~0.1 units in cellular environments can drastically change a cell's behavior and functions [118], this contribution becomes significant. Of course, this also indicates, as predicted, that other factors also contribute to the pH differential and regulatory processes in tumor cells. It should also be emphasized that this is an approximate estimation and relies on several assumptions that may not fully reflect the tumor cell at any given time. However, it allows us to consider the maximum effect that these CA-proton sequestration processes may have on tumor cell pH regulation.

It is also important to note that CAs catalyzes a reversible reaction and this reaction depends on substrate/product availability. Hypothesis 2 takes this bi-directionality into account, hence the buffering capacity of extracellular CAs. For instance, at pH values lower than ~6.8, CA IX has been shown to favor its dehydration reaction ($H^+ + H_2CO_3^- \rightarrow CO_2 + H_2O$) over hydration ($CO_2 + H_2O \rightarrow H^+ + H_2CO_3^-$), while the opposite is expected to happen at much higher pH [33]. This evidence supports the notion that membrane bound CAs, especially CA IX (pKa of Zn-H$_2$O ~6.3), work not only to prevent the unfavorable acidification of the tumor microenvironment, as a result of the metabolic switch, but to "adjust" the pH_e to favor tumor growth, progression, and eventually metastasis. The role of cytosolic CAs in both hypotheses remains similar. Cytosolic CAs will utilize available substrates in the form of CO_2, water, HCO_3^-, and protons, allowing them to buffer the intracellular environment to maintain the slightly alkaline pH that is observed.

It is possible that each of these hypotheses in defining the role of CAs in the differential pH environment of a tumor cell have validity. For instance, the fate of extracellular CO_2 converted to HCO_3^- can be rationally explained by Hypothesis 1, particularly at the initiation of tumor growth where hypoxia is not an issue, as this explains the location of CO_2 and HCO_3^- both in the intra- and extracellular environment. Alternatively, the process of proton sequestration and the role of extracellular CA activity in reducing and maintaining pH_e is more rationally explained by the Hypothesis 2, particularly with the support from recent experimental evidence [33,51,113]. Of course, more experiments will be needed to address the strength and weaknesses of these hypotheses in explaining the role of CA in tumor cells, which will drive a more accurate model of the function of these enzymes in the microenvironment. This becomes more important as the effort toward creating CA targeting anti-cancer therapies increases and as more compounds progress through clinical trials.

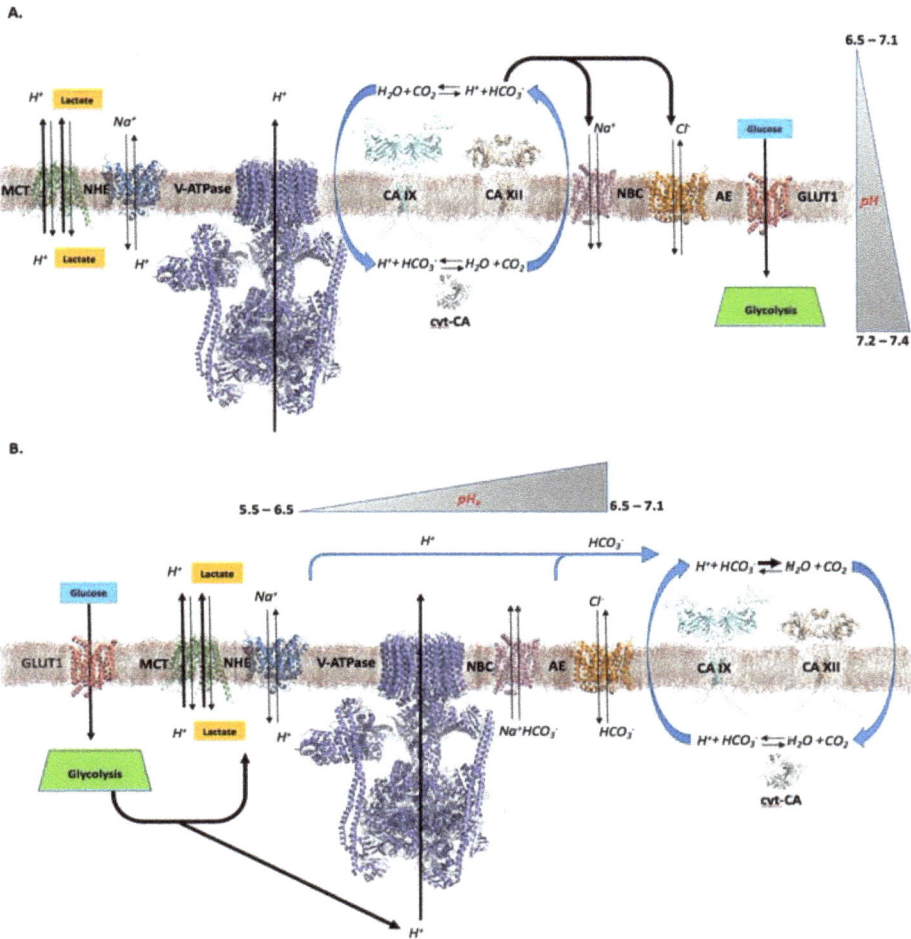

Figure 1. Illustrations of Hypothesis 1 (**A**) and Hypothesis 2 (**B**) to model the functional role of CA in the hypoxic tumor microenvironment. (**A**) Hypothesis 1: CA IX and XII (cyan and tan, respectively) are located on the extracellular membrane and adjacent to transporters involved in pH regulation. Here, CA IX and XII act in conjugation with cyt-CAs (grey) to cycle substrates of water, CO_2, HCO_3^-, and protons, to maintain the differential pH microenvironment. (**B**) Hypothesis 2: Extracellular CA IX and XII act to raise pH_e that has been reduced to levels < 6.0 due to excess expulsion of glycolysis byproducts. In this case, HCO_3^- dehydration and proton sequestration are catalytically favored by CA IX and XII. Alternatively, cyt-CAs act to convert excess CO_2 to HCO_3^- to buffer pH_i and provide substrates to be transported to the extracellular surface to be utilized by CA IX and XII. Transporters shown are MCT (green), NHE (blue), V-ATPase (purple), NBC (pink), AE (orange), and GLUT1 (red). Structural models were generated using PyMol [119]. Models of CA IX and XII were generated using PDBs 5DVX [51] and 1JCZ [120], respectively. PDB 3J9T was utilized to model V-ATPase [121], and PDB 4YZF [122] was utilized to generate models of AE, NBC, and NHE. For modeling of MCT and GLUT1, PDBs 1PW4 [123] and 4PYP [124] were used, respectively. To represent cyt-CAs, PDB 3KS3 [125] was utilized.

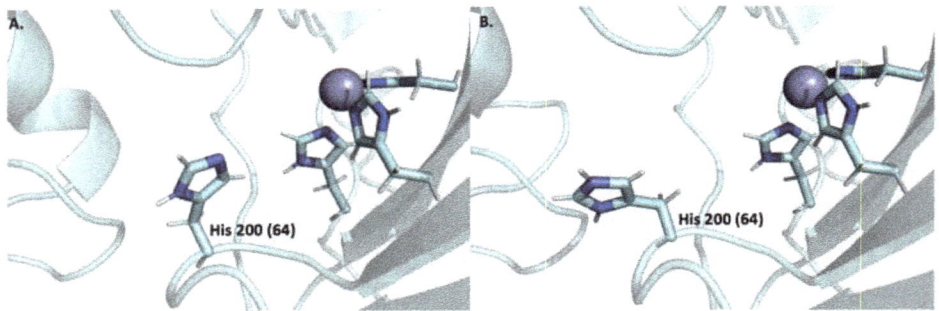

Figure 2. Positions and protonation states of CA IX catalytic His 200 (64 with CA II numbering) that has been implicated in proton sequestration in the acidic tumor microenvironment. (**A**) Deprotonated His 200 (PDB: 4Q49) in an "in" position at neutral pH = 7.5. (**B**) Protonated His 200 (PDB: 4Y0J) in an "out" position at acidic pH (pH = 6.0) representative of a "proton sequestered" state as in the acidic tumor microenvironment. Neutron crystallography structures of CA II were used to determine protonation states and positions and were previously determined by Fisher and McKenna [126].

3. Cytosolic CA Isoforms: Expression, Distribution, and Function

The eight active cytosolic CA (cyt-CA) isoforms include CA I, II, III, VII, VIII, X, XI, and XIII [3,48]. Because there is limited knowledge of the roles CA-RPs play in terms of normal or tumor physiology, they will not be further discussed in this review. The cyt-CA isoforms have >50% primary sequence identity (Table 1). The greatest number of conserved residues are between CAs I and II, both of which are highly expressed in red blood cells (Table 1) [127–129]. The cyt-CAs are ubiquitously distributed in human tissue and show diverse functional roles despite having conserved enzymatic activity. Expression of cyt-CAs has been observed in red blood cells, kidneys, skeletal muscle, adipose tissue, colon liver, brain, and neurons [39,127–132]. Primarily, functions of cyt-CAs include maintaining physiological pH of the blood through production of HCO_3^-, often through interactions with transporters to facilitate efficient HCO_3^-/proton flux together they are called transport metabolons [133–135]. In addition, cyt-CAs contribute to maintaining normal cell metabolism and also show involvement in neuronal excitement and signaling pathways [135]. Interestingly, specific cyt-CA function correlates with tissue distribution.

3.1. Expression and Function of CA I and CA II in Normal Cells

High levels of cyt-CAs I and II expression are found in red blood cells and are necessary for maintaining physiological pH of the blood. It is suggested that CA II activity is the more dominant because of its greater expression and increased catalytic efficiency (in vitro) compared to CA I, although the intracellular environment might dictate how these enzymes behave in vivo (Table 2). In addition to red blood cells, both CA I and II are expressed in the GI tract, lungs, bone marrow, the eye, and also in the extracellular exosome-enriched fraction from normal human urine [136–139]. On its own, CA II expression has been observed in kidney, liver, brain, salivary gland, testis, and minimal expression observed in breast tissue [137–140]. CA II function is essential for bone resorption and osteoclast differentiation, and for regulation of fluid secretion into the anterior chamber of the eye. Recent evidence has implicated CA II activity in association with and activation of several ion transporters, which suggests that CA II acts as mediator of certain metabolic pathways by providing additional substrates for the transporters to balance cytosolic pH. As a result of these interactions, CA II is associated with several diseases including glaucoma, renal tubular acidosis, cerebral calcification, cardiomyocyte hypertrophy, growth retardation and osteoporosis [31,130–132]. Abnormal levels of CA I expression in the blood are a marker for hemolytic anemia [131]. Taken together, these show that in comparison to CA I, CA II expression is abundant in normal tissues and consequently has well-established and important physiological functions.

3.2. Expression and Function of CA I and CA II in Tumor Cells

The potential of CA I as a tumor-associated isoform has not been extensively studied. However, according to multiple genomic databases including the cBioPortal of the cancer genome atlas (TCGA) and the human protein atlas, medium to high levels of CA1 mRNA was detected in acute myeloid leukemia, colorectal cancer, and renal carcinoma patients, based on RNA sequencing [136,139,141,142]. Immunohistochemical (IHC) staining of malignant tissues showed strong cytoplasmic and nuclear CA I staining in a few lymphomas and medium to high staining in renal cancer, melanomas, and stomach cancers [136,138]. Recent studies have also shown that CA I contributes to mammary microcalcification, tumorigenesis, and migration in breast cancer [143]. Furthermore, high CA I expression was observed in the sera of stage I non-small cell lung cancers (NSCLC), suggesting that CA I can be used as a potential biomarker for early detection of NSCLC [144].

CA I has also been shown to be upregulated in human pancreatic cancer (PDAC) where its expression correlated with tumor de-differentiation. Biopsies of patients with PSA positive gray-zone levels (serum prostate-specific antigen levels ranging from 4 to 10 ng/mL is considered a diagnostic gray zone for detecting prostate cancer) also tested positive for plasma CA I [145]. This suggests that CA I can potentially serve as a plasma biomarker, and the combination of PSA and CA I detection may have great advantages for diagnosing prostate cancer in patients with gray-zone PSA levels [145]. The most common alterations in cancer for the CA1 gene are amplifications, as high as 40%, in neuroendocrine prostate cancer, prostate adenocarcinomas and metastatic cancers [141,142]. Recently, CA1 gene amplification was detected in approximately 25% of breast cancer studies [141,142]. Other gene alternations include mutations (missense and truncations) and gene deletions. The most frequent mutations were observed in the mRNA of breast, lung and melanoma patients (Figure 3A) [141,142]. However, it is currently unknown if these mutations result in changes in gene expression or affect activity.

Unlike CA1, detection of CA2 mRNA expression using RNA sequencing showed a more widespread upregulation in cancers. These include but are not limited to prostate, melanomas, bladder, thyroid, breast, lung, liver, pancreas, gliomas (with the highest expression observed in glioblastomas), renal cell carcinomas, and head and neck cancers [141,142]. Like CA1, the most common gene alterations in CA2 were amplifications especially in neuroendocrine prostate cancer, breast cancer, prostate adenocarcinomas and metastatic cancers followed by mutations (Figure 3B) [141,142]. This infers an increase in expression and activity, although this has not been measured. IHC showed strong cytosolic staining in gastric, pancreatic, and cervical cancers, and medium staining in breast, renal, and liver cancers [136–138,140]. As a biomarker, low CA II protein expression is often associated with tumor aggressiveness and poor prognosis in some cancers including pancreatic ductal adenocarcinomas (PDAC), colorectal, gastric and gastrointestinal stromal cancers [35,146–148]. Therefore, this isoform can be considered both a diagnostic and an independent prognostic factor for favorable outcome and overall survival in the aforementioned cancers. However, in other cancers (such as astrocytomas, oligodendrogliomas, melanomas, pulmonary endocrine tumors, and breast cancer) CA II upregulation is associated with poor prognosis, tumor progression, and metastasis [146,149–151]. Thus both CAs I and II may be potential targets for the treatment of many cell-type specific cancers, Because we are in the early stages of developing targeting strategies, care should be taken not to overestimate the role of CA I and CA II in the regulation of pH and tumor growth, nor as prognosticators or targets, until these strategies, are further documented.

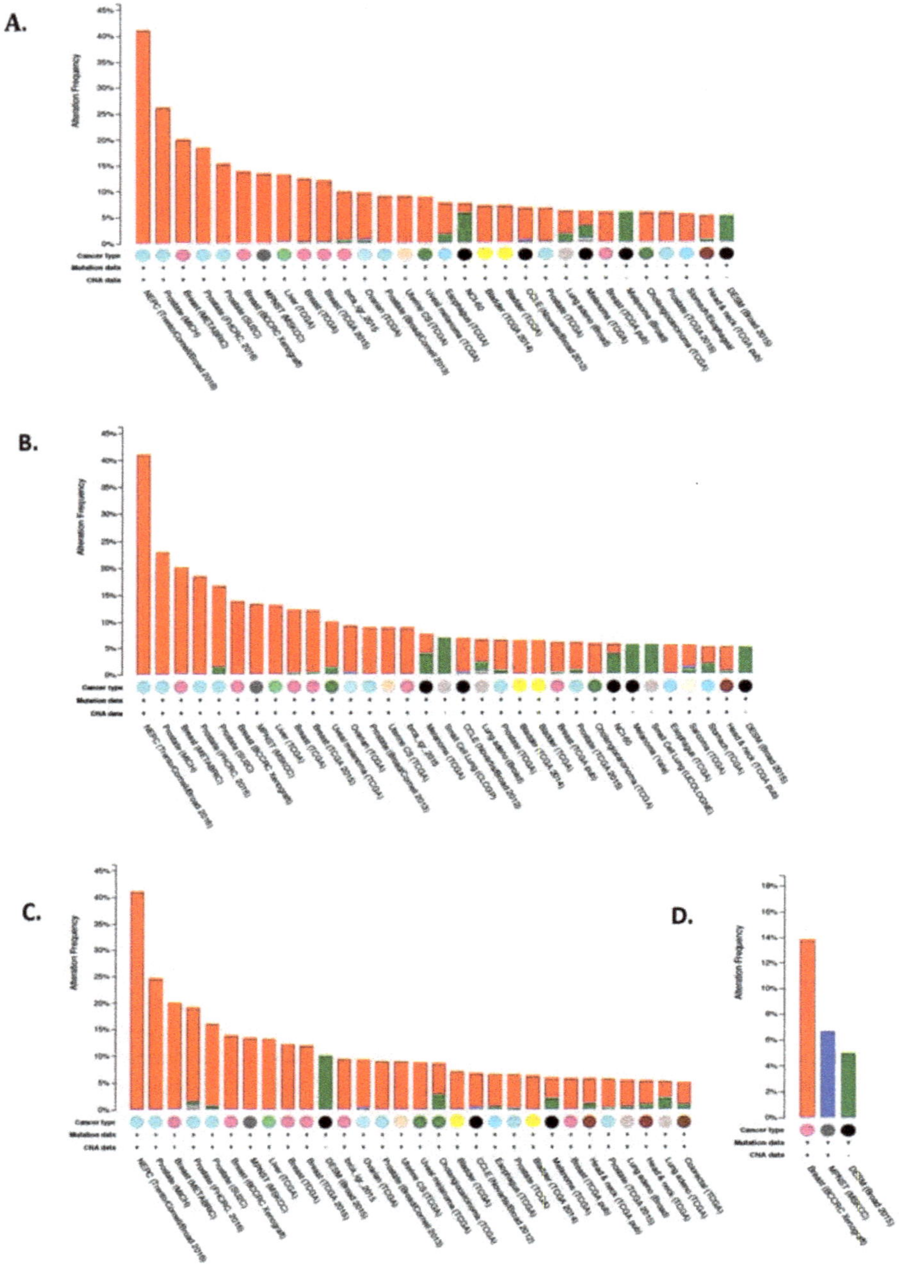

Figure 3. Summary of cross-cancer alterations for CA isoforms. (**A**) CA I, (**B**) CA II, (**C**) CA III, and (**D**) CA VII. Gene amplifications are shown in red, mutations in green, deletions in blue, and multiple alterations in gray. The minimum percent altered samples threshold was set to 5%. The data and results were obtained from the cBioPortal of the human genome atlas (TCGA) [141,142]. Experimental details of each study referenced can also be found in the TCGA database.

3.3. Expression and Function of CA III and CA VII in Normal Cells

In normal cells, CA III expression has been detected in both skeletal muscles and adipose tissues (both white and brown fat), with medium expression also observed in the breast [136,137,152]. CA III displays a 200-fold decrease in catalytic activity compared to CA II and is considered the slowest isoform in terms of its catalytic activity (Table 2) [48]. This difference in activity has led to the hypothesis that CA III might serve different physiological roles unrelated to its primary catalytic function. These include gene regulation, adipogenesis, metabolism, and protection in response to oxidative stress. Recent studies have also shown high CA III expression in osteocytes where its expression is regulated by parathyroid hormone both in vitro and in vivo. In that capacity, it functions to protect osteoclasts from hypoxia and oxidative stress [153]. CA VII, which is one of the least characterized CA family members, is expressed primarily in the colon, liver, skeletal muscle and brain. It has the highest esterase activity among the CA family members. CA VII, like CA III, has been suggested to play a role as an oxygen free radical scavenger, because of the presence of two reactive cysteines that can be glucothionylated. CA VII has also been implicated in neural excitation, seizures and in that regard may represent a drug target for treatment of seizures and neuropathic pain [3,4].

3.4. Expression and Function of CA III and CA VII in Tumor Cells

There are few published studies that have focused on the expression and function of CA III and CA VII in tumors. However, according to the RNA sequencing and IHC data deposited to the TCGA and human protein atlas respectively, CA3 mRNA transcripts were observed in multiple cancers, with the highest median expression in glioblastomas and thyroid cancers [135–139,141,142]. Although no CA III mutations in the aforementioned cancers were described, mutations were observed in lung, melanomas, and head and neck cancers and the most common mutations in these cancers include missense mutations (Figure 3C). Since two of the aforementioned cancer types are high among smokers, these mutations may be attributed to smoking. Ultraviolet damage from the sun may also occur in melanoma patients, although still speculative at this point. Furthermore, IHC showed strong cytoplasmic and nuclear staining in one study of renal cancer along with a few basal cell carcinomas of the skin [136,139]. Like CAs I and II, the most frequent alterations observed were gene amplifications, which occurred in prostate and breast cancers [141,142].

The median expression for CA7 mRNA transcript in tumors is much lower than the previously mentioned isoforms, with the highest expression observed in thyroid carcinoma, colorectal adenocarcinoma, and lower-grade brain gliomas [141,142]. In the latter, CA VII upregulation may act as a marker for poor prognosis. The most frequent gene alterations are also different from the previously mention cytosolic isoforms, because they are low in abundance and are more varied from tumor to tumor (Figure 3D). The highest alteration, which is CA7 amplification, was observed in breast cancer patient xenografts followed by deletions in malignant peripheral nerve sheath tumors and then mutations in desmoplastic melanoma (Figure 3D) [141,142]. IHC staining of CA VII showed weak to moderate cytoplasmic and occasional nuclear expression [136]. However, a few cases of ovarian and gastric cancers exhibited strong staining [136,139].

A study published by Kuo et al. discovered reduced levels of CA I, II and III in human hepatocellular carcinoma (HCC) compared to adjacent normal tissue. In 2008, a study showed that CA III expression promotes the transformation and invasive capacity of hepatoma cells through the focal adhesion kinase (FAK) signaling pathway [154]. In this study, it was hypothesized that CA III is re-expressed in later stages of metastatic progression of HCC, and it might have an important influence in the development of metastasis in liver cancer [154]. Since then, no other studies have been published specifically looking at CA III expression and function in cancer. A recent study has, however, shown that CA VII expression has some prognostic value in colorectal carcinoma (CRC) [155]. CA VII expression was frequently downregulated in CRC tissues at both the mRNA and protein levels. Decreased expression of CA VII was significantly correlated with poor differentiation, positive lymph node metastasis, advance TNM (T-refers to the tumor size, N-refers to 'node' status and M-refers to 'metastasis') stage and unfavorable clinical outcome [155]. This suggests that CA VII can be used an independent prognostic indicator for patients with early stage CRC, and CA III may serve as a therapeutic target in the treatment of metastatic liver.

3.5. Expression and Function of CA XIII in Normal and Tumor Cells

Human CA XIII isoform was first identified and characterized in 2004 [28,156]. It is considered a slow isoform in terms of its catalytic activity (1.5×10^5) comparable to CA I (Table 2). CA XIII expression is observed in several tissues including kidney, brain, lymph nodes, thyroid, liver, GI tract, skin, adipose, soft tissue, and in both male and female reproductive organs [39,136,138]. It has been hypothesized that CA XIII plays a role in pH regulation of reproductive processes including maintenance of sperm mobility and the normal fertilization process. To date, no direct proof for a significant physiological function for CA XIII has been reported. However, downregulation of CA XIII has been seen in cases of colorectal cancer and the lowest signal was detected in carcinoma samples, although the clinical significance of these observations is yet to be determined [39,136,139].

Nonetheless, the downregulation of cytosolic CA I, II and XIII in colorectal cancer may result from reduced levels of a common transcription factor or loss of the closely linked CA1, CA2 and CA13 alleles on chromosome 8. According to IHC data submitted in the human protein atlas, most cancer tissues showed weak to moderate CA XIII immunoreactivity. However, strong staining was observed in renal and pancreatic cancers. Most thyroid cancers, several gliomas, gastric, and liver cancers showed weak to moderate cytoplasmic staining, while melanomas, lung, and skin cancers were only weakly positive [136,138]. Similar to the IHC data, CA13 mRNA expression was most highly upregulated in thyroid, RCC, lung, pancreas but additionally colorectal and testicular germ line cancers [141,142]. The most common alteration, however, was gene amplification, which was observed in prostate and breast cancers (Figure 4A). Although not much attention has been given to CA XIII with regard to cancer because of its high expression in normal cells, it could have some prognostic and diagnostic value.

4. Mitochondrial CA Isoforms: Expression, Distribution, and Function

The first mitochondrial α-CA isoform discovered and isolated from guinea pig liver was CA V [25]. Two different transcripts of this isoform were later identified and coined CA VA and CA VB [26,27,157]. These two transcripts have 59% primary sequence identity and 184 conserved residues (Table 1). This number is slightly lower than that observed between CA isoforms I and II. Although, the tissue-specific distribution pattern between the two is significantly different, CA VA and CA VB are the only two isoforms exclusively expressed in the mitochondrial matrix of hepatocytes and adipocytes, respectively [157]. Interestingly the human ortholog for CA5B, which has broad tissue expression, has been mapped to chromosome Xp22.1, while CA5A was mapped to 16q24 [25,26]. Their roles within specific tissues include ureagenesis and lipogenesis but may also serve as mediators in several other metabolic pathways as discussed below. These functions suggest that both enzymes could be considered as anti-obesity and anti-diabetic drug targets, studies of which are currently being pursued [158–161].

4.1. Expression and Function of CA VA and CA VB in Normal Cells

Mitochondrial isoform CA VA has been shown to be directly associated with ureagenesis [3,4,162]. CA VA produces bicarbonate, which is a substrate for carbamoyl phosphate synthetase I in the synthesis of carbamoyl phosphate, the rate-limiting step of ureagenesis [162]. Bicarbonate production by CA VA can also drive other biosynthetic reactions like that of pyruvate carboxylase, which mediates an important anepleurotic step in the Krebs cycle from which substrates can be drawn for biosynthetic reactions, including gluconeogenesis in the liver. This indicates that CA VA can act as a key mediator in several metabolic pathways of the liver, the only organ in which it is normally expressed [163–165]. Conversely, CA VB has broad tissue distribution even though it is the only CA isoform found in the mitochondria of adipocytes [166]. High expression of this isoform is also observed in the mitochondria of adrenal glands, tonsils, lymph nodes, spleen, liver, colon, and testis, and medium expression observed in other organs including the brain, lungs, muscles, GI tract, breast, and skin. CA VB functions within the adipocytes like CA VA: it stimulates pyruvate carboxylase activity, thereby increasing substrate levels for drawing off citrate for cytoplasmic transport [166,167]. Deficiencies resulting from alterations in CA5A, which decrease its enzymatic activity, causes hyperammonemia in early childhood [168].

4.2. Expression and Function of CA VA and CA VB in Tumor Cells

Although few studies have shown expression of CA VA and CA VB in cancer, the TCGA database shows high expression of CA5A in an RNA sequence study performed with liver hepatocellular carcinoma patient samples [141,142]. The highest gene alterations observed were amplifications, as seen in breast cancer patient xenografts (~17%) and pancreatic cancer (~6%), and deletions in prostate adenocarcinomas and metastatic cancer (~10%) (Figure 4B). No information is currently available for IHC detection of CA VA in tumors, either because none was observed or the studies are yet to be performed. CA5B mRNA upregulation in cancer is more widespread than CA5A with highest expression observed in acute myeloid leukemia, prostate, and renal cell carcinomas [141,142]. Once again, the most common alterations are amplification; in prostate and breast cancers, mutations were detected, including frameshift and missense mutations (Figure 2C).

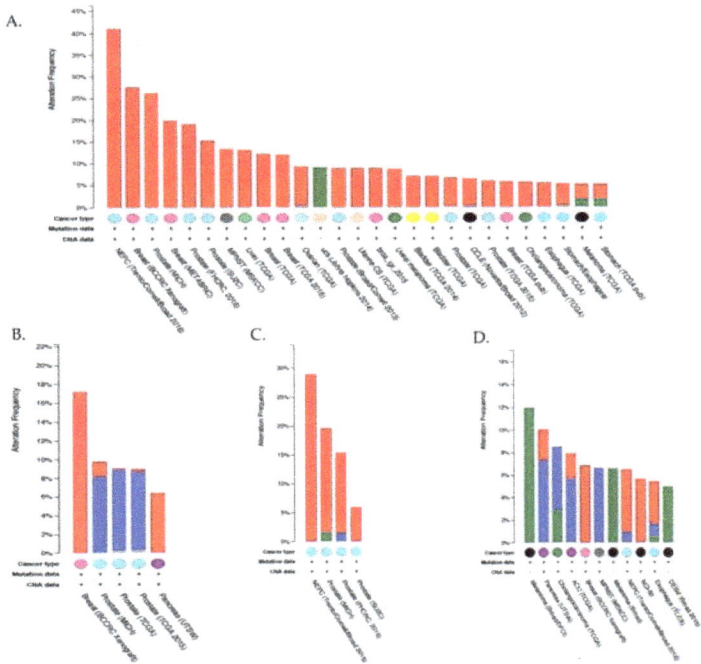

Figure 4. Summary of cross-cancer alterations for CA isoforms. (**A**) Cytosolic isoform CA XIII, (**B,C**) mitochondrial isoforms CA VA, and CA VB, respectively, and (**D**) secreted isoform CA VI. Gene amplifications are shown in red, mutations in green, deletions in blue, and multiple alterations in gray. The minimum percent altered samples threshold was set to 5%. The data and results were obtained from the cBioPortal of the TCGA database [141,142]. Experimental details of each study referenced can also be found in the TCGA database.

Most malignant cells exhibit moderate reactivity for CA VB in IHC staining [136–139]. A few cases of prostate and breast cancers showed strong staining. Most ovarian, skin, renal, urothelial, and gastric carcinomas were either weakly stained or negative. Because both isoforms are involved in metabolic processes, specifically lipogenesis, and tumors have been shown to upregulate lipogenesis, they could be important cancer targets. However, one might argue that since these isoforms are only found in the mitochondria of cells and mitochondrial dysfunction is one of the initiating factors of tumorigenesis, targeting mitochondrial CAs may not be as effective. It is also important to note that few studies have shown dysfunction in the Krebs cycle, specifically, as a result of mitochondrial dysregulation in the process of tumorigenesis. Citrate, an important precursor of fatty acid biosynthesis and lipogenesis, is an intermediate formed during this cycle in the mitochondria. While inhibition of the mitochondrial CAs alone might be effective, combining these inhibitors with drugs that directly target glycolysis, lipogenesis, and/or pH regulation may prove more potent for cancer therapy. These ideas must, however, be explored in more comprehensive and conclusive studies.

5. Secreted CA Isoforms: Expression, Distribution, and Function

CA VI is the only secreted isoform among the human CA family members [14,169]. It is distinct in terms of primary sequence and the number of conserved residues among the α-CAs (Table 1). The lowest percent sequence identity of 24.4% observed was with CA VB and the highest of 39% was surprisingly observed with CA IX (Table 1). CA VI has moderate catalytic activity and it is mostly expressed in salivary and mammary glands (Table 2). The physiological role of CA VI is still unclear, although it has been suggested that it is required for pH homeostasis of the mouth and taste perception [170–172].

Expression and Function of CA VI in Normal and Tumor Cells

As the only secreted isoform, CA VI is also known as gustin [173]. It has been found in tears, milk, respiratory airways, human serum, epithelial lining of the alimentary canal, enamel organs, and most significantly in human saliva [174–177]. Although the physiological role of CA VI has not been established, it may be required for oral homeostasis to regulate against acidic environments [170]. Maintenance of proper pH levels in the saliva protects against enamel erosion and acid neutralization in biofilms and prevents dental caries [171,178–180]. Inhibition of CA VI was shown to cause taste perversion and sometimes complete loss of taste [173,181]. Exposure to high levels of exogenous zinc reversed the observed phenotype [173,181,182]. These studies suggest that CA VI plays a key role in taste perception and in maintaining proper salivary functions by regulating pH [181]. CA VI expression in cancer is limited both at mRNA and protein levels [136,141,142]. The most common gene alterations observed were missense mutations in melanoma followed by both deletions and amplifications (Figure 4D). No studies to date have linked CA VI to tumorigenesis, cancer progression, or metastasis.

6. Membrane Associated CA Isoforms: Expression, Distribution, and Function

The membrane-associated CAs include the transmembrane isoforms: CA IX, XII, and XIV, and GPI-anchored isoform IV [3,4,48]. As might be expected, CA IV has the lowest sequence identity compared to the other membrane-associated isoforms, with an average of ~30% (Table 1). Surprisingly, CA XII and XIV are the most similar in terms of primary sequence and number of conserved residues, closely followed by CA IX and XIV. CA IX and XII have only 101 conserved residues, which equates to ~39% in primary sequence identity (Table 1). CA IV and CA IX are considered the fastest membrane-associated isoforms, with identical K_{cat} values of $1.1 \times 10^6 \text{ s}^{-1}$, similar to that of CA II (Table 2) [183]. The expression and roles of the membrane-associated isoforms also varies from the kidneys, where they are necessary for bicarbonate reabsorption and normal kidney function, to the lungs, prostate, ovaries, GI tract, breast, and brain. They are also involved in hyperactivity of the heart and pH regulation/balance in retina, muscles, and erythrocytes [3,33,51,137].

More recently, CA IV and CA IX expression have been observed in cells that support wound repair [184]. Further, Membrane isoforms IX and XII have been implicated in tumorigenesis, cancer progression, and metastasis [36–38,41,110].

6.1. Expression and Function of CA IV and CA XIV in Normal Cells

High expression of GPI-anchored CA IV has been observed in the bone marrow, GI tract, liver, and gallbladder, whereas low expression is observed in the pancreas, kidney, brain, adipose, and soft tissues [17,185–190]. In the kidney, its function is necessary for bicarbonate reabsorption and normal kidney function [191]. Like CA II, CA IV has been shown to interact with various transporters [167,192]. These interactions increase the activity of bicarbonate transport and are required for maintaining appropriate pH balance within the environment of the retina and retinal pigment epithelium [193,194]. However, its mutant forms have been shown to be responsible for an autosomal dominant form of retinitis pigmentosa causing rod and cone photoreceptor degeneration [193,194]. CA IV has also been detected in tissue regeneration using mouse skin wound models [184]. These studies performed by Barker et al. showed increased CA IV mRNA during the period of wound hypoxia in keratinocytes, which form structures beneath the migrating epidermis [184]. In this setting, CA IV is suggested to contribute to wound healing by providing an acidic environment in which the migrating dermis and neutrophils can survive [184]. CA XIV mRNA shows strong expression in the brain, muscles, seminal vesicles, and retina [195,196]. IHC staining showed medium expression in different parts of the brain, muscles, and skin. Low CA XIV expression was observed in the liver, GI tract, seminal vesicles, and cervix [137,138,140]. Strong luminal correlation between CA IV and CA XIV suggest functional overlap between the enzymes [196]. CA XIV has also been shown to interact with bicarbonate transporters and has been implicated in acid–base balance in muscles and erythrocytes in response to chronic hypoxia, hyperactivity of the heart, and pH regulation in the retina [194,197,198].

6.2. Expression and Function of CA IV and CA XIV in Tumor Cells

The tumor-associated potential of isoforms CA IV and CA XIV, like many other α-CAs, has not been extensively studied. CA XIV mRNA has been shown to be upregulated in many cancers, being most often observed in melanomas, gliomas, liver, and uterine cancers [139,141,142]. It is the most altered gene in cancer among all the membrane-associated isoforms, with about >5% alterations, most of which are amplifications, based on 30 independent studies deposited in the TCGA (Figure 5B). CA XIV is seen in prostate, breast, pancreatic, lung, liver, bladder, and ovarian cancers. CA IV mRNA expression in cancer is much lower than CA XIV, but nonetheless can be observed in gliomas, renal cell carcinomas, thyroid cancers, and melanomas [139,141,142]. The most common gene alteration observed was amplification and the most common mutation was missense mutations (Figure 5A). However, CA IV IHC staining showed that cancer tissues were essentially negative. IHC detection of CA XIV showed moderate membranous staining in the majority of melanomas, along with a few hepatocellular carcinomas and pancreatic cancers [136,140]. Further studies specifically linking CA XIV to tumorigenesis and progression of the aforementioned and other cancers are yet to be explored and/or published.

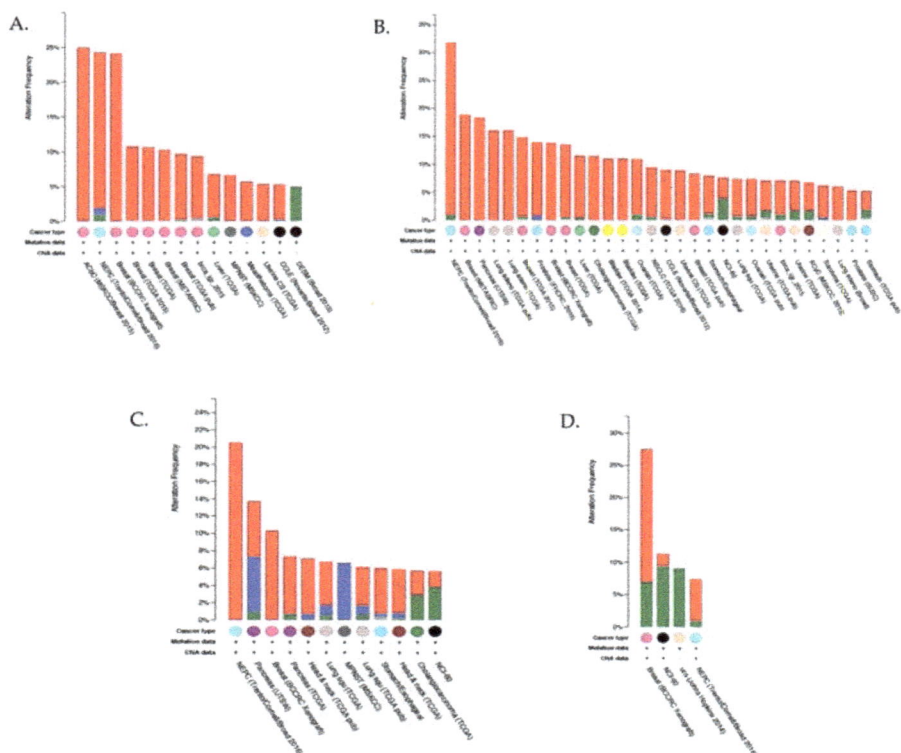

Figure 5. Summary of cross-cancer alterations for CA isoforms. (**A**) GPI-anchored CA IV, (**B–D**) transmembrane isoforms CA XIV, CA IX, and CA XIV, respectively. Gene amplifications are shown in red, mutations in green, deletions in blue, and multiple alterations in gray. The minimum percent altered samples threshold was set to 5%. The data and results were obtained from the cBioPortal of TCGA. Experimental details of each study referenced can also be found in the TCGA database.

6.3. Expression and Function of CA IX and CA XII in Normal Cells

CA IX was originally discovered by Pastorekova et al. as a component of MaTu quasi-viral agent [20]. Two years later, it was cloned and characterized as a tumor-associated member of the carbonic anhydrase family [199]. We now know that CA IX expression in normal adult tissue is limited to the outer shield of hair follicles, the epidermis of the skin during wound healing, and the GI tract [184,200]. In the GI tract CA IX is specifically found in the stomach and epithelial tissues of the gut, particularly the basolateral surfaces of the crypt, enterocytes of the duodenum, jejunum, and ileum of the small intestine [200]. In the large intestine, CA IX expression is restricted to the base of the glands in the cecum and colon [200,201]. Its function at these sites includes carbon dioxide and bicarbonate transport, acid–base balance, and signal transduction. Because of the presence of an "exofacial" proteoglycan-like domain in CA IX, which functions independently of its catalytic activity, it is also thought to contribute to the assembly and maturation of focal adhesion contacts during initial cell spreading [50,56,202,203]. In contrast, expression of CA XII has been observed in many normal tissues with high expression reported in the appendix, pancreas, colon, rectum, kidney, prostate, intestine, and activated lymphocytes. Moderate expression is observed in the esophagus, oral mucosa, urinary bladder, breast, vagina, cervix, endometrium, and skin, and low expression is seen in the stomach, seminal vesicles, and fallopian tubes [38,120,200,204,205]. Its biological functions in normal tissue range from facilitating bicarbonate transport in cells to maintenance of an internal steady state concentration of chloride ions within an organism [206,207]. Similar to CA IX, CA XII is also involved in chemical reactions and pathways involving small molecule metabolism. Furthermore, it has been postulated that CA XII may be important for normal kidney function [191,205].

6.4. Expression and Function of CA IX and CA XII in Tumor Cells

Both CA IX and CA XII are often regarded as tumor-associated CAs and thus are the most studied in the tumor setting. As we have discussed in previous sections, both CA IX and XII act in conjunction with cyt-CAs to establish the differential pH in the cellular microenvironment observed in hypoxic tumors [56,110,167]. CA IX has, however, garnered more attention because its limited expression in normal cells and upregulation in many aggressive cancers compared to CA XII. This suggests that CA IX is a more "druggable" target. Cancers associated with CA IX expression include brain, breast, bladder, cervix, colon, colorectal, head and neck, pancreas, kidney, lung, ovaries, stomach oral cavity, and T-cell lymphomas [52,68,136,139,208–210]. CA IX expression, especially in hypoxic tumors, is modulated by hypoxia-inducible factor 1 (HIF-1) in response to low oxygen levels and increased cell density [211]. CA IX expression can also be regulated both in chronic and mild hypoxic conditions by components of the mitogen-activated protein kinase (MAPK) pathway [212,213]. Studies have demonstrated that the MAPK pathway controls the CA9 promoter, using both the HIF-1-dependent and -independent signals, working as a downstream mediator of CA9 transcriptional response to both hypoxia and high cell density [213]. This is important since activating mutations of the MAPK and phosphatidylinositol-3 kinase (PI3K) pathways, which occur in many tumor types, may consequently upregulate CA9 gene expression and influence intratumoral distribution of the CA IX protein. Furthermore, there is evidence of CA IX expression at the leading edge of cells, displaying a migratory phenotype where it may serve as a modulator of tumor aggression [214].

Regulation of CA IX activity has been observed through phosphorylation of its intracellular domain at three putative sites. Phosphorylation at Thr 443 by protein kinase A (PKA) in response to cyclic adenosine monophosphate (cAMP) activates CA IX activity, under hypoxic conditions [215]. Epidermal growth factor (EGF) induced phosphorylation at the same position mediates cross talk between CA IX and PI3K to activate Akt kinase, although the impact on CA IX activity is unclear [216]. On the other hand, phosphorylation of CA IX at Ser 448 inhibits enzyme activity. The signal transduction activity of CA IX is activated when the tyrosine residue at position 449 is phosphorylated [215,216]. CA IX has also been shown to be a key modulator of tumor growth, survival and migration, tumorigenesis, pH control, cell adhesion, and proliferation, and is the basis of studies used to establish mechanisms proposed in both Hypothesis 1 and 2 presented in earlier sections (Figure 1) [33,52,202,217,218]. CA IX expression is regulated by tumor hypoxia and has not only been established as a prognostic indicator for a variety of cancers but also as an anticancer target. Overexpression of CA IX in most cancers is associated with poor prognosis, chemotherapeutic resistance, and poor clinical outcome [219–222].

Like CA IX, CA XII expression has also been shown to be upregulated in many tumor types. However, its expression is also abundant in normal tissues. The most abundant cancer-related CA12 expression, according to RNA sequencing datasets deposited in the TCGA database, was observed in renal cell carcinomas, colorectal, breast, bladder glioblastomas, and head and neck cancers [141,142]. The most common gene alterations include amplifications and mutations in breast cancer patient xenografts, mutations in NCI-60 cell lines, and uterine carcinomas (Figure 5D). Several cases of breast, renal, urothelial, skin, lung, endometrial, and cervical cancers also displayed moderate to strong membranous IHC staining, with additional positivity in a few cases [136,139]. Furthermore, high expression of CA XII has also been observed in glioblastomas, astrocytomas, and T-cell lymphomas [136]. Despite all the evidence that suggested CA XII protein has potential as a prognosticator, it has still not been fully established as a prognostic marker. This may be due to the fact that in some cancers it is a marker of "good prognosis," including in lung, cervical, and breast cancer, whereas its expression has no prognostic value in other types such as brain cancer [40,223–225]. Finally, in cancers such as colorectal, oral squamous carcinoma, and some kidney cancers, CA XII upregulation is associated with poor prognosis [205,226]. Recent studies have shown increased CA XII expression on the surface of chemoresistant cells, suggesting its potential as a therapeutic target to overcome chemoresistance in cancer cells [227]. Furthermore, CA XII has been shown to facilitate cancer cell survival and promote tumor cell migration, invasion, and maintenance of cancer cell stemness [36,53,205].

7. Conclusions

One current research interest in the field of CA drug discovery is enzymes' role in cancer, either as a prognostic marker or as a potential drug target. To date, experimental evidence has indicated an important relationship between pH regulation and tumor cell proliferation and survival. Many of these studies show direct involvement of CA in these processes, marking them as important avenues for pH-disruptive anti-cancer treatments. This is particular true in tumors that have undergone hypoxic transition, where CAs establish pH gradients that favor cancer cell growth at the expense of adjacent normal cell (and patient) survival. Of the 15 human CAs, however, attention has only been directed toward two isoforms, CA IX and XII. Of course, this has been fruitful as inhibitors are now in clinical trials that specifically target each isoform. Despite this achievement, the roles of the remaining CAs have only recently been considered in cancers. This area of research has strong potential as many of the human isoforms, beyond CA IX and XII, have shown an association with tumorigenesis in primary cancers. With much still to understand in terms of the "combined role" of CAs in different cancer types, it will be important moving forward for researchers to consider the effects beside those of CA IX and XII. As the search for viable and novel treatments against the most aggressive and detrimental cancers continues, priority should be placed on deciphering the roles of cancer-related CAs.

Acknowledgments: This work was financed by the National Institutes of Health, CA165284 (SCF), minority supplement CA165284-03S1 (MYM) and the Cancer Biology Dissertation Award (MYM).

Author Contributions: All four authors contributed to the writing and editing of this review.

Conflicts of Interest: The authors declare no conflict of interest.

References

1. Stadie, W.C.; O'Brien, H. The catalytic hydration of carbondioxide and dehydration of carbonic acid by enzyme isolated from red blood cells. *J. Biol. Chem.* **1933**, *103*, 521–529.
2. Supuran, C.T. Structure and function of carbonic anhydrases. *Biochem. J.* **2016**, *473*, 2023–2032. [CrossRef] [PubMed]
3. Frost, S.C.; McKenna, R. (Eds.) *Carbonic Anhydrase: Mechanism, Regulation, Links to Disease, and Industrial Applications, Subcellular Biochemistry*; Springer: Berlin/Heidelberg, Germany, 2014.
4. Supuran, C.T. Carbonic anhydrases—An overview. *Curr. Pharm. Des.* **2008**, *14*, 603–614. [CrossRef] [PubMed]
5. Del Prete, S.; Vullo, D.; Fisher, G.M.; Andrews, K.T.; Poulsen, S.-A.; Capasso, C.; Supuran, C.T. Discovery of a new family of carbonic anhydrases in the malaria pathogen *Plasmodium falciparum*—the η-carbonic anhydrases. *Bioorg. Med. Chem. Lett.* **2014**, *24*, 4389–4396. [CrossRef] [PubMed]
6. Nyman, P.; Lindskog, S. Amino acid composition of various forms of bovine and human erythrocyte carbonic anhydrase. *Biochim. Biophys. Acta* **1964**, *85*, 141–151. [CrossRef]
7. Andersson, B.; Nyman, P.O.; Strid, L. Amino acid sequence of human erythrocyte carbonic anhydrase B. *Biochem. Biophys. Res. Commun.* **1972**, *48*, 670–677. [CrossRef]
8. Lindskog, S. Purification and properties of bovine erythrocyte carbonic anhydrase. *Biochim. Biophys. Acta* **1960**, *39*, 218–226. [CrossRef]
9. Nyman, P.O. Purification and properties of carbonic anhydrase from human erythrocytes. *Biochim. Biophys. Acta* **1961**, *52*, 1–12. [CrossRef]

10. Liljas, A.; Kannan, K.K.; Bergstén, P.C.; Waara, I.; Fridborg, K.; Strandberg, B.; Carlbom, U.; Järup, L.; Lövgren, S.; Petef, M. Crystal structure of human carbonic anhydrase C. *Nat. New Biol.* **1972**, *235*, 131–137. [CrossRef] [PubMed]

11. Lin, K.T.; Deutsch, H.F. Human carbonic anhydrases. XI. The complete primary structure of carbonic anhydrase B. *J. Biol. Chem.* **1973**, *248*, 1885–1893. [PubMed]

12. Balckburn, M.N.; Chirgwin, J.M.; Gordon, J.T.; Thomas, K.D.; Parsons, T.F.; Register, A.M.; Schnackerz, K.D.; Noltmann, E.A. Pseudoisoenzymes of rabbit muscle phosphoglucose isomerase. *J. Biol. Chem.* **1972**, *247*, 1170–1179.

13. Fernley, R.T.; Wright, R.D.; Coghlan, J.P. A novel carbonic anhydrase from the ovine parotid gland. *FEBS Lett.* **1979**, *105*, 299–302. [CrossRef]

14. Murakami, H.; Sly, W.S. Purification and characterization of human salivary carbonic anhydrase. *J. Biol. Chem.* **1987**, *262*, 1382–1388. [PubMed]

15. Whitney, P.L.; Briggle, T.V. Membrane-associated carbonic anhydrase purified from bovine lung. *J. Biol. Chem.* **1982**, *257*, 12056–12059. [PubMed]

16. Wistrand, P.J. Properties of membrane-bound carbonic anhydrase. *Ann. N. Y. Acad. Sci.* **1984**, *429*, 195–206. [CrossRef] [PubMed]

17. Carter, N.D.; Fryer, A.; Grant, A.G.; Hume, R.; Strange, R.G.; Wistrand, P.J. Membrane specific carbonic anhydrase (CAIV) expression in human tissues. *Biochim. Biophys. Acta* **1990**, *1026*, 113–116. [CrossRef]

18. Zhu, X.L.; Sly, W.S. Carbonic anhydrase IV from human lung. Purification, characterization, and comparison with membrane carbonic anhydrase from human kidney. *J. Biol. Chem.* **1990**, *265*, 8795–8801. [PubMed]

19. Montgomery, J.C.; Venta, P.J.; Eddy, R.L.; Fukushima, Y.-S.; Shows, T.B.; Tashian, R.E. Characterization of the human gene for a newly discovered carbonic anhydrase, CA VII, and its localization to chromosome 16. *Genomics* **1991**, *11*, 835–848. [CrossRef]

20. Pastoreková, S.; Závadová, Z.; Kostál, M.; Babusíková, O.; Závada, J. A novel quasi-viral agent, MaTu, is a two-component system. *Virology* **1992**, *187*, 620–626. [CrossRef]

21. Opavský, R.; Pastoreková, S.; Zelník, V.; Gibadulinová, A.; Stanbridge, E.J.; Závada, J.; Kettmann, R.; Pastorek, J. Human MN/CA9 gene, a novel member of the carbonic anhydrase family: Structure and exon to protein domain relationships. *Genomics* **1996**, *33*, 480–487. [CrossRef] [PubMed]

22. Türeci, O.; Sahin, U.; Vollmar, E.; Siemer, S.; Göttert, E.; Seitz, G.; Parkkila, A.K.; Shah, G.N.; Grubb, J.H.; Pfreundschuh, M.; et al. Human carbonic anhydrase XII: CDNA cloning, expression, and chromosomal localization of a carbonic anhydrase gene that is overexpressed in some renal cell cancers. *Proc. Natl. Acad. Sci. USA* **1998**, *95*, 7608–7613. [CrossRef] [PubMed]

23. Fujikawa-Adachi, K.; Nishimori, I.; Taguchi, T.; Onishi, S. Human carbonic anhydrase XIV (CA14): CDNA cloning, mRNA expression, and mapping to chromosome 1. *Genomics* **1999**, *61*, 74–81. [CrossRef] [PubMed]

24. Mori, K.; Ogawa, Y.; Ebihara, K.; Tamura, N.; Tashiro, K.; Kuwahara, T.; Mukoyama, M.; Sugawara, A.; Ozaki, S.; Tanaka, I.; et al. Isolation and characterization of CA XIV, a novel membrane-bound carbonic anhydrase from mouse kidney. *J. Biol. Chem.* **1999**, *274*, 15701–15705. [CrossRef] [PubMed]

25. Nagao, Y.; Platero, J.S.; Waheed, A.; Sly, W.S. Human mitochondrial carbonic anhydrase: CDNA cloning, expression, subcellular localization, and mapping to chromosome 16. *Proc. Natl. Acad. Sci. USA* **1993**, *90*, 7623–7627. [CrossRef] [PubMed]

26. Fujikawa-Adachi, K.; Nishimori, I.; Taguchi, T.; Onishi, S. Human mitochondrial carbonic anhydrase VB: CDNA cloning, mRNA expression, subcellular localization, and mapping to chromosome X. *J. Biol. Chem.* **1999**, *274*, 21228–21233. [CrossRef] [PubMed]

27. Idrees, D.; Kumar, S.; Rehman, S.A.A.; Gourinath, S.; Islam, A.; Ahmad, F.; Imtaiyaz Hassan, M. Cloning, expression, purification and characterization of human mitochondrial carbonic anhydrase VA. *3 Biotech* **2016**, *6*. [CrossRef] [PubMed]

28. Lehtonen, J.; Shen, B.; Vihinen, M.; Casini, A.; Scozzafava, A.; Supuran, C.T.; Parkkila, A.-K.; Saarnio, J.; Kivelä, A.J.; Waheed, A.; et al. Characterization of CA XIII, a novel member of the carbonic anhydrase isozyme family. *J. Biol. Chem.* **2004**, *279*, 2719–2727. [CrossRef] [PubMed]

29. Ozensoy Guler, O.; Capasso, C.; Supuran, C.T. A magnificent enzyme superfamily: Carbonic anhydrases, their purification and characterization. *J. Enzyme Inhib. Med. Chem.* **2015**, 1–6. [CrossRef] [PubMed]

30. Supuran, C.T. Carbonic anhydrases: Novel therapeutic applications for inhibitors and activators. *Nat. Rev. Drug Discov.* **2008**, *7*, 168–181. [CrossRef] [PubMed]

31. Thiry, A.; Dogné, J.-M.; Supuran, C.T.; Masereel, B. Carbonic anhydrase inhibitors as anticonvulsant agents. *Curr. Top. Med. Chem.* **2007**, *7*, 855–864. [CrossRef] [PubMed]
32. Supuran, C.T. How many carbonic anhydrase inhibition mechanisms exist? *J. Enzyme Inhib. Med. Chem.* **2016**, *31*, 345–360. [CrossRef] [PubMed]
33. Li, Y.; Tu, C.; Wang, H.; Silverman, D.N.; Frost, S.C. Catalysis and pH control by membrane-associated carbonic anhydrase IX in MDA-MB-231 breast cancer cells. *J. Biol. Chem.* **2011**, *286*, 15789–15796. [CrossRef] [PubMed]
34. Sly, W.S. The membrane carbonic anhydrases: From CO_2 transport to tumor markers. *EXS* **2000**, 95–104.
35. Nordfors, K.; Haapasalo, J.; Korja, M.; Niemelä, A.; Laine, J.; Parkkila, A.-K.; Pastorekova, S.; Pastorek, J.; Waheed, A.; Sly, W.S.; et al. The tumour-associated carbonic anhydrases CA II, CA IX and CA XII in a group of medulloblastomas and supratentorial primitive neuroectodermal tumours: An association of CA IX with poor prognosis. *BMC Cancer* **2010**, *10*, 148. [CrossRef] [PubMed]
36. Lounnas, N.; Rosilio, C.; Nebout, M.; Mary, D.; Griessinger, E.; Neffati, Z.; Chiche, J.; Spits, H.; Hagenbeek, T.J.; Asnafi, V.; et al. Pharmacological inhibition of carbonic anhydrase XII interferes with cell proliferation and induces cell apoptosis in T-cell lymphomas. *Cancer Lett.* **2013**, *333*, 76–88. [CrossRef] [PubMed]
37. Saarnio, J.; Parkkila, S.; Parkkila, A.K.; Haukipuro, K.; Pastoreková, S.; Pastorek, J.; Kairaluoma, M.I.; Karttunen, T.J. Immunohistochemical study of colorectal tumors for expression of a novel transmembrane carbonic anhydrase, MN/CA IX, with potential value as a marker of cell proliferation. *Am. J. Pathol.* **1998**, *153*, 279–285. [CrossRef]
38. Kivelä, A.J.; Parkkila, S.; Saarnio, J.; Karttunen, T.J.; Kivelä, J.; Parkkila, A.K.; Pastoreková, S.; Pastorek, J.; Waheed, A.; Sly, W.S.; et al. Expression of transmembrane carbonic anhydrase isoenzymes IX and XII in normal human pancreas and pancreatic tumours. *Histochem. Cell Biol.* **2000**, *114*, 197–204. [PubMed]
39. Kummola, L.; Hämäläinen, J.M.; Kivelä, J.; Kivelä, A.J.; Saarnio, J.; Karttunen, T.; Parkkila, S. Expression of a novel carbonic anhydrase, CA XIII, in normal and neoplastic colorectal mucosa. *BMC Cancer* **2005**, *5*, 41. [CrossRef] [PubMed]
40. Barnett, D.H.; Sheng, S.; Charn, T.H.; Waheed, A.; Sly, W.S.; Lin, C.-Y.; Liu, E.T.; Katzenellenbogen, B.S. Estrogen receptor regulation of carbonic anhydrase XII through a distal enhancer in breast cancer. *Cancer Res.* **2008**, *68*, 3505–3515. [CrossRef] [PubMed]
41. Chen, L.Q.; Howison, C.M.; Spier, C.; Stopeck, A.T.; Malm, S.W.; Pagel, M.D.; Baker, A.F. Assessment of carbonic anhydrase IX expression and extracellular pH in B-cell lymphoma cell line models. *Leuk. Lymphoma* **2015**, *56*, 1432–1439. [CrossRef] [PubMed]
42. Savile, C.K.; Lalonde, J.J. Biotechnology for the acceleration of carbon dioxide capture and sequestration. *Curr. Opin. Biotechnol.* **2011**, *22*, 818–823. [CrossRef] [PubMed]
43. Bond, G.M.; Stringer, J.; Brandvold, D.K.; Simsek, F.A.; Medina, M.-G.; Egeland, G. Development of integrated system for biomimetic CO_2 sequestration using the enzyme carbonic anhydrase. *Energy Fuels* **2001**, *15*, 309–316. [CrossRef]
44. Lee, S.-W.; Park, S.-B.; Jeong, S.-K.; Lim, K.-S.; Lee, S.-H.; Trachtenberg, M.C. On carbon dioxide storage based on biomineralization strategies. *Micron* **2010**, *41*, 273–282. [CrossRef] [PubMed]
45. Kaar, J.L.; Oh, H.-I.; Russell, A.J.; Federspiel, W.J. Towards improved artificial lungs through bio-catalysis. *Biomaterials* **2007**, *28*, 3131–3139. [CrossRef] [PubMed]
46. Sreenivasan, R.; Bassett, E.K.; Hoganson, D.M.; Vacanti, J.P.; Gleason, K.K. Ultra-thin, gas permeable free-standing and composite membranes for microfluidic lung assist devices. *Biomaterials* **2011**, *32*, 3883–3889. [CrossRef] [PubMed]
47. Stadermann, M.; Baxamusa, S.H.; Aracne-Ruddle, C.; Chea, M.; Li, S.; Youngblood, K.; Suratwala, T. Fabrication of Large-area Free-standing Ultrathin Polymer Films. *J. Vis. Exp.* **2015**. [CrossRef] [PubMed]
48. Pinard, M.A.; Mahon, B.; McKenna, R. Probing the surface of human carbonic anhydrase for clues towards the design of isoform specific inhibitors. *BioMed Res. Int.* **2015**, *2015*, 1–15. [CrossRef] [PubMed]
49. Mboge, M.Y.; McKenna, R.; Frost, S.C. Advances in anti-cancer drug development targeting carbonic anhydrase IX and XII. In *Topics in Anti-Cancer Research*; Bentham Science Publishers: Sharjah, United Arab Emirates, 2016; Volume 5, pp. 3–42.
50. Mahon, B.P.; Pinard, M.A.; McKenna, R. Targeting carbonic anhydrase IX activity and expression. *Molecules* **2015**, *20*, 2323–2348. [CrossRef] [PubMed]

51. Mahon, B.P.; Bhatt, A.; Socorro, L.; Driscoll, J.M.; Okoh, C.; Lomelino, C.L.; Mboge, M.Y.; Kurian, J.J.; Tu, C.; Agbandje-McKenna, M.; et al. The structure of carbonic anhydrase IX is adapted for low-pH catalysis. *Biochemistry* **2016**, *55*, 4642–4653. [CrossRef] [PubMed]
52. Yang, J.-S.; Lin, C.-W.; Chuang, C.-Y.; Su, S.-C.; Lin, S.-H.; Yang, S.-F. Carbonic anhydrase IX overexpression regulates the migration and progression in oral squamous cell carcinoma. *Tumour Biol. J. Int. Soc. Oncodev. Biol. Med.* **2015**, *36*, 9517–9524. [CrossRef] [PubMed]
53. Hsieh, M.-J.; Chen, K.-S.; Chiou, H.-L.; Hsieh, Y.-S. Carbonic anhydrase XII promotes invasion and migration ability of MDA-MB-231 breast cancer cells through the p38 MAPK signaling pathway. *Eur. J. Cell Biol.* **2010**, *89*, 598–606. [CrossRef] [PubMed]
54. Pacchiano, F.; Carta, F.; McDonald, P.C.; Lou, Y.; Vullo, D.; Scozzafava, A.; Dedhar, S.; Supuran, C.T. Ureido-substituted benzenesulfonamides potently inhibit carbonic anhydrase IX and show antimetastatic activity in a model of breast cancer metastasis. *J. Med. Chem.* **2011**, *54*, 1896–1902. [CrossRef] [PubMed]
55. Winum, J.-Y.; Carta, F.; Ward, C.; Mullen, P.; Harrison, D.; Langdon, S.P.; Cecchi, A.; Scozzafava, A.; Kunkler, I.; Supuran, C.T. Ureido-substituted sulfamates show potent carbonic anhydrase IX inhibitory and antiproliferative activities against breast cancer cell lines. *Bioorg. Med. Chem. Lett.* **2012**, *22*, 4681–4685. [CrossRef] [PubMed]
56. Mahon, B.P.; McKenna, R. Regulation and role of carbonic anhydrase IX and use as a biomarker and therapeutic target in cancer. *Res. Trends Curr. Top. Biochem. Res.* **2013**, *15*, 1–21.
57. Supuran, C.T. Carbonic Anhydrase inhibition and the management of hypoxic tumors. *Metabolites* **2017**, *7*, 48. [CrossRef] [PubMed]
58. Mulders, P.; Bleumer, I.; Debruyne, F.; Oosterwijk, E. Specific monoclonal antibody-based immunotherapy by targeting the RCC-associated antigen carbonic anhydrase-IX(G250/MN). *Urol. Ausg A* **2004**, *43*, S146–S147. [CrossRef] [PubMed]
59. Davis, I.D.; Wiseman, G.A.; Lee, F.-T.; Gansen, D.N.; Hopkins, W.; Papenfuss, A.T.; Liu, Z.; Moynihan, T.J.; Croghan, G.A.; Adjei, A.A.; et al. A phase I multiple dose, dose escalation study of cG250 monoclonal antibody in patients with advanced renal cell carcinoma. *Cancer Immun.* **2007**, *7*, 13. [PubMed]
60. Stillebroer, A.B.; Boerman, O.C.; Desar, I.M.E.; Boers-Sonderen, M.J.; van Herpen, C.M.L.; Langenhuijsen, J.F.; Smith-Jones, P.M.; Oosterwijk, E.; Oyen, W.J.G.; Mulders, P.F.A. Phase 1 radioimmunotherapy study with lutetium 177-labeled anti-carbonic anhydrase IX monoclonal antibody girentuximab in patients with advanced renal cell carcinoma. *Eur. Urol.* **2013**, *64*, 478–485. [CrossRef] [PubMed]
61. Muselaers, C.H.J.; Boers-Sonderen, M.J.; van Oostenbrugge, T.J.; Boerman, O.C.; Desar, I.M.E.; Stillebroer, A.B.; Mulder, S.F.; van Herpen, C.M.L.; Langenhuijsen, J.F.; Oosterwijk, E.; et al. Phase 2 study of lutetium 177-labeled anti-carbonic anhydrase IX monoclonal antibody girentuximab in patients with advanced renal cell carcinoma. *Eur. Urol.* **2016**, *69*, 767–770. [CrossRef] [PubMed]
62. Dubois, L.J.; Niemans, R.; van Kuijk, S.J.A.; Panth, K.M.; Parvathaneni, N.-K.; Peeters, S.G.J.A.; Zegers, C.M.L.; Rekers, N.H.; van Gisbergen, M.W.; Biemans, R.; et al. New ways to image and target tumour hypoxia and its molecular responses. *Radiother. Oncol.* **2015**, *116*, 352–357. [CrossRef] [PubMed]
63. Li, J.; Zhang, G.; Wang, X.; Li, X.-F. Is carbonic anhydrase IX a validated target for molecular imaging of cancer and hypoxia? *Future Oncol.* **2015**, *11*, 1531–1541. [CrossRef] [PubMed]
64. Ahlskog, J.K.J.; Schliemann, C.; Mårlind, J.; Qureshi, U.; Ammar, A.; Pedley, R.B.; Neri, D. Human monoclonal antibodies targeting carbonic anhydrase IX for the molecular imaging of hypoxic regions in solid tumours. *Br. J. Cancer* **2009**, *101*, 645–657. [CrossRef] [PubMed]
65. Chen, F.; Zhuang, X.; Lin, L.; Yu, P.; Wang, Y.; Shi, Y.; Hu, G.; Sun, Y. New horizons in tumor microenvironment biology: Challenges and opportunities. *BMC Med.* **2015**, *13*. [CrossRef] [PubMed]
66. Whiteside, T.L. The tumor microenvironment and its role in promoting tumor growth. *Oncogene* **2008**, *27*, 5904–5912. [CrossRef] [PubMed]
67. Brown, J.M. Tumor hypoxia in cancer therapy. *Methods Enzymol.* **2007**, *435*, 297–321. [CrossRef] [PubMed]
68. Jamali, S.; Klier, M.; Ames, S.; Felipe Barros, L.; McKenna, R.; Deitmer, J.W.; Becker, H.M. Hypoxia-induced carbonic anhydrase IX facilitates lactate flux in human breast cancer cells by non-catalytic function. *Sci. Rep.* **2015**, *5*. [CrossRef] [PubMed]
69. Warburg, O. The chemical constitution of respiratory ferment. *Science* **1928**, *68*, 437–443. [CrossRef] [PubMed]
70. Warburg, O.; Wind, F.; Negelein, E. The metabolism of tumors in the body. *J. Gen. Physiol.* **1927**, *8*, 519–530. [CrossRef] [PubMed]

71. Gatenby, R.A.; Gillies, R.J. Why do cancers have high aerobic glycolysis? *Nat. Rev. Cancer* **2004**, *4*, 891–899. [CrossRef] [PubMed]

72. Newsholme, P.; Lima, M.M.R.; Procopio, J.; Pithon-Curi, T.C.; Doi, S.Q.; Bazotte, R.B.; Curi, R. Glutamine and glutamate as vital metabolites. *Braz. J. Med. Biol. Res.* **2003**, *36*, 153–163. [CrossRef] [PubMed]

73. Brekke, E.; Morken, T.S.; Walls, A.B.; Waagepetersen, H.; Schousboe, A.; Sonnewald, U. Anaplerosis for glutamate synthesis in the neonate and in adulthood. In *The Glutamate/GABA-Glutamine Cycle*; Springer International Publishing: Basel, Switzerland, 2016; Volume 13, pp. 43–58.

74. DeBerardinis, R.J.; Mancuso, A.; Daikhin, E.; Nissim, I.; Yudkoff, M.; Wehrli, S.; Thompson, C.B. Beyond aerobic glycolysis: Transformed cells can engage in glutamine metabolism that exceeds the requirement for protein and nucleotide synthesis. *Proc. Natl. Acad. Sci. USA* **2007**, *104*, 19345–19350. [CrossRef] [PubMed]

75. Currie, E.; Schulze, A.; Zechner, R.; Walther, T.C.; Farese, R.V. Cellular fatty acid metabolism and cancer. *Cell Metab.* **2013**, *18*, 153–161. [CrossRef] [PubMed]

76. Pavlova, N.N.; Thompson, C.B. The emerging hallmarks of cancer metabolism. *Cell Metab.* **2016**, *23*, 27–47. [CrossRef] [PubMed]

77. Gillies, R.J.; Raghunand, N.; Karczmar, G.S.; Bhujwalla, Z.M. MRI of the tumor microenvironment. *J. Magn. Reson. Imaging JMRI* **2002**, *16*, 430–450. [CrossRef] [PubMed]

78. Webb, B.A.; Chimenti, M.; Jacobson, M.P.; Barber, D.L. Dysregulated pH: A perfect storm for cancer progression. *Nat. Rev. Cancer* **2011**, *11*, 671–677. [CrossRef] [PubMed]

79. Stüwe, L.; Müller, M.; Fabian, A.; Waning, J.; Mally, S.; Noël, J.; Schwab, A.; Stock, C. pH dependence of melanoma cell migration: Protons extruded by NHE1 dominate protons of the bulk solution. *J. Physiol.* **2007**, *585*, 351–360. [CrossRef] [PubMed]

80. Balkwill, F.R.; Capasso, M.; Hagemann, T. The tumor microenvironment at a glance. *J. Cell Sci.* **2012**, *125*, 5591–5596. [CrossRef] [PubMed]

81. Wilson, W.R.; Hay, M.P. Targeting hypoxia in cancer therapy. *Nat. Rev. Cancer* **2011**, *11*, 393–410. [CrossRef] [PubMed]

82. Dietl, K.; Renner, K.; Dettmer, K.; Timischl, B.; Eberhart, K.; Dorn, C.; Hellerbrand, C.; Kastenberger, M.; Kunz-Schughart, L.A.; Oefner, P.J.; et al. Lactic acid and acidification inhibit TNF secretion and glycolysis of human monocytes. *J. Immunol.* **2010**, *184*, 1200–1209. [CrossRef] [PubMed]

83. Kuwata, F.; Suzuki, N.; Otsuka, K.; Taguchi, M.; Sasai, Y.; Wakino, H.; Ito, M.; Ebihara, S.; Suzuki, K. Enzymatic regulation of glycolysis and gluconeogenesis in rabbit periodontal ligament under various physiological pH conditions. *J. Nihon Univ. Sch. Dent.* **1991**, *33*, 81–90. [CrossRef] [PubMed]

84. Chan, F.K.-M.; Moriwaki, K.; De Rosa, M.J. Detection of necrosis by release of lactate dehydrogenase activity. In *Immune Homeostasis*; Humana Press: Totowa, NJ, USA, 2013; Volume 979, pp. 65–70.

85. Gray, J.A. Kinetics of enamel dissolution during formation of incipient caries-like lesions. *Arch. Oral Biol.* **1966**. [CrossRef]

86. Putney, L.K.; Barber, D.L. Expression profile of genes regulated by activity of the Na-H exchanger NHE1. *BMC Genom.* **2004**, *5*, 46. [CrossRef] [PubMed]

87. Chen, J.L.-Y.; Lucas, J.E.; Schroeder, T.; Mori, S.; Wu, J.; Nevins, J.; Dewhirst, M.; West, M.; Chi, J.-T. The genomic analysis of lactic acidosis and acidosis response in human cancers. *PLoS Genet.* **2008**, *4*, e1000293. [CrossRef] [PubMed]

88. Menard, L.; Maughan, D.; Vigoreaux, J. The structural and functional coordination of glycolytic enzymes in muscle: Evidence of a metabolon? *Biology* **2014**, *3*, 623–644. [CrossRef] [PubMed]

89. Campanella, M.E.; Chu, H.; Wandersee, N.J.; Peters, L.L.; Mohandas, N.; Gilligan, D.M.; Low, P.S. Characterization of glycolytic enzyme interactions with murine erythrocyte membranes in wild-type and membrane protein knockout mice. *Blood* **2008**, *112*, 3900–3906. [CrossRef] [PubMed]

90. Stock, C.; Schwab, A. Protons make tumor cells move like clockwork. *Pflugers Arch.* **2009**, *458*, 981–992. [CrossRef] [PubMed]

91. Wilhelm, S.M.; Shao, Z.H.; Housley, T.J.; Seperack, P.K.; Baumann, A.P.; Gunja-Smith, Z.; Woessner, J.F. Matrix metalloproteinase-3 (stromelysin-1). Identification as the cartilage acid metalloprotease and effect of pH on catalytic properties and calcium affinity. *J. Biol. Chem.* **1993**, *268*, 21906–21913. [PubMed]

92. Bourguignon, L.Y.W.; Singleton, P.A.; Diedrich, F.; Stern, R.; Gilad, E. CD44 interaction with Na$^+$-H$^+$ exchanger (NHE1) creates acidic microenvironments leading to hyaluronidase-2 and cathepsin B activation and breast tumor cell invasion. *J. Biol. Chem.* **2004**, *279*, 26991–27007. [CrossRef] [PubMed]

93. Lee, H.-S.; Bellin, R.M.; Walker, D.L.; Patel, B.; Powers, P.; Liu, H.; Garcia-Alvarez, B.; de Pereda, J.M.; Liddington, R.C.; Volkmann, N.; et al. Characterization of an actin-binding site within the talin FERM domain. *J. Mol. Biol.* **2004**, *343*, 771–784. [CrossRef] [PubMed]

94. Moseley, J.B.; Okada, K.; Balcer, H.I.; Kovar, D.R.; Pollard, T.D.; Goode, E.L. Twinfilin is an actin-filament-severing protein and promotes rapid turnover of actin structures in vivo. *J. Cell Sci.* **2006**, *119*, 1547–1557. [CrossRef] [PubMed]

95. Pope, B.J.; Zierler-Gould, K.M.; Kühne, R.; Weeds, A.G.; Ball, L.J. Solution structure of human cofilin: Actin binding, pH sensitivity, and relationship to actin-depolymerizing factor. *J. Biol. Chem.* **2004**, *279*, 4840–4848. [CrossRef] [PubMed]

96. Srivastava, J.; Barreiro, G.; Groscurth, S.; Gingras, A.R.; Goult, B.T.; Critchley, D.R.; Kelly, M.J.S.; Jacobson, M.P.; Barber, D.L. Structural model and functional significance of pH-dependent talin-actin binding for focal adhesion remodeling. *Proc. Natl. Acad. Sci. USA* **2008**, *105*, 14436–14441. [CrossRef] [PubMed]

97. Grey, M.J.; Tang, Y.; Alexov, E.; McKnight, C.J.; Raleigh, D.P.; Palmer, A.G. Characterizing a partially folded intermediate of the villin headpiece domain under non-denaturing conditions: Contribution of His41 to the pH-dependent stability of the N-terminal subdomain. *J. Mol. Biol.* **2006**, *355*, 1078–1094. [CrossRef] [PubMed]

98. McLachlan, G.D.; Cahill, S.M.; Girvin, M.E.; Almo, S.C. Acid-induced equilibrium folding intermediate of human platelet profilin. *Biochemistry* **2007**, *46*, 6931–6943. [CrossRef] [PubMed]

99. Lagadic-Gossmann, D.; Huc, L.; Lecureur, V. Alterations of intracellular pH homeostasis in apoptosis: Origins and roles. *Cell Death Differ.* **2004**, *11*, 953–961. [CrossRef] [PubMed]

100. Matsuyama, S.; Llopis, J.; Deveraux, Q.L.; Tsien, R.Y.; Reed, J.C. Changes in intramitochondrial and cytosolic pH: Early events that modulate caspase activation during apoptosis. *Nat. Cell Biol.* **2000**, *2*, 318–325. [CrossRef] [PubMed]

101. Pouysségur, J.; Franchi, A.; L'Allemain, G.; Paris, S. Cytoplasmic pH, a key determinant of growth factor-induced DNA synthesis in quiescent fibroblasts. *FEBS Lett.* **1985**, *190*, 115–119. [CrossRef]

102. Bower, J.J.; Vance, L.D.; Psioda, M.; Smith-Roe, S.L.; Simpson, D.A.; Ibrahim, J.G.; Hoadley, K.A.; Perou, C.M.; Kaufmann, W.K. Patterns of cell cycle checkpoint deregulation associated with intrinsic molecular subtypes of human breast cancer cells. *Npj Breast Cancer* **2017**, *3*. [CrossRef] [PubMed]

103. Khaled, A.R.; Kim, K.; Hofmeister, R.; Muegge, K.; Durum, S.K. Withdrawal of IL-7 induces Bax translocation from cytosol to mitochondria through a rise in intracellular pH. *Proc. Natl. Acad. Sci. USA* **1999**, *96*, 14476–14481. [CrossRef] [PubMed]

104. Swietach, P.; Vaughan-Jones, R.D.; Harris, A.L.; Hulikova, A. The chemistry, physiology and pathology of pH in cancer. *Philos. Trans. R. Soc. B Biol. Sci.* **2014**. [CrossRef] [PubMed]

105. Neri, D.; Supuran, C.T. Interfering with pH regulation in tumours as a therapeutic strategy. *Nat. Rev. Drug Discov.* **2011**, *10*, 767–777. [CrossRef] [PubMed]

106. Uda, N.R.; Seibert, V.; Stenner-Liewen, F.; Müller, P.; Herzig, P.; Gondi, G.; Zeidler, R.; van Dijk, M.; Zippelius, A.; Renner, C. Esterase activity of carbonic anhydrases serves as surrogate for selecting antibodies blocking hydratase activity. *J. Enzyme Inhib. Med. Chem.* **2015**, *30*, 955–960. [CrossRef] [PubMed]

107. Verpoorte, J.A.; Mehta, S.; Edsall, J.T. Esterase activities of human carbonic anhydrases B and C. *J. Biol. Chem.* **1967**, *242*, 4221–4229. [PubMed]

108. Lindskog, S.; Coleman, J.E. The catalytic mechanism of carbonic anhydrase. *Proc. Natl. Acad. Sci. USA* **1973**, *70*, 2505–2508. [CrossRef] [PubMed]

109. Al-Samir, S.; Papadopoulos, S.; Scheibe, R.J.; Meißner, J.D.; Cartron, J.-P.; Sly, W.S.; Alper, S.L.; Gros, G.; Endeward, V. Activity and distribution of intracellular carbonic anhydrase II and their effects on the transport activity of anion exchanger AE1/SLC4A1: Role of CAII in the function of AE1. *J. Physiol.* **2013**, *591*, 4963–4982. [CrossRef] [PubMed]

110. Benej, M.; Pastorekova, S.; Pastorek, J. Carbonic anhydrase IX: Regulation and role in cancer. *Subcell. Biochem.* **2014**, *75*, 199–219. [CrossRef] [PubMed]

111. Alterio, V.; Hilvo, M.; Di Fiore, A.; Supuran, C.T.; Pan, P.; Parkkila, S.; Scaloni, A.; Pastorek, J.; Pastorekova, S.; Pedone, C.; et al. Crystal structure of the catalytic domain of the tumor-associated human carbonic anhydrase IX. *Proc. Natl. Acad. Sci. USA* **2009**, *106*, 16233–16238. [CrossRef] [PubMed]

112. Widmann, M.; Trodler, P.; Pleiss, J. The isoelectric region of proteins: A systematic analysis. *PLoS ONE* **2010**, *5*, e10546. [CrossRef] [PubMed]

113. Klier, M.; Andes, F.T.; Deitmer, J.W.; Becker, H.M. Intracellular and extracellular carbonic anhydrases cooperate non-enzymatically to enhance activity of monocarboxylate transporters. *J. Biol. Chem.* **2014**, *289*, 2765–2775. [CrossRef] [PubMed]

114. Becker, H.M.; Klier, M.; Schüler, C.; McKenna, R.; Deitmer, J.W. Intramolecular proton shuttle supports not only catalytic but also noncatalytic function of carbonic anhydrase II. *Proc. Natl. Acad. Sci. USA* **2011**, *108*, 3071–3076. [CrossRef] [PubMed]

115. Svastova, E.; Witarski, W.; Csaderova, L.; Kosik, I.; Skvarkova, L.; Hulikova, A.; Zatovicova, M.; Barathova, M.; Kopacek, J.; Pastorek, J.; et al. Carbonic anhydrase IX interacts with bicarbonate transporters in lamellipodia and increases cell migration via its catalytic domain. *J. Biol. Chem.* **2012**, *287*, 3392–3402. [CrossRef] [PubMed]

116. Innocenti, A.; Vullo, D.; Scozzafava, A.; Casey, J.R.; Supuran, C.T. Carbonic anhydrase inhibitors. Interaction of isozymes I, II, IV, V, and IX with carboxylates. *Bioorg. Med. Chem. Lett.* **2005**, *15*, 573–578. [CrossRef] [PubMed]

117. Swietach, P.; Patiar, S.; Supuran, C.T.; Harris, A.L.; Vaughan-Jones, R.D. The role of carbonic anhydrase 9 in regulating extracellular and intracellular pH in three-dimensional tumor cell growths. *J. Biol. Chem.* **2009**, *284*, 20299–20310. [CrossRef] [PubMed]

118. Giffard, R.G.; Monyer, H.; Christine, C.W.; Choi, D.W. Acidosis reduces NMDA receptor activation, glutamate neurotoxicity, and oxygen-glucose deprivation neuronal injury in cortical cultures. *Brain Res.* **1990**, *506*, 339–342. [CrossRef]

119. The PyMOL Molecular Graphics System, Version 2.0 Schrödinger, LLC. Available online: https://pymol.org/2/ (accessed on 21 February 2018).

120. Whittington, D.A.; Waheed, A.; Ulmasov, B.; Shah, G.N.; Grubb, J.H.; Sly, W.S.; Christianson, D.W. Crystal structure of the dimeric extracellular domain of human carbonic anhydrase XII, a bitopic membrane protein overexpressed in certain cancer tumor cells. *Proc. Natl. Acad. Sci. USA* **2001**, *98*, 9545–9550. [CrossRef] [PubMed]

121. Zhao, J.; Benlekbir, S.; Rubinstein, J.L. Electron cryomicroscopy observation of rotational states in a eukaryotic V-ATPase. *Nature* **2015**, *521*, 241–245. [CrossRef] [PubMed]

122. Arakawa, T.; Kobayashi-Yurugi, T.; Alguel, Y.; Iwanari, H.; Hatae, H.; Iwata, M.; Abe, Y.; Hino, T.; Ikeda-Suno, C.; Kuma, H.; et al. Crystal structure of the anion exchanger domain of human erythrocyte band 3. *Science* **2015**, *350*, 680–684. [CrossRef] [PubMed]

123. Huang, Y. Structure and mechanism of the glycerol-3-phosphate transporter from escherichia coli. *Science* **2003**, *301*, 616–620. [CrossRef] [PubMed]

124. Deng, D.; Xu, C.; Sun, P.; Wu, J.; Yan, C.; Hu, M.; Yan, N. Crystal structure of the human glucose transporter GLUT1. *Nature* **2014**, *510*, 121–125. [CrossRef] [PubMed]

125. Avvaru, B.S.; Kim, C.U.; Sippel, K.H.; Gruner, S.M.; Agbandje-McKenna, M.; Silverman, D.N.; McKenna, R. A short, Strong hydrogen bond in the active site of human carbonic anhydrase II. *Biochemistry* **2010**, *49*, 249–251. [CrossRef] [PubMed]

126. Michalczyk, R.; Unkefer, C.J.; Bacik, J.-P.; Schrader, T.E.; Ostermann, A.; Kovalevsky, A.Y.; McKenna, R.; Fisher, S.Z. Joint neutron crystallographic and NMR solution studies of Tyr residue ionization and hydrogen bonding: Implications for enzyme-mediated proton transfer. *Proc. Natl. Acad. Sci. USA* **2015**, *112*, 5673–5678. [CrossRef] [PubMed]

127. Chiang, W.L.; Chu, S.C.; Lai, J.C.; Yang, S.F.; Chiou, H.L.; Hsieh, Y.S. Alternations in quantities and activities of erythrocyte cytosolic carbonic anhydrase isoenzymes in glucose-6-phosphate dehydrogenase-deficient individuals. *Clin. Chim. Acta Int. J. Clin. Chem.* **2001**, *314*, 195–201. [CrossRef]

128. Maren, T.H.; Swenson, E.R. A comparative study of the kinetics of the Bohr effect in vertebrates. *J. Physiol.* **1980**, *303*, 535–547. [CrossRef] [PubMed]

129. Swenson, E.R. Respiratory and renal roles of carbonic anhydrase in gas exchange and acid-base regulation. In *The Carbonic Anhydrases*; Birkhäuser Basel: Basel, Switzerland, 2000; pp. 281–341.

130. Brown, B.F.; Quon, A.; Dyck, J.R.B.; Casey, J.R. Carbonic anhydrase II promotes cardiomyocyte hypertrophy. *Can. J. Physiol. Pharmacol.* **2012**, *90*, 1599–1610. [CrossRef] [PubMed]

131. Kuo, W.-H.; Yang, S.-F.; Hsieh, Y.-S.; Tsai, C.-S.; Hwang, W.-L.; Chu, S.-C. Differential expression of carbonic anhydrase isoenzymes in various types of anemia. *Clin. Chim. Acta Int. J. Clin. Chem.* **2005**, *351*, 79–86. [CrossRef] [PubMed]

132. Gilmour, K.M. Perspectives on carbonic anhydrase. *Comp. Biochem. Physiol. A Mol. Integr. Physiol.* **2010**, *157*, 193–197. [CrossRef] [PubMed]

133. Becker, H.M.; Deitmer, J.W. Carbonic anhydrase II increases the activity of the human electrogenic Na$^+$/HCO$_3^-$ cotransporter. *J. Biol. Chem.* **2007**, *282*, 13508–13521. [CrossRef] [PubMed]

134. Becker, H.M.; Deitmer, J.W. Nonenzymatic proton handling by carbonic anhydrase II during H$^+$-lactate cotransport via monocarboxylate transporter 1. *J. Biol. Chem.* **2008**, *283*, 21655–21667. [CrossRef] [PubMed]

135. Stridh, M.H.; Alt, M.D.; Wittmann, S.; Heidtmann, H.; Aggarwal, M.; Riederer, B.; Seidler, U.; Wennemuth, G.; McKenna, R.; Deitmer, J.W.; et al. Lactate flux in astrocytes is enhanced by a non-catalytic action of carbonic anhydrase II: CAII enhances lactate transport in astrocytes. *J. Physiol.* **2012**, *590*, 2333–2351. [CrossRef] [PubMed]

136. Uhlén, M.; Björling, E.; Agaton, C.; Szigyarto, C.A.-K.; Amini, B.; Andersen, E.; Andersson, A.-C.; Angelidou, P.; Asplund, A.; Asplund, C.; et al. A human protein atlas for normal and cancer tissues based on antibody proteomics. *Mol. Cell. Proteom.* **2005**, *4*, 1920–1932. [CrossRef] [PubMed]

137. Uhlen, M.; Oksvold, P.; Fagerberg, L.; Lundberg, E.; Jonasson, K.; Forsberg, M.; Zwahlen, M.; Kampf, C.; Wester, K.; Hober, S.; et al. Towards a knowledge-based human protein atlas. *Nat. Biotechnol.* **2010**, *28*, 1248–1250. [CrossRef] [PubMed]

138. Uhlen, M.; Fagerberg, L.; Hallstrom, B.M.; Lindskog, C.; Oksvold, P.; Mardinoglu, A.; Sivertsson, A.; Kampf, C.; Sjostedt, E.; Asplund, A.; et al. Tissue-based map of the human proteome. *Science* **2015**, *347*. [CrossRef] [PubMed]

139. Uhlen, M.; Zhang, C.; Lee, S.; Sjöstedt, E.; Fagerberg, L.; Bidkhori, G.; Benfeitas, R.; Arif, M.; Liu, Z.; Edfors, F.; et al. A pathology atlas of the human cancer transcriptome. *Science* **2017**. [CrossRef] [PubMed]

140. Thul, P.J.; Åkesson, L.; Wiking, M.; Mahdessian, D.; Geladaki, A.; Ait Blal, H.; Alm, T.; Asplund, A.; Björk, L.; Breckels, L.M.; et al. A subcellular map of the human proteome. *Science* **2017**, *356*, eaal3321. [CrossRef] [PubMed]

141. Cerami, E.; Gao, J.; Dogrusoz, U.; Gross, B.E.; Sumer, S.O.; Aksoy, B.A.; Jacobsen, A.; Byrne, C.J.; Heuer, M.L.; Larsson, E.; et al. The cBio Cancer Genomics Portal: An open platform for exploring multidimensional cancer genomics data: Figure 1. *Cancer Discov.* **2012**, *2*, 401–404. [CrossRef] [PubMed]

142. Gao, J.; Aksoy, B.A.; Dogrusoz, U.; Dresdner, G.; Gross, B.; Sumer, S.O.; Sun, Y.; Jacobsen, A.; Sinha, R.; Larsson, E.; et al. Integrative Analysis of Complex Cancer Genomics and Clinical Profiles Using the cBioPortal. *Sci. Signal.* **2013**. [CrossRef] [PubMed]

143. Zheng, Y.; Xu, B.; Zhao, Y.; Gu, H.; Li, C.; Wang, Y.; Chang, X. CA1 contributes to microcalcification and tumourigenesis in breast cancer. *BMC Cancer* **2015**, *15*. [CrossRef] [PubMed]

144. Wang, D.; Lu, X.; Zhang, X.; Li, Z.; Li, C. Carbonic anhydrase 1 is a promising biomarker for early detection of non-small cell lung cancer. *Tumor Biol.* **2016**, *37*, 553–559. [CrossRef] [PubMed]

145. Takakura, M.; Yokomizo, A.; Tanaka, Y.; Kobayashi, M.; Jung, G.; Banno, M.; Sakuma, T.; Imada, K.; Oda, Y.; Kamita, M.; et al. Carbonic anhydrase I as a new plasma biomarker for prostate cancer. *ISRN Oncol.* **2012**, *2012*, 1–10. [CrossRef] [PubMed]

146. Zhou, R.; Huang, W.; Yao, Y.; Wang, Y.; Li, Z.; Shao, B.; Zhong, J.; Tang, M.; Liang, S.; Zhao, X.; et al. CA II, a potential biomarker by proteomic analysis, exerts significant inhibitory effect on the growth of colorectal cancer cells. *Int. J. Oncol.* **2013**, *43*, 611–621. [CrossRef] [PubMed]

147. Parkkila, S.; Lasota, J.; Fletcher, J.A.; Ou, W.; Kivelä, A.J.; Nuorva, K.; Parkkila, A.-K.; Ollikainen, J.; Sly, W.S.; Waheed, A.; et al. Carbonic anhydrase II. A novel biomarker for gastrointestinal stromal tumors. *Mod. Pathol.* **2010**, *23*, 743–750. [CrossRef] [PubMed]

148. Järvinen, P.; Kivelä, A.J.; Nummela, P.; Lepistö, A.; Ristimäki, A.; Parkkila, S. Carbonic anhydrase II: A novel biomarker for pseudomyxoma peritonei. *APMIS* **2017**, *125*, 207–212. [CrossRef] [PubMed]

149. Waterman, E.A.; Cross, N.A.; Lippitt, J.M.; Cross, S.S.; Rehman, I.; Holen, I.; Hamdy, F.C.; Eaton, C.L. The antibody MAB8051 directed against osteoprotegerin detects carbonic anhydrase II: Implications for association studies with human cancers. *Int. J. Cancer* **2007**, *121*, 1958–1966. [CrossRef] [PubMed]

150. Liu, L.-C. Overexpression of carbonic anhydrase II and Ki-67 proteins in prognosis of gastrointestinal stromal tumors. *World J. Gastroenterol.* **2013**, *19*, 2473. [CrossRef] [PubMed]

151. Liu, C.-M.; Lin, Y.-M.; Yeh, K.-T.; Chen, M.-K.; Chang, J.-H.; Chen, C.-J.; Chou, M.-Y.; Yang, S.-F.; Chien, M.-H. Expression of carbonic anhydrases I/II and the correlation to clinical aspects of oral squamous cell carcinoma analyzed using tissue microarray: CA I/II and correlation to OSCC. *J. Oral Pathol. Med.* **2012**. [CrossRef] [PubMed]

152. Harju, A.-K.; Bootorabi, F.; Kuuslahti, M.; Supuran, C.T.; Parkkila, S. Carbonic anhydrase III: A neglected isozyme is stepping into the limelight. *J. Enzyme Inhib. Med. Chem.* **2013**, *28*, 231–239. [CrossRef] [PubMed]

153. Shi, C.; Uda, Y.; Dedic, C.; Azab, E.; Sun, N.; Hussein, A.I.; Petty, C.A.; Fulzele, K.; Mitterberger-Vogt, M.C.; Zwerschke, W.; et al. Carbonic anhydrase III protects osteocytes from oxidative stress. *FASEB J.* **2018**, *32*, 440–452. [CrossRef] [PubMed]

154. Dai, H.-Y.; Hong, C.-C.; Liang, S.-C.; Yan, M.-D.; Lai, G.-M.; Cheng, A.-L.; Chuang, S.-E. Carbonic anhydrase III promotes transformation and invasion capability in hepatoma cells through FAK signaling pathway. *Mol. Carcinog.* **2008**, *47*, 956–963. [CrossRef] [PubMed]

155. Bootorabi, F.; Haapasalo, J.; Smith, E.; Haapasalo, H.; Parkkila, S. Carbonic anhydrase VII—A potential prognostic marker in gliomas. *Health* **2011**, *3*, 6–12. [CrossRef]

156. Hilvo, M.; Innocenti, A.; Monti, S.M.; De Simone, G.; Supuran, C.T.; Parkkila, S. Recent advances in research on the most novel carbonic anhydrases, CA XIII and XV. *Curr. Pharm. Des.* **2008**, *14*, 672–678. [PubMed]

157. Shah, G.N.; Hewett-Emmett, D.; Grubb, J.H.; Migas, M.C.; Fleming, R.E.; Waheed, A.; Sly, W.S. Mitochondrial carbonic anhydrase CA VB: Differences in tissue distribution and pattern of evolution from those of CA VA suggest distinct physiological roles. *Proc. Natl. Acad. Sci. USA* **2000**, *97*, 1677–1682. [CrossRef] [PubMed]

158. Nishimori, I.; Vullo, D.; Innocenti, A.; Scozzafava, A.; Mastrolorenzo, A.; Supuran, C.T. Carbonic Anhydrase Inhibitors. The Mitochondrial Isozyme VB as a new target for sulfonamide and sulfamate inhibitors. *J. Med. Chem.* **2005**, *48*, 7860–7866. [CrossRef] [PubMed]

159. Arechederra, R.L.; Waheed, A.; Sly, W.S.; Supuran, C.T.; Minteer, S.D. Effect of sulfonamides as carbonic anhydrase VA and VB inhibitors on mitochondrial metabolic energy conversion. *Bioorg. Med. Chem.* **2013**, *21*, 1544–1548. [CrossRef] [PubMed]

160. Shah, G.N.; Rubbelke, T.S.; Hendin, J.; Nguyen, H.; Waheed, A.; Shoemaker, J.D.; Sly, W.S. Targeted mutagenesis of mitochondrial carbonic anhydrases VA and VB implicates both enzymes in ammonia detoxification and glucose metabolism. *Proc. Natl. Acad. Sci. USA* **2013**, *110*, 7423–7428. [CrossRef] [PubMed]

161. Poulsen, S.-A.; Wilkinson, B.L.; Innocenti, A.; Vullo, D.; Supuran, C.T. Inhibition of human mitochondrial carbonic anhydrases VA and VB with para-(4-phenyltriazole-1-yl)-benzenesulfonamide derivatives. *Bioorg. Med. Chem. Lett.* **2008**, *18*, 4624–4627. [CrossRef] [PubMed]

162. Lusty, C.J. Carbamoylphosphate synthetase I of rat-liver mitochondria. Purification, properties, and polypeptide molecular weight. *Eur. J. Biochem.* **1978**, *85*, 373–383. [CrossRef] [PubMed]

163. Dodgson, S.J.; Forster, R.E.; Storey, B.T. The role of carbonic anhydrase in hepatocyte metabolism. *Ann. N. Y. Acad. Sci.* **1984**, *429*, 516–524. [CrossRef] [PubMed]

164. Dodgson, S.J.; Forster, R.E. Inhibition of CA V decreases glucose synthesis from pyruvate. *Arch. Biochem. Biophys.* **1986**, *251*, 198–204. [CrossRef]

165. Dodgson, S.J. Inhibition of mitochondrial carbonic anhydrase and ureagenesis: A discrepancy examined. *J. Appl. Physiol.* **1987**, *63*, 2134–2141. [CrossRef] [PubMed]

166. Hazen, S.A.; Waheed, A.; Sly, W.S.; LaNoue, K.F.; Lynch, C.J. Differentiation-dependent expression of CA V and the role of carbonic anhydrase isozymes in pyruvate carboxylation in adipocytes. *FASEB J.* **1996**, *10*, 481–490. [CrossRef] [PubMed]

167. Henry, R.P. Multiple roles of carbonic anhydrase in cellular transport and metabolism. *Annu. Rev. Physiol.* **1996**, *58*, 523–538. [CrossRef] [PubMed]

168. Van Karnebeek, C.D.; Sly, W.S.; Ross, C.J.; Salvarinova, R.; Yaplito-Lee, J.; Santra, S.; Shyr, C.; Horvath, G.A.; Eydoux, P.; Lehman, A.M.; et al. Mitochondrial Carbonic Anhydrase VA Deficiency Resulting from CA5A Alterations presents with hyperammonemia in early childhood. *Am. J. Hum. Genet.* **2014**, *94*, 453–461. [CrossRef] [PubMed]

169. Kivelä, J.; Parkkila, S.; Parkkila, A.-K.; Leinonen, J.; Rajaniemi, H. Salivary carbonic anhydrase isoenzyme VI. *J. Physiol.* **1999**, *520*, 315–320. [CrossRef] [PubMed]

170. Frasseto, F.; Parisotto, T.M.; Peres, R.C.R.; Marques, M.R.; Line, S.R.P.; Nobre Dos Santos, M. Relationship among salivary carbonic anhydrase VI activity and flow rate, biofilm pH and caries in primary dentition. *Caries Res.* **2012**, *46*, 194–200. [CrossRef] [PubMed]

171. Kimoto, M.; Kishino, M.; Yura, Y.; Ogawa, Y. A role of salivary carbonic anhydrase VI in dental plaque. *Arch. Oral Biol.* **2006**, *51*, 117–122. [CrossRef] [PubMed]

172. Feeney, E.L.; Hayes, J.E. Exploring associations between taste perception, oral anatomy and polymorphisms in the carbonic anhydrase (gustin) gene *CA6*. *Physiol. Behav.* **2014**, *128*, 148–154. [CrossRef] [PubMed]

173. Thatcher, B.J.; Doherty, A.E.; Orvisky, E.; Martin, B.M.; Henkin, R.I. Gustin from human parotid saliva is carbonic anhydrase VI. *Biochem. Biophys. Res. Commun.* **1998**, *250*, 635–641. [CrossRef] [PubMed]

174. Karhumaa, P.; Leinonen, J.; Parkkila, S.; Kaunisto, K.; Tapanainen, J.; Rajaniemi, H. The identification of secreted carbonic anhydrase VI as a constitutive glycoprotein of human and rat milk. *Proc. Natl. Acad. Sci. USA* **2001**, *98*, 11604–11608. [CrossRef] [PubMed]

175. Ogawa, Y.; Matsumoto, K.; Maeda, T.; Tamai, R.; Suzuki, T.; Sasano, H.; Fernley, R.T. Characterization of lacrimal gland carbonic anhydrase VI. *J. Histochem. Cytochem.* **2002**, *50*, 821–827. [CrossRef] [PubMed]

176. Leinonen, J.S.; Saari, K.A.; Seppänen, J.M.; Myllylä, H.M.; Rajaniemi, H.J. Immunohistochemical demonstration of carbonic anhydrase isoenzyme VI (CA VI) expression in rat lower airways and lung. *J. Histochem. Cytochem.* **2004**, *52*, 1107–1112. [CrossRef] [PubMed]

177. Kivelä, J.; Parkkila, S.; Waheed, A.; Parkkila, A.K.; Sly, W.S.; Rajaniemi, H. Secretory carbonic anhydrase isoenzyme (CA VI) in human serum. *Clin. Chem.* **1997**, *43*, 2318–2322. [PubMed]

178. Li, Z.-Q.; Hu, X.-P.; Zhou, J.-Y.; Xie, X.-D.; Zhang, J.-M. Genetic polymorphisms in the carbonic anhydrase VI gene and dental caries susceptibility. *Genet. Mol. Res.* **2015**, *14*, 5986–5993. [CrossRef] [PubMed]

179. Sengul, F.; Kilic, M.; Gurbuz, T.; Tasdemir, S. Carbonic anhydrase VI gene polymorphism rs2274327 relationship between salivary parameters and dental-oral health status in children. *Biochem. Genet.* **2016**, *54*, 467–475. [CrossRef] [PubMed]

180. Dowd, F.J. Saliva and dental caries. *Dent. Clin. N. Am.* **1999**, *43*, 579–597. [PubMed]

181. Shatzman, A.R.; Henkin, R.I. Gustin concentration changes relative to salivary zinc and taste in humans. *Proc. Natl. Acad. Sci. USA* **1981**, *78*, 3867–3871. [CrossRef] [PubMed]

182. Henkin, R.I.; Lippoldt, R.E.; Bilstad, J.; Edelhoch, H. A zinc protein isolated from human parotid saliva. *Proc. Natl. Acad. Sci. USA* **1975**, *72*, 488–492. [CrossRef] [PubMed]

183. Innocenti, A.; Pastorekova, S.; Pastorek, J.; Scozzafava, A.; Simone, G.D.; Supuran, C.T. The proteoglycan region of the tumor-associated carbonic anhydrase isoform IX acts as anintrinsic buffer optimizing CO_2 hydration at acidic pH values characteristic of solid tumors. *Bioorg. Med. Chem. Lett.* **2009**, *19*, 5825–5828. [CrossRef] [PubMed]

184. Barker, H.; Aaltonen, M.; Pan, P.; Vähätupa, M.; Kaipiainen, P.; May, U.; Prince, S.; Uusitalo-Järvinen, H.; Waheed, A.; Pastoreková, S.; et al. Role of carbonic anhydrases in skin wound healing. *Exp. Mol. Med.* **2017**, *49*, e334. [CrossRef] [PubMed]

185. Fujikawa-Adachi, K.; Nishimori, I.; Sakamoto, S.; Morita, M.; Onishi, S.; Yonezawa, S.; Hollingsworth, M.A. Identification of carbonic anhydrase IV and VI mRNA expression in human pancreas and salivary glands. *Pancreas* **1999**, *18*, 329–335. [CrossRef] [PubMed]

186. Wistrand, P.J.; Carter, N.D.; Conroy, C.W.; Mahieu, I. Carbonic anhydrase IV activity is localized on the exterior surface of human erythrocytes. *Acta Physiol. Scand.* **1999**, *165*, 211–218. [CrossRef] [PubMed]

187. Sender, S.; Decker, B.; Fenske, C.D.; Sly, W.S.; Carter, N.D.; Gros, G. Localization of carbonic anhydrase IV in rat and human heart muscle. *J. Histochem. Cytochem.* **1998**, *46*, 855–861. [CrossRef] [PubMed]

188. Parkkila, S.; Parkkila, A.K.; Juvonen, T.; Waheed, A.; Sly, W.S.; Saarnio, J.; Kaunisto, K.; Kellokumpu, S.; Rajaniemi, H. Membrane-bound carbonic anhydrase IV is expressed in the luminal plasma membrane of the human gallbladder epithelium. *Hepatology* **1996**, *24*, 1104–1108. [CrossRef] [PubMed]

189. Sender, S.; Gros, G.; Waheed, A.; Hageman, G.S.; Sly, W.S. Immunohistochemical localization of carbonic anhydrase IV in capillaries of rat and human skeletal muscle. *J. Histochem. Cytochem.* **1994**, *42*, 1229–1236. [CrossRef] [PubMed]

190. Fleming, R.E.; Parkkila, S.; Parkkila, A.K.; Rajaniemi, H.; Waheed, A.; Sly, W.S. Carbonic anhydrase IV expression in rat and human gastrointestinal tract regional, cellular, and subcellular localization. *J. Clin. Investig.* **1995**, *96*, 2907–2913. [CrossRef] [PubMed]

191. Purkerson, J.M.; Schwartz, G.J. The role of carbonic anhydrases in renal physiology. *Kidney Int.* **2007**, *71*, 103–115. [CrossRef] [PubMed]

192. Alvarez, B.V.; Loiselle, F.B.; Supuran, C.T.; Schwartz, G.J.; Casey, J.R. Direct extracellular interaction between carbonic anhydrase IV and the human NBC1 sodium/bicarbonate co-transporter. *Biochemistry* **2003**, *42*, 12321–12329. [CrossRef] [PubMed]
193. Alvarez, B.V.; Vithana, E.N.; Yang, Z.; Koh, A.H.; Yeung, K.; Yong, V.; Shandro, H.J.; Chen, Y.; Kolatkar, P.; Palasingam, P.; et al. Identification and characterization of a novel mutation in the carbonic anhydrase IV gene that causes retinitis pigmentosa. *Invest. Ophthalmol. Vis. Sci.* **2007**, *48*, 3459–3468. [CrossRef] [PubMed]
194. Yang, Z.; Alvarez, B.V.; Chakarova, C.; Jiang, L.; Karan, G.; Frederick, J.M.; Zhao, Y.; Sauvé, Y.; Li, X.; Zrenner, E.; et al. Mutant carbonic anhydrase 4 impairs pH regulation and causes retinal photoreceptor degeneration. *Hum. Mol. Genet.* **2005**, *14*, 255–265. [CrossRef] [PubMed]
195. Parkkila, S.; Parkkila, A.K.; Rajaniemi, H.; Shah, G.N.; Grubb, J.H.; Waheed, A.; Sly, W.S. Expression of membrane-associated carbonic anhydrase XIV on neurons and axons in mouse and human brain. *Proc. Natl. Acad. Sci. USA* **2001**, *98*, 1918–1923. [CrossRef] [PubMed]
196. Kaunisto, K.; Parkkila, S.; Rajaniemi, H.; Waheed, A.; Grubb, J.; Sly, W.S. Carbonic anhydrase XIV: Luminal expression suggests key role in renal acidification. *Kidney Int.* **2002**, *61*, 2111–2118. [CrossRef] [PubMed]
197. Juel, C.; Lundby, C.; Sander, M.; Calbet, J.A.L.; van Hall, G. Human skeletal muscle and erythrocyte proteins involved in acid-base homeostasis: Adaptations to chronic hypoxia. *J. Physiol.* **2003**, *548*, 639–648. [CrossRef] [PubMed]
198. Vargas, L.A.; Alvarez, B.V. Carbonic anhydrase XIV in the normal and hypertrophic myocardium. *J. Mol. Cell. Cardiol.* **2012**, *52*, 741–752. [CrossRef] [PubMed]
199. Pastorek, J.; Pastoreková, S.; Callebaut, I.; Mornon, J.P.; Zelník, V.; Opavský, R.; Zat'ovicová, M.; Liao, S.; Portetelle, D.; Stanbridge, E.J. Cloning and characterization of MN, a human tumor-associated protein with a domain homologous to carbonic anhydrase and a putative helix-loop-helix DNA binding segment. *Oncogene* **1994**, *9*, 2877–2888. [PubMed]
200. Liao, S.-Y.; Lerman, M.I.; Stanbridge, E.J. Expression of transmembrane carbonic anhydrases, CAIX and CAXII, in human development. *BMC Dev. Biol.* **2009**, *9*, 22. [CrossRef] [PubMed]
201. Ivanov, S.; Shu-Yuan, L.; Ivanova, A.; Danilkovitch-Miagkova, A.; Tarasova, N.; Weirich, G.; Merrill, M.J.; Proescholdt, M.A.; Oldfield, E.H.; Lee, J.; et al. Expression of hypoxia inducible cell surface transmembrane carbonic anhydrases in human cancer. *Am. J. Pathol.* **2001**, *158*, 905–919. [CrossRef]
202. Radvak, P.; Repic, M.; Svastova, E.; Takacova, M.; Csaderova, L.; Strnad, H.; Pastorek, J.; Pastorekova, S.; Kopacek, J. Suppression of carbonic anhydrase IX leads to aberrant focal adhesion and decreased invasion of tumor cells. *Oncol. Rep.* **2013**, *29*, 1147–1153. [CrossRef] [PubMed]
203. Csaderova, L.; Debreova, M.; Radvak, P.; Stano, M.; Vrestiakova, M.; Kopacek, J.; Pastorekova, S.; Svastova, E. The effect of carbonic anhydrase IX on focal contacts during cell spreading and migration. *Front. Physiol.* **2013**, *4*. [CrossRef] [PubMed]
204. Karhumaa, P.; Kaunisto, K.; Parkkila, S.; Waheed, A.; Pastoreková, S.; Pastorek, J.; Sly, W.S.; Rajaniemi, H. Expression of the transmembrane carbonic anhydrases, CA IX and CA XII, in the human male excurrent ducts. *Mol. Hum. Reprod.* **2001**, *7*, 611–616. [CrossRef] [PubMed]
205. Parkkila, S.; Parkkila, A.K.; Saarnio, J.; Kivelä, J.; Karttunen, T.J.; Kaunisto, K.; Waheed, A.; Sly, W.S.; Türeci, O.; Virtanen, I.; et al. Expression of the membrane-associated carbonic anhydrase isozyme XII in the human kidney and renal tumors. *J. Histochem. Cytochem.* **2000**, *48*, 1601–1608. [CrossRef] [PubMed]
206. Lee, M.; Vecchio-Pagán, B.; Sharma, N.; Waheed, A.; Li, X.; Raraigh, K.S.; Robbins, S.; Han, S.T.; Franca, A.L.; Pellicore, M.J.; et al. Loss of carbonic anhydrase XII function in individuals with elevated sweat chloride concentration and pulmonary airway disease. *Hum. Mol. Genet.* **2016**, *25*, 1923–1933. [CrossRef] [PubMed]
207. Feldshtein, M.; Elkrinawi, S.; Yerushalmi, B.; Marcus, B.; Vullo, D.; Romi, H.; Ofir, R.; Landau, D.; Sivan, S.; Supuran, C.T.; et al. Hyperchlorhidrosis caused by homozygous mutation in CA12, encoding carbonic anhydrase XII. *Am. J. Hum. Genet.* **2010**, *87*, 713–720. [CrossRef] [PubMed]
208. Capkova, L.; Koubkova, L.; Kodet, R. Expression of carbonic anhydrase IX (CAIX) in malignant mesothelioma. An immunohistochemical and immunocytochemical study. *Neoplasma* **2014**, *61*, 161–169. [CrossRef] [PubMed]
209. Yang, J.-S.; Chen, M.-K.; Yang, S.-F.; Chang, Y.-C.; Su, S.-C.; Chiou, H.-L.; Chien, M.-H.; Lin, C.-W. Increased expression of carbonic anhydrase IX in oral submucous fibrosis and oral squamous cell carcinoma. *Clin. Chem. Lab. Med.* **2014**, *52*, 1367–1377. [CrossRef] [PubMed]

210. Jomrich, G.; Jesch, B.; Birner, P.; Schwameis, K.; Paireder, M.; Asari, R.; Schoppmann, S.F. Stromal expression of carbonic anhydrase IX in esophageal cancer. *Clin. Transl. Oncol.* **2014**, *16*, 966–972. [CrossRef] [PubMed]

211. Wykoff, C.C.; Beasley, N.J.; Watson, P.H.; Turner, K.J.; Pastorek, J.; Sibtain, A.; Wilson, G.D.; Turley, H.; Talks, K.L.; Maxwell, P.H.; et al. Hypoxia-inducible expression of tumor-associated carbonic anhydrases. *Cancer Res.* **2000**, *60*, 7075–7083. [PubMed]

212. Kopacek, J.; Barathova, M.; Dequiedt, F.; Sepelakova, J.; Kettmann, R.; Pastorek, J.; Pastorekova, S. MAPK pathway contributes to density- and hypoxia-induced expression of the tumor-associated carbonic anhydrase IX. *Biochim. Biophys. Acta BBA—Gene Struct. Expr.* **2005**, *1729*, 41–49. [CrossRef] [PubMed]

213. Kaluz, S.; Kaluzová, M.; Chrastina, A.; Olive, P.L.; Pastoreková, S.; Pastorek, J.; Lerman, M.I.; Stanbridge, E.J. Lowered oxygen tension induces expression of the hypoxia marker MN/carbonic anhydrase IX in the absence of hypoxia-inducible factor 1 alpha stabilization: A role for phosphatidylinositol 3'-kinase. *Cancer Res.* **2002**, *62*, 4469–4477. [PubMed]

214. Swayampakula, M.; McDonald, P.C.; Vallejo, M.; Coyaud, E.; Chafe, S.C.; Westerback, A.; Venkateswaran, G.; Shankar, J.; Gao, G.; Laurent, E.M.N.; et al. The interactome of metabolic enzyme carbonic anhydrase IX reveals novel roles in tumor cell migration and invadopodia/MMP14-mediated invasion. *Oncogene* **2017**, *36*, 6244–6261. [CrossRef] [PubMed]

215. Ditte, P.; Dequiedt, F.; Svastova, E.; Hulikova, A.; Ohradanova-Repic, A.; Zatovicova, M.; Csaderova, L.; Kopacek, J.; Supuran, C.T.; Pastorekova, S.; et al. Phosphorylation of carbonic anhydrase IX controls its ability to mediate extracellular acidification in hypoxic tumors. *Cancer Res.* **2011**, *71*, 7553–7567. [CrossRef] [PubMed]

216. Dorai, T.; Sawczuk, I.S.; Pastorek, J.; Wiernik, P.H.; Dutcher, J.P. The role of carbonic anhydrase IX overexpression in kidney cancer. *Eur. J. Cancer* **2005**, *41*, 2935–2947. [CrossRef] [PubMed]

217. Nasu, K.; Yamaguchi, K.; Takanashi, T.; Tamai, K.; Sato, I.; Ine, S.; Sasaki, O.; Satoh, K.; Tanaka, N.; Tanaka, Y.; et al. Crucial role of carbonic anhydrase IX in tumorigenicity of xenotransplanted adult T-cell leukemia-derived cells. *Cancer Sci.* **2017**, *108*, 435–443. [CrossRef] [PubMed]

218. Li, Y.; Dong, M.; Sheng, W.; Huang, L. Roles of carbonic anhydrase IX in development of pancreatic cancer. *Pathol. Oncol. Res.* **2016**, *22*, 277–286. [CrossRef] [PubMed]

219. Li, G.; Feng, G.; Zhao, A.; Péoc'h, M.; Cottier, M.; Mottet, N. CA9 as a biomarker in preoperative biopsy of small solid renal masses for diagnosis of clear cell renal cell carcinoma. *Biomarkers* **2017**, *22*, 123–126. [CrossRef] [PubMed]

220. Smith, A.D.; Truong, M.; Bristow, R.; Yip, P.; Milosevic, M.F.; Joshua, A.M. The utility of serum CA9 for prognostication in prostate cancer. *Anticancer Res.* **2016**, *36*, 4489–4492. [CrossRef] [PubMed]

221. De Martino, M.; Lucca, I.; Mbeutcha, A.; Wiener, H.G.; Haitel, A.; Susani, M.; Shariat, S.F.; Klatte, T. Carbonic anhydrase IX as a diagnostic urinary marker for urothelial bladder cancer. *Eur. Urol.* **2015**, *68*, 552–554. [CrossRef] [PubMed]

222. Huang, W.-J.; Jeng, Y.-M.; Lai, H.-S.; Fong, I.-U.; Sheu, F.-Y. B.; Lai, P.-L.; Yuan, R.-H. Expression of hypoxic marker carbonic anhydrase IX predicts poor prognosis in resectable hepatocellular carcinoma. *PLoS ONE* **2015**, *10*, e0119181. [CrossRef] [PubMed]

223. Ilie, M.I.; Hofman, V.; Ortholan, C.; Ammadi, R.E.; Bonnetaud, C.; Havet, K.; Venissac, N.; Mouroux, J.; Mazure, N.M.; Pouysségur, J.; et al. Overexpression of carbonic anhydrase XII in tissues from resectable non-small cell lung cancers is a biomarker of good prognosis. *Int. J. Cancer* **2011**, *128*, 1614–1623. [CrossRef] [PubMed]

224. Kobayashi, M.; Matsumoto, T.; Ryuge, S.; Yanagita, K.; Nagashio, R.; Kawakami, Y.; Goshima, N.; Jiang, S.-X.; Saegusa, M.; Iyoda, A.; et al. CAXII is a sero-diagnostic marker for lung cancer. *PLoS ONE* **2012**, *7*, e33952. [CrossRef] [PubMed]

225. Yoo, C.W.; Nam, B.-H.; Kim, J.-Y.; Shin, H.-J.; Lim, H.; Lee, S.; Lee, S.-K.; Lim, M.-C.; Song, Y.-J. Carbonic anhydrase XII expression is associated with histologic grade of cervical cancer and superior radiotherapy outcome. *Radiat. Oncol. Lond. Engl.* **2010**, *5*, 101. [CrossRef] [PubMed]

226. Chien, M.-H.; Ying, T.-H.; Hsieh, Y.-H.; Lin, C.-H.; Shih, C.-H.; Wei, L.-H.; Yang, S.-F. Tumor-associated carbonic anhydrase XII is linked to the growth of primary oral squamous cell carcinoma and its poor prognosis. *Oral Oncol.* **2012**, *48*, 417–423. [CrossRef] [PubMed]
227. Kopecka, J.; Campia, I.; Jacobs, A.; Frei, A.P.; Ghigo, D.; Wollscheid, B.; Riganti, C. Carbonic anhydrase XII is a new therapeutic target to overcome chemoresistance in cancer cells. *Oncotarget* **2015**, *6*, 6776–6793. [CrossRef] [PubMed]

metabolites

MDPI

Review

Carbonic Anhydrase IX (CAIX), Cancer, and Radiation Responsiveness

Carol Ward [1,2,*] , James Meehan [1,3], Mark Gray [1,2], Ian H. Kunkler [1], Simon P. Langdon [1] and David J. Argyle [2]

1   Cancer Research UK Edinburgh Centre and Division of Pathology Laboratory, Institute of Genetics and Molecular Medicine, University of Edinburgh, Edinburgh EH4 2XU, UK; j.meehan@hw.ac uk (J.M.); s9900757@sms.ed.ac.uk (M.G.); I.Kunkler@ed.ac.uk (I.H.K.); simon.langdon@ed.ac.uk (S.P.L.)
2   Royal (Dick) School of Veterinary Studies and Roslin Institute, The University of Edinburgh, Easter Bush, Midlothian EH25 9RG, UK; david.argyle@roslin.ed.ac.uk
3   Institute of Sensors, Signals and Systems, School of Engineering and Physical Sciences, Heriot-Watt University, Edinburgh EH14 4AS, UK
*   Correspondence: drcward@hotmail.co.uk; Tel.: +44-131-537-1763

Received: 12 January 2018; Accepted: 7 February 2018; Published: 10 February 2018

Abstract: Carbonic anhydrase IX has been under intensive investigation as a therapeutic target in cancer. Studies demonstrate that this enzyme has a key role in pH regulation in cancer cells, allowing these cells to adapt to the adverse conditions of the tumour microenviroment. Novel CAIX inhibitors have shown efficacy in both *in vitro* and *in vivo* pre-clinical cancer models, adversely affecting cell viability, tumour formation, migration, invasion, and metastatic growth when used alone. In co-treatments, CAIX inhibitors may enhance the effects of anti-angiogenic drugs or chemotherapy agents. Research suggests that these inhibitors may also increase the response of tumours to radiotherapy. Although many of the anti-tumour effects of CAIX inhibition may be dependent on its role in pH regulation, recent work has shown that CAIX interacts with several of the signalling pathways involved in the cellular response to radiation, suggesting that pH-independent mechanisms may also be an important basis of its role in tumour progression. Here, we discuss these pH-independent interactions in the context of the ability of CAIX to modulate the responsiveness of cancer to radiation.

Keywords: carbonic anhydrase IX; cancer; hypoxia; radiation; resistance

1. Introduction

During growth, many solid tumours develop areas of low oxygen tension, or hypoxia, caused by malformation of the tumour vasculature and the increasing distance of tumour cells from the capillary bed. In tissues, O_2 concentrations of 2–9% are typical, O_2 concentrations $\leq$2% O_2 are defined as hypoxic, and $\leq$0.02% are defined as severely hypoxic [1]. The diffusion distance of O_2 from capillaries is approximately 100–200 μm; tumour cells situated further than this become hypoxic, as oxygen gradients develop in the tumour [2,3]. Circulation in the tumour is often cyclic, causing periods of acute or chronic hypoxia [4]. Tumour pH also falls with increased distance from blood vessels, with decreases from 7.4 to 6.0 measured around 300 μm from the vasculature [5]. However, the intracellular pH (pH_i) of tumour cells is maintained between 7.0 and 7.4, by the actions of pH regulating proteins [6]. Poor perfusion also inhibits the removal of waste metabolites from the tumour, allowing acidosis to develop within the tumour microenvironment (TME). Hypoxia and acidosis cause major problems in cancer treatment, contributing to increased levels of resistance to both radiotherapy and chemotherapy [7].

2. Survival Strategies

To survive in these adverse conditions, cancer cells must adapt or die [8]. As one of the mechanisms of adaptation, cells in hypoxic conditions activate the transcription factor hypoxia inducible factor-1 (HIF-1), consisting of a heterodimer constructed from α and β subunits. In normal cellular O_2 concentrations, the α subunit is rapidly degraded. The oxygen-dependent activation of prolyl hydroxylases causes hydroxylation of two proline residues (402 and 564) on HIF-1α, allowing interaction with an E3 ubiquitin ligase, VHL (the Von Hippel-Lindau factor), which targets this subunit for destruction in the proteasome [9]. This does not occur in hypoxic conditions and instead HIF-1α is stabilised and interacts with the HIF-1β subunit forming HIF-1, which activates gene transcription after nuclear translocation. Hypoxia-independent mechanisms for HIF-1α stabilisation additionally occur [10–12]. Other proteins also regulate HIF-1 activation, for example, factor inhibiting hypoxia inducible factor-1 (FIH-1), which prevents full activation of HIF-1 in moderate hypoxia. This protein maintains its activity in low O_2 concentrations [13,14] by impairing the interaction between the C-terminal transactivation domain of HIF-1α and its co-activator proteins, causing partial activation of HIF-1 [15]. FIH-1 itself is inhibited in severe hypoxia, or by membrane type-1 matrix metalloproteinase/(MMP14), allowing full HIF-1 activation [13,16].

HIF-1 regulates the expression of genes involved in glycolysis, angiogenesis, pH regulation, migration, epithelial-mesenchymal transition (EMT), and invasion [17–19], many of which aid cancer progression. For example, EMT involves E-cadherin loss, which allows cancer cells to disperse and develop a migratory and invasive phenotype, and is also linked to increased resistance to chemotherapy and radiotherapy [20]. Hypoxia via HIF-1 causes E-cadherin loss by stimulating the lysyl oxidase (LOX)-Snail pathway [21]. LOX inhibition decreases the motility and invasiveness of cancer cells in hypoxia and also reduces metastasis *in vivo* [22]. Hypoxia also interferes with the homologous recombination, non-homologous end-joining, and mismatch repair DNA pathways, and inhibits the G1/S cell cycle checkpoint. This increases DNA errors and causes chromosomal instability [6,21].

Cancer cells use aerobic glycolysis for energy and to provide components for cell growth and proliferation, even in normoxic conditions, causing higher rates of glycolysis and increased production of CO_2, H^+, and lactate [8,23]. These metabolic by-products must be removed from the cell to prevent the pH_i becoming acidic, and thus maintain a slightly alkaline pH_i consistent with survival. Early studies using D_2O in yeast demonstrated that active transport mechanisms are likely to be dependent on protons, since deuterons could not substitute for protons in these processes [24], and further illustrated the role of alkaline pH_i in transformation, tumorigenicity, and proliferation [25,26]. Tumour cells can maintain their pH_i through increased expression and activation of pH regulatory proteins, some of which are HIF1-dependent, such as monocarboxylate transporter 4 (MCT4), which exports lactate and H^+ from tumour cells, or carbonic anhydrase IX (CAIX), an enzyme that accelerates the conversion of CO_2 and H_2O to HCO_3^- and H^+ [18,27–29]. HCO_3^- is transported back into tumour cells via HCO_3^- transporters and used to buffer pH_i [28,30]. The role of CAIX is shown in Figure 1.

As a result of low O_2 levels, hypoxic cancer cells are required to undergo lactic acid fermentation for the production of energy, a process that leads to the production of H^+ ions. If these H^+ ions are allowed to build up in the cytoplasm, they can lead to changes in pH, which can be detrimental to the cell. The metabolic acids generated within the cell can react with HCO_3^-, leading to the production of H_2O and CO_2. Membrane-permeant CO_2 is a form in which much acid is removed from cancer cells. CAIX facilitates CO_2 diffusion out of the cell by catalysing the extracellular hydration of CO_2, leading to the production of H^+ and HCO_3^-. CAIX therefore maintains a steeper efflux gradient for CO_2, leading to a more alkaline intracellular pH, while also causing the acidification of the extracellular milieu.

Carbonic anhydrases (CAs) are ubiquitous metalloenzymes that catalyse the reversible formation of HCO_3^- and H^+ ions from H_2O and CO_2 [31]. At least 16 different isoforms of CAs have been isolated from mammals and differ in terms of cellular location, activity, and tissue locations. One CA, CAVI, is secreted, two (CAVA and VB) are found in the mitochondria, five are cytosolic (CAs I, II, III, VII, and XIII), and five are found on membranes (CA IV, IX, XII, XIV, and XV); of these, CAIX and CAXII have been shown to play an important role in cancer progression [32,33]. Some CAs have been shown to operate as part of transport 'metabolons' to increase the effectiveness of HCO_3^-- and H^+-transporters [34–37]. This contributes to the maintenance of an alkaline pH_i in tumour cells and an acidic pH_e in the TME, which supports tumour growth, invasion, metastasis, and resistance to both chemotherapy and radiotherapy [5,6,38–40]. For example, in tumours, the most invasive regions are those exhibiting the lowest pH, which causes activation and increases expression of proteinases and metalloproteases that degrade components of the extracellular matrix (ECM), facilitating invasion and migration [5,41–43]. Alkaline pH_i causes resistance to apoptotic stimuli because caspase activation occurs in acidic pH_i conditions [44]; it also increases both DNA synthesis and cell proliferation, allowing tumour growth and progression [6,8,45,46]. Figure 2 illustrates the expression of CAIX, proliferation, the hypoxic region, and apoptotic staining in human 3D breast cancer models.

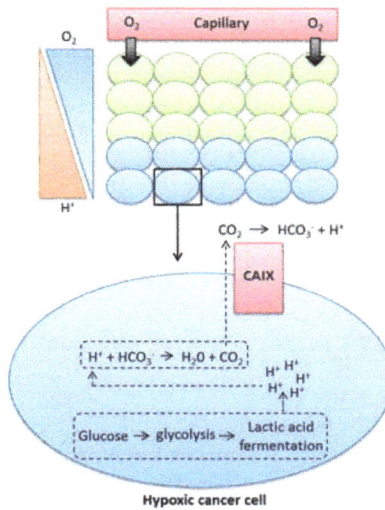

Figure 1. Contribution of CAIX to the movement of glycolytic protons from inside the cytoplasm to the extracellular milieu.

In Figure 2, the illustrations on the left demonstrate overlapping staining for CAIX and the hypoxic marker hypoxyprobe-1 in 3D spheroid cultures of HBL-100 human breast cancer cells. The right-hand illustrations demonstrate staining for CAIX, Ki67 (a proliferation marker), and caspase-3 (apoptosis) in a xenograft model of MDA-MB-231 human breast cancer cells.

Multiple studies have demonstrated a role for CAIX in pH regulation in cancer cells [28,47–50]. Targeting CAIX is proposed as a logical strategy for anti-cancer therapy, since it is an extracellular target, mainly associated with malignant growth, and is largely absent from most healthy tissue, with the exception of the gastro-intestinal tract and stomach [32,38,47,51,52]. CAIX staining in tumours is associated with poor prognosis and progression in several types of cancer, and in a series of lymph node-positive breast tumours it was found to correlate with metastasis [40,53–58]. Knockdown of CAIX in murine models leads to few phenotypic abnormalities other than gastric hyperplasia, inferring limited toxicity issues in normal tissue [59,60].

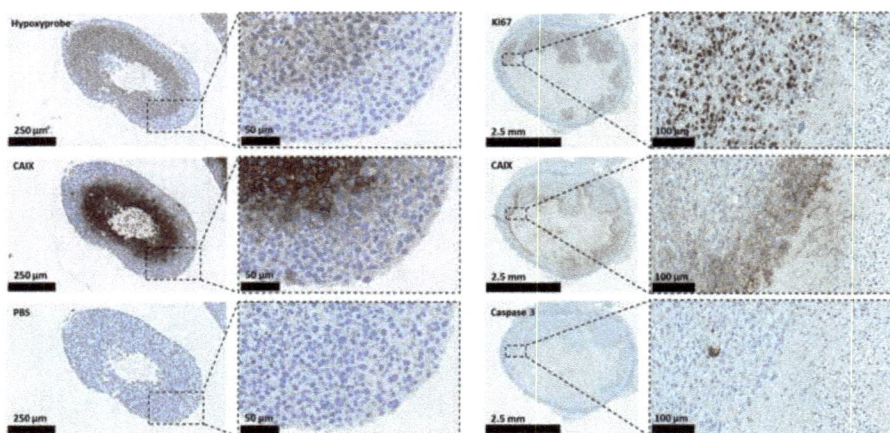

Figure 2. The identification of hypoxic areas and the expression of CAIX, proliferation, and apoptotic markers in 3D human breast cancer models.

3. CAIX Inhibition as a Cancer Therapy

The effectiveness of CAIX inhibition as an anti-cancer strategy has been demonstrated in many pre-clinical studies using various cancer models. CAIX inhibitors negatively affect cancer cell viability and migration, as well as collagen destruction and invasion, and hinder both tumour formation and metastic growth in murine models, suggesting that increased expression and activity of CAIX in a cancer will adversely affect progression and prognosis. Many groups have now validated a crucial role for CAIX in growth, migration, invasion, and metastasis of tumours [40,52,61–68].

In vitro investigations have demonstrated that inhibition of CAIX using siRNA or CAIX inhibitors decreased the invasiveness of renal and ovarian cancer cells, while also reducing the amount of cells invading from human breast carcinoma spheroids [40,66,69,70]. A novel class of sulfamate CAIX inhibitors reduced the invasion, proliferation, and migration of human breast cancer cells, and also exhibited the capacity to reverse established invasion in a model consisting of breast tumour tissue from naïve biopsies [40,63,66,71]. CAIX knockdown significantly reduced the proliferation and survival of cancer cells under both normoxic and hypoxic conditions [40,66,72].

In vivo studies have shown that knockdown of CAIX can decrease tumour volume in both breast and colon cancer xenografts. Additive results could be obtained in co-treatments alongside antiangiogenic therapy. Knockdown also inhibited lung metastasis in breast cancer models [52,63,64]. CAIX overexpression studies in a colon cancer model demonstrated increased rates of tumour growth and expression of Ki-67, a marker of proliferation [64]. CAIX inhibitors also slowed tumour growth in breast cancer xenografts by decreasing proliferation and increasing cell death [40]. One of these inhibitors also exhibited anti-metastatic effects in a xenograft model of human breast cancer [73]. Interestingly, the use of CAIX inhibitors in *in vivo* systems did not lead to any reports of non-specific toxicity [40,69,73,74]. The development of novel CAIX inhibitors and clinical trials has been reviewed recently [8,28–30,75].

4. Other Possible Functions for CAIX in Cancer Progression

Although the main function of CAIX in cancer is as a regulator of pH_i, several studies show that other possible mechanisms may be linked to this enzyme, thereby expanding its role in tumour progression. A recent study found novel roles for CAIX in tumour cell migration and MMP14-mediated invasion [67]. Interactions were identified with $\beta 1$ integrins, metabolic transporters, integrin-associated protein CD98hc, and MMP14. CAIX appears to associate with MMP14 through phosphorylation sites in the intracellular domain of CAIX and can increase the degradation of collagen by MMP14 through providing the H^+ the protease needs for its catalytic activity [67].

Interestingly, another recent report has linked MMP14 to the invasive capacity of breast cancer cell lines [66]. This study showed that although HIF-1α levels increased in hypoxic conditions, the expression of CAIX was variable between cell lines, and was only markedly upregulated in MCF-7 cells, a non-invasive cell line, after exposure to chronic hypoxia. HIF-1α gene expression can be constrained by FIH-1, which in turn can be inhibited by MMP14 [16]. Although MCF-7 cells expressed high FIH-1 levels, they lacked the active form of MMP14, suggesting that in this cell line, FIH-1 is able to prevent full HIF-1 activation in acute hypoxia, but not in chronic hypoxia [66]. This is in agreement with prior studies showing that FIH-1 suppression increases CAIX expression in hepatoma and osteosarcoma cell lines [14]. In the MCF-7 models, FIH-1 knockdown increased CAIX expression in hypoxic cells [66]. Taken together, these results suggest a strong link between hypoxia, HIF-1, FIH-1, CAIX, and MMP14 in the regulation of cancer cell invasive potential [14,16,66,67].

CAIX can decrease binding of E-cadherin to the cytoskeleton and affect cell adhesion [76] while also increasing the metastatic potential of tumour cells by effects on the activity of Rho-GTPase [77]. It similarly interacts with DKK1 protein, and thus the Rho/ROCK pathway, activating paxillin and stimulating migration [77]. CAIX can be phosphorylated at Thr-443 by Protein Kinase A during hypoxia, causing activation of CAIX and facilitating migration via increased transcription of proteins involved in cytoskeletal organisation and EMT [77–79]. Overexpression of CAIX also modulates Rho/ROCK signalling (which is pH sensitive), activates paxillin, increases focal adhesion turnover, cell migration, and activation of the FAK/PI3K/mTOR/p70S6K signalling pathway [77,80]. Conversely, CAIX inhibition impedes ROCK1 and decreases invasion [68,81].

5. Radiation

Radiotherapy is used to treat approximately 50% of cancer patients, either alone or in concert with chemotherapy or surgery [82]. It aims to eliminate cancer cells from the primary tumour, regional lymph nodes, or oligometastatic disease whilst limiting normal tissue damage. Radiation responses depend on the production of free radicals and intermediate ions that cause DNA damage in the form of single-strand (SSBs) or double-strand (DSBs) breaks in DNA. DBSs are the most effective in terms of inflicting cell damage and activate the DNA damage response (DDR) pathway that regulates whether the cell repairs DNA or undergoes cell death [83,84].

DSBs in DNA are detected by ataxia telangiectasia mutated (ATM) and ataxia telangiectasia and Rad3-related (ATR) kinases, which activate signalling pathways that stimulate cell cycle checkpoints and DNA repair. H2AX, once phosphorylated by ATM, recruits DNA repair proteins to the damaged area, and cyclins and cyclin-dependent kinases (CDKs) at G1/S and G2/M interphases delay cell division while DNA is repaired. The damage is restored by either homologous recombination or nonhomologous end joining (NHEJ) repair pathways. If DNA is not repaired, the damage results in cell death.

However, some tumours may either acquire or possess intrinsic radioresistance, and a major clinical advantage would be achieved in cancer treatment if new approaches to sensitizing these tumours to radiotherapy were developed [84,85]. Studies in the 1950s acertained the role of hypoxia in radioresistance and, conversely, the role that O_2 plays in radiation responses [86–88]. Maximal cell kill in response to radiotherapy needs O_2 to form free radicals that damage DNA, and stabilise or 'fix' radiation-induced DNA damage [89]. This causes changes in the DNA that cannot be repaired, leading to cell death if the cell tries to undergo cell division [1,90]. Hypoxic cells can be 2 to 3 times more resistant to the same dose of radiation, because fewer DSBs are stabilised [88]. Another factor is the decrease in cell proliferation caused by hypoxia, as DNA damage is higher in rapidly dividing cells [5]. The phase of the cell cycle also affects radiation responses, with cells in G2/M and G1 phases the most radiosensitive and those in the S phase more radioresistant [91,92]. Increased acidification decreases the effectiveness of radiation; cells cultured in acidic media are more resistant to radiation, with acidic pH_e shown to reduce fixation of radiation-induced DNA damage, inhibit radiation-induced apoptosis, and delay G2/M-phase arrest allowing more time for treated cells to repair DNA damage, thus increasing radioresistance [7,93–99].

6. CAIX Inhibition and Radiation

Studies have shown that CAIX can influence the response of cancers to radiation [53,100]. The knockdown of hypoxia-induced CAIX, or CAIX and CAXII together, sensitised tumour cells to radiation by decreasing the number of cells in the radioresistant S phase in both *in vitro* and *in vivo* models. This knockdown caused decreased intracellular pH values, which were found to correlate with enhanced cell death, suggesting that active CAIX is protecting cells against radiation by preserving an alkaline pH_i, since ectopic CAIX expression increased cell survival after radiation treatment [29,62]. Recent studies have also shown the ability of CAIX inhibitors to sensitise renal cell carcinoma to radiation by increasing radiation-induced apoptosis [101], which is one mechanism known to be involved in the therapeutic effect of radiotherapy [102]. CAIX inhibitors similarly enhanced radiation sensitivity when used in combination in a breast cancer model; proteomic studies indicated that co-treatment increased expression of pro-apoptotic proteins and reduced expression of anti-apoptotic proteins [66]. Other pre-clinical data using xenograft models have shown that tumours expressing high levels of CAIX were less responsive to radiation, but that CAIX inhibitors could significantly increase radiosensitivity [103,104]. For example, the co-treatment of CAIX inhibitors with radiation in a colon HT29 mouse xenograft model demonstrated an improved therapeutic effect [103]. Although this was not apparent in *in vitro* studies, a novel class of sulfamate CAIX inhibitors enhanced the effects of radiation in a colorectal model, both *in vitro* and *in vivo* [103,104]. The use of isotopic substitution experiments could give insight into whether CAIX inhibition induces radiation sensitivity by increasing intracellular H^+ concentrations [24].

7. CAIX and Radiation Responses, and Other Mechanisms

Although most research suggests that CAIX influences cancer responses via pH regulation, this enzyme can also interact with other mechanisms involved in cellular reactions to radiation, suggesting that additional factors may be involved in its ability to radiosensitize cancer cells. Radiation triggers several signalling cascades known to be involved in cell survival such as the PI3K/AKT and ERK pathways; this occurs through activation of the epidermal growth factor receptor (EGFR) [105,106]. EGFR can influence radiation responses by binding to DNA-dependent protein kinases (DNA-PK), enhancing their activity and thus DNA repair [106–108]. Radiation induces nuclear accumulation of EGFR, where it is involved in the relaxation of chromatin, allowing DNA repair proteins access and thus enhancing resistance to radiation [109]. Decreased expression of EGFR or AKT has been shown to increase radiation sensitivity in human cancer cells [110]. Activation of EGFR by epidermal growth factor causes phosphorylation of CAIX on Tyr449, which in turn can activate the PI3K/AKT pathway by interacting with the p85 regulatory subunit [111]. This suggests that radiation itself may activate survival pathways through a mechanism that at least partially involves CAIX. PI3K/Akt is one of the pathways stimulated by radiation that is known to be involved in the inhibition of cell death via apoptosis; further, several studies have also linked overexpression and activation of EGFR with radiation resistance in cancer [112–118]. EGFR inhibitors can sensitise cancer cells to radiation both *in vitro* and *in vivo*, with positive results also observed in a Phase III trial in head and neck cancer [119,120]. Whether this response to radiation is in part through the prevention of CAIX phosphorlyation and activation of PI3K/AKT is currently unclear.

NF-κB activity is stimulated by hypoxia and acidic pH [121–125]. It is also activated by radiation and has a key role in radiation resistance and cell survival [126–130]. CAIX can interact with the NF-κB signalling pathway via a mechanism involving β1-integrin. Expression of β1 integrin is increased in various cancers, where it is involved in cell survival, proliferation, apoptosis, invasion, metastasis, and resistance to both chemotherapy and radiotherapy [131–135]. Studies have shown that cancer cells can be sensitized to radiotherapy by targeting β1 integrins [136,137]. The radioresistance mediated by β1 integrin is regulated by NF-κB, which increases β1-integrin expression, but conversely, inhibition of β1-integrin can inhibit the transcriptional activity of NF-κB [138]. A recent study demonstrated an interaction between CAIX and β1 integrin in tumour cells [67]; therefore this is another possible mode of interaction between CAIX and radiation responses. The downmodulation of β1 integrin can synergistically inhibit Akt-mediated survival in breast cancer cell lines and enhance radiotherapy in breast cancer xenografts [134,136,139]. A further study has shown that CAIX is required for the activation of NF-κB in hypoxia and can, via this interaction, stimulate the production of G-CSF to promote movement of granulocytic myeloid-derived suppressor cells (MDSC) to the metastatic niche of the lung [140]. Interestingly, G-CSF is strongly linked with protection from radiation damage [141,142], and high expression of the receptor is associated with poor response to radiotherapy in rectal cancer [143].

Signal transducer and activator of transcription 3 (Stat3) is overexpressed in many cancers; STAT3 has been shown to be involved in the regulation of CAIX expression in several studies [144,145]. Inhibition of STAT3 has been found to increase radiation sensitivity in cancer cells, and to inhibit radiation-induced progression in glioma [146–148]. IL-6 promotes tumour growth and invasion through STAT3 activation [140], and has likewise been linked to radiation resistance [149,150]. IL-6 is also an NF-κB responsive gene [151]. Taken together, it is therefore possible that CAIX is part of an IL-6-STAT3-NF-κB signalling axis involved in radiation resistance as illustrated in Figure 3. These interactions appear to be independent of the pH-regulating roles of CAIX.

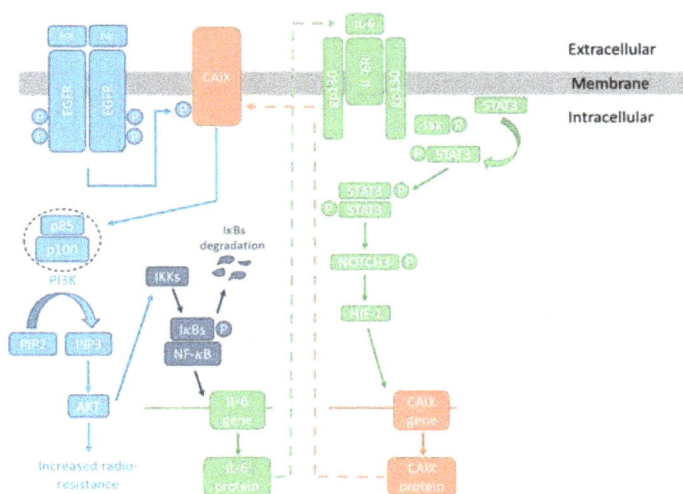

Figure 3. The CAIX/NF-κB/IL-6 signalling node and radioresistance.

EGF induces phosphorylation of CAIX via the EGFR, allowing interaction with the p85 regulatory subunit of PI3K and activation of survival pathways. AKT can in turn activate I-κB kinase (IKK) and the NF-κB transcription factor, causing the production of IL-6. IL-6 can, via STAT3 and HIF, increase CAIX expression. Overexpression of CAIX, EGFR, STAT3, and IL-6, and activation of NF-κB, EGFR, and STAT3, have all been linked with radiation resistance.

The increased presence of lactate in the TME has been linked to both radiation and chemoresistance [36,152,153], which may be due to the antioxidant effects of lactate in the case of radioresistance [154]. The secretion of excess lactic acid from the cell is regulated by MCTs, such as the HIF1-inducible MCT4, which operates almost solely to export lactate, or MCT1, which also transports other monocarboxylates [18,155–157]. Both MCT1 and 4 co-transport H$^+$ with lactate [18]. MCT1 expression is increased in various cancers such as ovarian, prostate, breast, and colorectal cancers, where it correlates with progression and poor patient prognosis [158–160]. In xenografts of lung, colorectal, or small cell lung cancer, MCT1 inhibitors decreased lactate secretion and increased radiosensitivity [161,162].

The MCT1/4 accessory molecule CD147 is required for plasma membrane expression of these transporters; if targeted, it can decrease expression of these MCTs and inhibit tumour growth in an *in vivo* model [163]; therefore, CD147 should also be linked to radiation responses. Studies show that CD147 can promote radioresistance in hepatocellular carcinoma cells *in vitro* and *in vivo*, and it has been linked to radioresistance in cervical cancer [164–166]. Interestingly, CD147 has also been demonstrated to interact with integrin β1 in hepatocellular carcinoma, causing activation of the FAK pathway and increasing malignancy of these cells.

Could CAIX inhibition affect lactate secretion from cancer cells and thus sensitise resistant tissues to radiation? In an interesting study, it was found that the increased lactate efflux from hypoxic breast cancer cells was not due to amplified expression of MCTs, but to a hypoxia-induced upregulation of CAIX, via a mechanism that was independent of its catalytic activity [37]. Other studies have also demonstrated that CAs have effects on cells that are not dependent on the catalytic activity of these enzymes. For example, it has been demonstrated that both the cytosolic CAII and CAIV can enhance the activity of monocarboxylate transporters 1 and 4 in a non-enzymatic manner, thus increasing lactate flux [34,36,167,168]. Jamali et al. showed that knock-down of CAIX decreased lactate flux by approximately 50%. This study proposed that CAIX may function as a 'proton-collecting/distributing antenna' that accelerates proton transfer and requires an extracellular location for CAIX, and that could facilitate both MCT1 and MCT4 activity. They further suggested that this collaboration between MCTs and CAIX would not be inhibited by compounds that specifically target CAIX catalytic activity [37].

Recently, lactate has been shown to induce the expression of CAIX in normoxic cancer cells both *in vitro* and *in vivo* in a mechanism involving both HIF-1 and specificity protein (SP-1) transcription factors [169]. The major mechanism of CAIX increase appeared to be through redox-dependent stabilisation of HIF-1α. Again, this suggests a possible signalling loop in which hypoxic cells produce lactate that can increase CAIX expression via HIF, and which in turn allows increased lactate export via CAIX/MCT co-operation and thus increased radiation resistance.

Hypoxia induces dedifferentiation of cancer cells to become phenotypically more stem cell-like [170]. It has been proposed that CAIX may be involved in this process, as CAIX expression has been shown to correlate well with that of CD44, a breast cancer stem cell marker [171–173]. Radioresistance is considered an inherent characteristic of cancer stem cells [174–176]. It has been suggested that such cells may repair DNA more effectively after radiation, since they express high levels of genes associated with DNA damage repair [177–180]. Inhibition of CAIX depletes the number of breast cancer stem cells in tumour hypoxic subvolumes, and therefore CAIX inhibitors may be useful to treat the radioresistant cancer stem cell population. Furthermore, it has been inferred that CAIX is required to maintain the stemness phenotype within the hypoxic niche of breast tumors [65]. This may be due to the possible mechanisms outlined above, or to the effect of CAIX on the acidic TME, since extracellular acidosis has also been linked to the development of 'stemness' [181]. Interestingly, increased lactate concentrations can also cause cancer cells to develop a cancer stem cell phenotype [182]. In pancreatic cancer stem cells, STAT3 is activated, but STAT3 inhibition decreases both radioresistance and stem cell numbers in pancreatic cancer [148]. Therefore, since both lactate and STAT3 activation can increase expression of CAIX [144,145,169], it is certainly possible that the effect of CAIX inhibitors on radiation sensitivity of cancer stem cells is due to the interactions of CAIX with lactate and STAT3.

8. Conclusions

CAIX is an attractive target for the treatment of cancer [8,31–33]. Data suggests that CAIX inhibition is a therapeutic strategy that could interfere with cancer cell proliferation, migration, and invasion, while *in vivo* studies demonstrate that metastatic growth could also be limited. While evidence indicates that the effectiveness of this inhibition is through interference with pH regulation in cancer cells, recent studies show that CAIX can interact with many other signalling pathways and mechanisms known to be active in cancer cells, many of which appear to influence the response of cancer cells to radiation. These pathways are not mutually exclusive, and sensitivity to radiation could be determined by additive or synergistic interactions between pH-dependent and independent mechanisms, which suggests that CAIX may have many important roles in cancer cells that could potentially be exploited therapeutically, particularly by radiation oncologists.

Acknowledgments: This work was supported by funding from the UK Engineering and Physical Sciences Research Council, through the IMPACT programme Grant (EP/K-34510/1). No funds were received for covering the costs to publish in open access.

Author Contributions: C.W., S.P.L., I.H.K., and D.J.A. conceptualized and designed the structure of the article; C.W. wrote the paper; J.M. and M.G. performed the immunohistochemistry and designed the figures; C.W., J.M., M.G., I.H.K., and D.J.A. were involved in the critical reading and editing of the manuscript. All authors have read and approved the final version of the manuscript.

Conflicts of Interest: The authors declare no conflict of interest.

References

1. Bertout, J.A.; Patel, S.A.; Simon, M.C. The impact of O$_2$ availability of human cancer. *Nat. Rev. Cancer* **2008**, *8*, 967–975. [CrossRef] [PubMed]

2. Secomb, T.W.; Dewhirst, M.W.; Pries, A.R. Structural adaptation of normal and tumour vascular networks. *Basic Clin. Pharmacol. Toxicol.* **2012**, *110*, 63–69. [CrossRef] [PubMed]

3. Vaupel, P.; Mayer, A. Hypoxia in tumors: Pathogenesis-related classification, characterization of hypoxia subtypes, and associated biological and clinical implications. *Adv. Exp. Med. Biol.* **2014**, *812*, 19–24. [CrossRef] [PubMed]

4. Michiels, C.; Tellier, C.; Feron, O. Cycling hypoxia: A key feature of the tumor microenvironment. *Biochim. Biophys. Acta* **2016**, *1866*, 76–86. [CrossRef] [PubMed]

5. Gatenby, R.A.; Smallbone, K.; Maini, P.K.; Rose, F.; Averill, J.; Nagle, R.B.; Worrall, L.; Gillies, R.J. Cellular adaptations to hypoxia and acidosis during somatic evolution of breast cancer. *Br. J. Cancer* **2007**, *97*, 646–653. [CrossRef] [PubMed]

6. Webb, B.A.; Chimenti, M.; Jacobson, M.P.; Barber, D.L. Dysregulated pH: A perfect storm for cancer progression. *Nat. Rev. Cancer* **2011**, *11*, 671–677. [CrossRef] [PubMed]

7. Vaupel, P. Tumor microenvironmental physiology and its implications for radiation oncology. *Semin. Radiat. Oncol.* **2004**, *14*, 198–206. [CrossRef] [PubMed]

8. Ward, C.; Langdon, S.P.; Mullen, P.; Harris, A.L.; Harrison, D.J.; Supuran, C.T.; Kunkler, I.H. New strategies for targeting the hypoxic tumour microenvironment in breast cancer. *Cancer Treat. Rev.* **2013**, *39*, 171–179. [CrossRef] [PubMed]

9. Wang, G.L.; Jiang, B.; Rue, E.A.; Semenza, G.L. Hypoxia-inducible factor 1 is a basic-helix-loop-helix-PAS heterodimer regulated by cellular O$_2$ tension. *Proc. Natl. Acad. Sci. USA* **1995**, *92*, 5510–5514. [CrossRef] [PubMed]

10. Doe, M.R.; Ascano, J.M.; Kaur, M.; Cole, M.D. Myc posttranslationally induces HIF1 protein and target gene expression in normal and cancer cells. *Cancer Res.* **2012**, *72*, 949–957. [CrossRef] [PubMed]

11. Semenza, G.L. Targeting HIF-1 for cancer therapy. *Nat. Rev. Cancer* **2003**, *3*, 721–732. [CrossRef] [PubMed]

12. Lu, H.; Forbes, R.A.; Verma, A. Hypoxia-inducible factor 1 activation by aerobic glycolysis implicates the Warburg effect in carcinogenesis. *J. Biol. Chem.* **2002**, *277*, 23111–23115. [CrossRef] [PubMed]

13. Tian, Y.M.; Yeoh, K.K.; Lee, M.K.; Eriksson, T.; Kessler, B.M.; Kramer, H.B.; Edelmann, M.J.; Willam, C.; Pugh, C.W.; Schofield, C.J.; et al. Differential sensitivity of hypoxia inducible factor hydroxylation sites to hypoxia and hydroxylase inhibitors. *J. Biol. Chem.* **2011**, *286*, 13041–13051. [CrossRef] [PubMed]

14. Stolze, I.P.; Tian, Y.M.; Appelhoff, R.J.; Turley, H.; Wykoff, C.C.; Gleadle, J.M.; Ratcliffe, P.J. Genetic analysis of the role of the asparaginyl hydroxylase factor inhibiting hypoxia-inducible factor (FIH) in regulating hypoxia-inducible factor (HIF) transcriptional target genes. *J. Biol. Chem.* **2004**, *279*, 42719–42725. [CrossRef] [PubMed]

15. Dayan, F.; Mazure, N.M.; Brahimi-Horn., M.C.; Pouyssegur, J. A dialogue between the hypoxia-inducible factor and the tumor microenvironment. *Cancer Microenviron.* **2008**, *1*, 53–68. [CrossRef] [PubMed]

16. Sakamoto, T.; Seiki, M. A membrane protease regulates energy production in macrophages by activating hypoxia-inducible factor-1 via a non-proteolytic mechanism. *J. Biol. Chem.* **2010**, *285*, 29951–29964. [CrossRef] [PubMed]

17. Wykoff, C.C.; Beasley, N.J.; Watson, P.H.; Turner, K.J.; Pastorek, J.; Sibtain, A.; Wilson, G.D.; Turley, H.; Talks, K.L.; Maxwell, P.H.; et al. Hypoxia-inducible expression of tumor-associated carbonic anhydrases. *Cancer Res.* **2000**, *60*, 7075–7083. [PubMed]

18. Ullah, M.S.; Davies, A.J.; Halestrap, A.P. The plasma membrane lactate transporter MCT4, but not MCT1, is up-regulated by hypoxia through a HIF-1α-dependent mechanism. *J. Biol. Chem.* **2006**, *281*, 9030–9037. [CrossRef] [PubMed]

19. Potter, C.; Harris, A.L. Hypoxia inducible carbonic anhydrase IX, marker of tumour hypoxia, survival pathway and therapy target. *Cell Cycle* **2004**, *3*, 164–167. [CrossRef] [PubMed]

20. Marie-Egyptienne, D.T.; Lohse, I.; Hill, R.P. Cancer stem cells, the epithelial to mesenchymal transition (EMT) and radioresistance: Potential role of hypoxia. *Cancer Lett.* **2013**, *341*, 63–72. [CrossRef] [PubMed]

21. Bristow, R.G.; Hill, R.P. Hypoxia and metabolism. Hypoxia, DNA repair and genetic instability. *Nat. Rev. Cancer* **2008**, *8*, 180–192. [CrossRef] [PubMed]

22. Chen, H.H.; Su, W.C.; Lin, P.W.; Guo, H.R.; Lee, W.Y. Hypoxia-inducible factor-1α correlates with MET and metastasis in node-negative breast cancer. *Breast Cancer Res. Treat.* **2007**, *103*, 167–175. [CrossRef] [PubMed]

23. Hanahan, D.; Weinberg, R.A. Hallmarks of cancer: The next generation. *Cell* **2011**, *144*, 646–674. [CrossRef] [PubMed]

24. Kotyk, A.; Dvorakova, M.; Koryta, J. Deuterons cannot replace protons in active transport processes in yeast. *FEBS Lett.* **1990**, *264*, 203–205. [CrossRef]

25. Perona, R.; Serrano, R. Increased pH and tumorigenicity of fibroblasts expressing a yeast proton pump. *Nature* **1988**, *334*, 438–440. [CrossRef] [PubMed]

26. Perona, R.; Portillo, F.; Giraldez, F.; Serrano, R. Transformation and pH homeostasis of fibroblasts expressing yeast H⁺-ATPase containing site-directed mutations. *Mol. Cell. Biol.* **1990**, *10*, 4110–4115.27. [CrossRef] [PubMed]

27. Pinheiro, C.; Longatto-Filho, A.; Azevedo-Silva, J.; Casal, M.; Schmitt, F.C.; Baltazar, F. Role of monocarboxylate transporters in human cancers: State of the art. *J. Bioenerg. Biomembr.* **2012**, *44*, 127–139. [CrossRef] [PubMed]

28. Swietach, P.; Hulikova, A.; Vaughan-Jones, R.D.; Harris, A.L. New insights into the physiological role of carbonic anhydrase IX in tumour pH regulation. *Oncogene* **2010**, *29*, 6509–6521. [CrossRef] [PubMed]

29. Chiche, J.; Brahimi-Horn, M.C.; Pouyssegur, J. Tumour hypoxia induces a metabolic shift causing acidosis: A common feature in cancer. *J. Cell. Mol. Med.* **2010**, *4*, 771–794. [CrossRef] [PubMed]

30. Parks, S.K.; Chiche, J.; Pouyssegur, J. pH control mechanisms of tumor survival and growth. *J. Cell Physiol.* **2011**, *226*, 299–308. [CrossRef] [PubMed]

31. Supuran, C.T. Carbonic anhydrase inhibitors. *Bioorg. Med. Chem. Lett.* **2010**, *20*, 3467–3474. [CrossRef] [PubMed]

32. Neri, D.; Supuran, C.T. Interfering with pH regulation in tumors as a therapeutic strategy. *Nat. Rev. Drug Discov.* **2011**, *10*, 767–777. [CrossRef] [PubMed]

33. Supuran, C.T. Carbonic anhydrases: Novel therapeutic applications for inhibitors and activators. *Nat. Rev. Drug Discov.* **2008**, *7*, 168–181. [CrossRef] [PubMed]

34. Becker, H.M.; Klier, M.; Schüler, C.; McKenna, R.; Deitmer, J.W. Intramolecular proton shuttle supports not only catalytic but also noncatalytic function of carbonic anhydrase II. *Proc. Natl. Acad. Sci. USA* **2011**, *108*, 3071–3076. [CrossRef] [PubMed]

35. Deitmer, J.W.; Becker, H.M. Transport metabolons with carbonic anhydrases. *Front. Physiol.* **2013**, *4*, 291. [CrossRef] [PubMed]

36. Klier, M.; Andes, F.T.; Deitmer, J.W.; Becker, H.M. Intracellular and extracellular carbonic anhydrases cooperate non-enzymaticaly to enhance activity of monocarboxylate transporters. *J. Biol. Chem.* **2014**, *289*, 2765–2775. [CrossRef] [PubMed]
37. Jamali, S.; Klier, M.; Ames, S.; Felipe Barros, L.; McKenna, R.; Deiter, J.W.; Becker, H.M. Hypoxia-induced carbonic anhydrase IX facilitates lactate flux in human breast cancer cells by non-catalytic function. *Sci. Rep.* **2015**, *5*, 13605. [CrossRef] [PubMed]
38. Parks, S.K.; Chiche, J.; Pouyssegur, J. Disrupting proton dynamics and energy metabolism for cancer therapy. *Nat. Rev. Cancer* **2013**, *13*, 611–623. [CrossRef] [PubMed]
39. Gillies, R.J.; Verduzco, D.; Gatenby, R.A. Evolutionary dynamics of carcinogenesis and why targeted therapy does not work. *Nat. Rev. Cancer* **2012**, *12*, 487–493. [CrossRef] [PubMed]
40. Ward, C.; Meehan, J.; Mullen, P.; Supuran, C.; Dixon, J.M.; Thomas, J.S.; Winum, J.Y.; Lambin, P.; Dubois, L.; Pavathaneni, N.K.; et al. Evaluation of carbonic anhydrase IX as a therapeutic target for inhibition of breast cancer invasion and metastasis using a series of in vitro breast cancer models. *Oncotarget* **2015**, *6*, 24856–24870. [CrossRef] [PubMed]
41. Rofstad, E.K.; Mathiesen, B.; Kindem, K.; Galappathi, K. Acidic extracellular pH promotes experimental metastasis of human melanoma cells in athymic nude mice. *Cancer Res.* **2006**, *66*, 6699–6707. [CrossRef] [PubMed]
42. Kato, Y.; Ozawa, S.; Tsukuda, M.; Kubota, E.; Miyazaki, K.; St-Pierre, Y.; Hata, R. Acidic extracellular pH increases calcium influx-triggered phospholipase D activity along with acidic spingomyelinase activation to induce matrix metalloproteinase-9 expression in mouse metastatic melanoma. *FEBS J.* **2007**, *274*, 3171–3183. [CrossRef] [PubMed]
43. Gatenby, R.A.; Gawlinski, E.T.; Gmitro, A.F.; Kaylor, B.; Gillies, R.J. Acid-mediated tumor invasion; a multidisciplinary study. *Cancer Res.* **2006**, *66*, 5216–5223. [CrossRef] [PubMed]
44. Matsuyama, S.; Llopis, J.; Deveraux, Q.L.; Tsien, R.Y.; Reed, J.C. Changes in intramitochondrial and cytosolic pH: Early events that modulate caspase activation during apoptosis. *Nat. Cell Biol.* **2000**, *2*, 318–325. [CrossRef] [PubMed]
45. Lee, C.H.; Cragoe, E.J., Jr.; Edwards, A.M. Control of hepatocyte DNA synthesis by intracellular pH and its role in the action of tumor promoters. *J. Cell. Physiol.* **2003**, *195*, 61–69. [CrossRef] [PubMed]
46. Schreiber, R. Ca^{2+} signalling, intracellular pH and cell volume in cell proliferation. *J. Membr. Biol.* **2005**, *205*, 129–137. [CrossRef] [PubMed]
47. Svastova, E.; Hulikova, A.; Rafajoba, M.; Zat'ovivova, M.; Gibadulinova, A.; Casini, A.; Cecchi, A.; Scozzafava, A.; Supuran, C.T.; Pastorek, J.; et al. Hypoxia activates the capacity of tumor-associated carbonic anhydrase IX to acidify extracellular pH. *FEBS Lett.* **2004**, *577*, 439–445. [CrossRef] [PubMed]
48. Swietach, P.; Vaughan-Jones, R.D.; Harris, A.L.; Hulikova, A. The chemistry, physiology and pathology of pH in cancer. *Philos. Trans. R. Soc. Lond. Biol. Sci.* **2014**, *369*, 20130099. [CrossRef] [PubMed]
49. Hulikova, A.; Vaughan-Jones, R.D.; Swietach, P. Dual role of CO_2/HCO_3^- buffer in the regulation of intracellular pH of three-dimensional tumor growths. *J. Biol. Chem.* **2011**, *286*, 13815–13826. [CrossRef] [PubMed]
50. Pouysségur, J.; Dayan, F.; Mazure, N.M. Hypoxia signalling in cancer and approaches to enforce tumour regression. *Nature* **2006**, *441*, 437–443. [CrossRef] [PubMed]
51. Pastoreková, S.; Parkkila, S.; Parkkila, A.K.; Opavsky, R.; Zelnik, V.; Saarnio, J.; Pastorek, J. Carbonic anhydrase IX, MN/CA IX: Analysis of stomach complementary DNA sequence and expression in human and rat alimentary tracts. *Gastroenterology* **1997**, *112*, 398–408. [CrossRef] [PubMed]
52. Chiche, J.; Ilc, K.; Laferriere, J.; Trottier, E.; Dayan, F.; Mazure, N.M.; Brahimi-Horn, M.C.; Pouysségur, J. Hypoxia-inducible carbonic anhydrase IX and XII promote tumor cell growth by counteracting acidosis through the regulation of the intracellular pH. *Cancer Res.* **2009**, *69*, 358–368. [CrossRef] [PubMed]
53. Koukourakis, M.I.; Giatromanolaki, A.; Sivridis, E.; Simopoulos, K.; Pastorek, J.; Wykoff, C.C.; Gatter, K.C.; Harris, A.L. Hypoxia-regulated carbonic anhydrase-9 (CA9) relates to poor vascularization and the resistance of squamous head and neck cancer to chemoradiotherapy. *Clin. Cancer Res.* **2001**, *11*, 3399–3403.
54. Korkeila, E.; Talvinen, K.; Jaakkola, P.M.; Minn, H.; Syrjanen, K.; Sundstrom, J.; Pyrhonen, S. Expression of carbonic anhydrase IX suggests poor outcome in rectal cancer. *Br. J. Cancer* **2009**, *100*, 874–880. [CrossRef] [PubMed]

55. Tan, E.Y.; Yan, M.; Campo, L.; Han, C.; Takano, E.; Turley, H.; Candiloro, I.; Pezzella, F.; Gatter, K.C.; Millar, E.K.; et al. The key hypoxia regulated gene CAIX is upregulated in basal-like breast tumours and is associated with resistance to chemotherapy. *Br. J. Cancer* **2009**, *100*, 405–411. [CrossRef] [PubMed]

56. Chia, S.K.; Wykoff, C.C.; Watson, P.H.; Han, C.; Leek, R.D.; Pastorek, J.; Gatter, K.C.; Ratcliffe, P.; Harris, A.L. Prognostic significance of a novel hypoxia-regulated marker, carbonic anhydrase IX, invasive breast carcinoma. *J. Clin. Oncol.* **2001**, *19*, 3660–3668. [CrossRef] [PubMed]

57. Bartosova, M.; Parkkila, S.; Pohlodek, K.; Karttunen, T.J.; Galbavy, S.; Mucha, V.; Harris, A.L.; Pastorek, J.; Pastorekova, S. Expression of carbonic anhydrase IX in breast is associated with malignant tissues and is related to overexpression of c-erbB2. *J. Pathol.* **2002**, *197*, 314–321. [CrossRef] [PubMed]

58. Generali, D.; Fox, S.B.; Berruti, A.; Brizzi, M.P.; Campo, L.; Bonardi, S.; Wigfield, S.M.; Bruzzi, P.; Bersiga, A.; Allevi, G.; et al. Role of carbonic anhydrase IX expression in prediction of the efficacy and outcome of primary epirubicin/tamoxifen therapy for breast cancer. *Endocr. Relat. Cancer* **2006**, *13*, 921–930. [CrossRef] [PubMed]

59. Gut, M.O.; Parkkila, S.; Vernerová, Z.; Rohde, E.; Závada, J.; Höcker, M.; Pastorek, J.; Karttunen, T.; Gibadulinová, A.; Závadová, Z.; et al. Gastric hyperplasia in mice with targeted disruption of the carbonic anhydrase gene Car9. *Gastroenterology* **2002**, *123*, 1889–1903. [CrossRef] [PubMed]

60. Leppilampi, M.; Karttunen, J.; Kivela, J.; Gut, M.O.; Pastorekova, S.; Pastorek, J.; Parkkila, S. Gastric pit cell hyperplasia and glandular atrophy in carbonic anhydrase IX knockout mice: Studies on two strains C57/BL6 and BALB/C. *Transgenic Res.* **2005**, *14*, 655–663. [CrossRef] [PubMed]

61. Morris, J.C.; Chiche, J.; Grellier, C.; Lopez, M.; Bornaghi, L.F.; Maresca, A.; Supuran, C.T.; Pouysségur, J.; Poulsen, S.A. Targeting hypoxic tumor cell viability with carbohydrate-based carbonic anhydrase IX and XII inhibitors. *J. Med. Chem.* **2011**, *54*, 6905–6918. [CrossRef] [PubMed]

62. Doyen, J.; Parks, S.K.; Marcie, S.; Pouyssegur, J.; Chiche, J. Knock-down of hypoxia-induced carbonic anhydrases IX and XII radiosensitizes tumor cells by increasing intracellular acidosis. *Front. Oncol.* **2013**, *2*, 199.2013. [CrossRef] [PubMed]

63. Lou, Y.; McDonald, P.C.; Oloumi, A.; Chia, S.; Ostlund, C.; Ahamdi, A.; Kyle, A.; Auf dem Keller, U.; Leung, S.; Huntsman, D.; et al. Targeting tumor hypoxia: Suppression of breast tumor growth and metastasis by novel carbonic anhydrase IX inhibitors. *Cancer Res.* **2011**, *71*, 3364–3376. [CrossRef] [PubMed]

64. McIntyre, A.; Patiar, S.; Wigfield, S.; Li, J.; Ledaki, I.; Turley, H.; Leek, R.; Snell, C.; Gatter, K.; Sly, W.S.; et al. Carbonic anhydrase IX promotes tumor growth and necrosis in vivo and inhibition enhances anti-VEGF therapy. *Clin. Cancer Res.* **2012**, *18*, 3100–3111. [CrossRef] [PubMed]

65. Lock, F.E.; McDonald, P.C.; Lou, Y.; Serrano, I.; Chafe, S.C.; Ostlund, C.; Aparicio, S.; Winum, J.Y.; Supuran, C.T.; Dedhar, S. Targeting carbonic anhydrase IX depletes breast cancer stem cells within the hypoxic niche. *Oncogene* **2013**, *32*, 5210–5219. [CrossRef] [PubMed]

66. Meehan, J.; Ward, C.; Turnbull, A.; Bukowski-Wills, J.; Finch, A.J.; Jarman, E.J.; Xintaropoulou, C.; Martinez-Perez, C.; Gray, M.; Pearson, M.; et al. Inhibition of pH regulation as a therapeutic strategy in hypoxic human breast cancer cells. *Oncotarget* **2017**, *8*, 42857–42875. [CrossRef] [PubMed]

67. Swayampakula, M.; McDonald, P.; Vallejo, M.; Coyaud, E.; Chafe, S.C.; Westerbeck, A.; Venkateswaran, G.; Shankar, J.; Gao, G.; Laurent, E.M.N.; et al. The interactome of metabolic enzyme carbonic anhydrase IX reveals novel roles in tumor cell migration and invadopodia/MMP14-mediated invasion. *Oncogene* **2017**, *36*, 6244–6261. [CrossRef] [PubMed]

68. Radvak, P.; Repic, M.; Svastova, E.; Takacova, M.; Csaderova, L.; Strnad, H.; Pastorek, J.; Pastorekova, S.; Kopacek, J. Suppression of carbonic anhydrase IX leads to aberrant focal adhesion and decreased invasion of tumor cells. *Oncol. Rep.* **2013**, *29*, 1147–1153. [CrossRef] [PubMed]

69. Sansone, P.; Storci, G.; Tavolari, S.; Guarnieri, T.; Giovannini, C.; Taffurelli, M.; Ceccarelli, C.; Santini, D.; Paterini, P.; Marcu, K.B.; et al. IL-6 triggers malignant features in mammospheres from human ductal breast carcinoma and normal mammary gland. *J. Clin. Investig.* **2007**, *117*, 3988–4002. [CrossRef] [PubMed]

70. Parkkila, S.; Rajaniemi, H.; Parkkila, A.K.; Kivela, J.; Waheed, A.; Pastorekova, S.; Pastorek, J.; Sly, W.S. Carbonic anhydrase inhibitor suppresses invasion of renal cancer cells in vitro. *Proc. Natl. Acad. Sci. USA* **2000**, *97*, 2220–2224. [CrossRef] [PubMed]

71. Winum, J.Y.; Carta, F.; Ward, C.; Mullen, P.; Harrison, D.; Langdon, S.P.; Cecchi, A.; Scozzafava, A.; Kunkler, I.; Supuran, C.T. Ureido-substituted sulfamates show potent carbonic anhydrase IX inhibitory and antiproliferative activities against breast cancer cell lines. *Bioorg. Med. Chem. Lett.* **2012**, *22*, 4681–4685. [CrossRef] [PubMed]

72. Robertson, N.; Potter, C.; Harris, A.L. Role of carbonic anhydrase IX in human tumor cell growth, survival and invasion. *Cancer Res.* **2004**, *64*, 6160–6165. [CrossRef] [PubMed]

73. Gieling, R.G.; Babur, M.; Mamnani, L.; Burrows, N.; Telfer, B.A.; Carta, F.; Winum, J.Y.; Scozzafava, A.; Supuran, C.T.; Williams, K.J. Antimetastatic effect of sulfamate carbonic anhydrase IX inhibitors in breast carcinoma xenografts. *J. Med. Chem.* **2012**, *55*, 5591–6000. [CrossRef] [PubMed]

74. Pacchiano, F.; Carta, F.; McDonald, P.C.; Lou, Y.; Vullo, D.; Scozzafava, A.; Dedhar, S.; Supuran, C.T. Ureido-substituted benzenesulfonamides potently inhibit carbonic anhydrase IX and show antimetastatic activity in a model of breast cancer metastasis. *J. Med. Chem.* **2011**, *54*, 1896–1902. [CrossRef] [PubMed]

75. Supuran, C.T. Carbonic Anhydrase Inhibition and the Management of Hypoxic Tumors. *Metabolites* **2017**, *7*, 48. [CrossRef]

76. Svastová, E.; Zilka, N.; Zaťovicová, M.; Gibadulinová, A.; Ciampor, F.; Pastorek, J.; Pastoreková, S. Carbonic anhydrase IX reduces E-cadherin-mediated adhesion of MDCK cells via interaction with catenin. *Exp. Cell Res.* **2003**, *290*, 332–345. [CrossRef]

77. Shin, H.J.; Rho, S.B.; Jung, D.; Han, I.O.; Oh, E.S.; Kim, J.Y. Carbonic anhydrase IX (CA9) modulates tumor-associated cell migration and invasion. *J. Cell Sci.* **2011**, *124*, 1077–1087. [CrossRef] [PubMed]

78. Hulikova, A.; Zatovicova, M.; Svastova, E.; Ditte, P.; Brasseur, R.; Kettmann, R.; Supuran, C.T.; Kopacek, J.; Pastorek, J.; Pastorekova, S. Intact intracellular tail is critical for proper functioning of the tumour-associated, hypoxia-regulated carbonic anhydrase IX. *FEBS Lett.* **2009**, *583*, 3563–3568. [CrossRef] [PubMed]

79. Ditte, P.; Dequiedt, F.; Svastova, E.; Hulikova, A.; Ohradanova-Repic, A.; Zatovicova, M.; Csaderova, L.; Kopacek, J.; Supuran, C.T.; Pastorekova, S.; et al. Phosphorylation of carbonic anhydrase IX controls its ability to mediate extracellular acidification in hypoxic tumors. *Cancer Res.* **2011**, *71*, 7558–7567. [CrossRef] [PubMed]

80. Worthylake, R.A.; Burridge, K. RhoA and ROCK promote migration by limiting membrane protrusions. *J. Biol. Chem.* **2003**, *278*, 13578–13584. [CrossRef] [PubMed]

81. Csaderova, L.; Debreova, M.; Radvak, P.; Stano, M.; Vrestiakova, M.; Kopacek, J.; Pastorekova, S.; Svastova, E. The effect of carbonic anhydrase IX on focal contacts during cell spreading and migration. *Front. Physiol.* **2013**, *4*, 271. [CrossRef] [PubMed]

82. Thariat, J.; Hannoun-Levi, J.M.; Sun Myint, A.; Vuong, T.; Gérard, J.P. Past, present, and future of radiotherapy for the benefit of patients. *Nat. Rev. Clin. Oncol.* **2013**, *10*, 52–60. [CrossRef] [PubMed]

83. Kavanagh, J.N.; Redmond, K.M.; Schettino, G.; Prise, K.M. DNA double strand break repair: A radiation perspective. *Antioxid. Redox Signal.* **2013**, *18*, 2458–2472. [CrossRef] [PubMed]

84. Raleigh, D.R.; Haas-Kogan, D.A. Molecular targets and mechanisms of radiosensitization using DNA damage response pathways. *Future Oncol.* **2013**, *9*, 219–233. [CrossRef] [PubMed]

85. Curtin, N.J. DNA repair dysregulation from cancer driver to therapeutic target. *Nat. Rev. Cancer* **2012**, *12*, 801–817. [CrossRef] [PubMed]

86. Deschner, E.E.; Gray, L.H. Influence of oxygen tension on X-ray-induced chromosomal damage in Ehrlich ascites tumor cells irradiated in vitro and in vivo. *Radiat. Res.* **1959**, *11*, 115–146. [CrossRef] [PubMed]

87. Dewey, D.L. Effect of oxygen and nitric oxide on the radio-sensitivity of human cells in tissue culture. *Nature* **1960**, *186*, 780–782. [CrossRef] [PubMed]

88. Gray, L.H. The initiation and development of cellular damage by ionizing radiations; the thirty-second Silvanus Thompson Memorial Lecture. *Br. J. Radiol.* **1953**, *26*, 609–618. [CrossRef] [PubMed]

89. Brown, J.M. Tumor hypoxia in cancer therapy. *Methods Enzymol.* **2007**, *435*, 297–321. [CrossRef] [PubMed]

90. Eriksson, D.; Stigbrand, T. Radiation-induced cell death mechanisms. *Tumour Biol.* **2010**, *31*, 363–372. [CrossRef] [PubMed]

91. Hwang, H.S.; Davis, T.W.; Houghton, J.A.; Kinsella, T.J. Radiosensitivity of thymidylate synthase-deficient human tumor cells is affected by progression through the G1 restriction point into S-phase: Implications for fluoropyrimidine radiosensitization. *Cancer Res.* **2000**, *60*, 92–100. [PubMed]

92. Pawlik, T.M.; Keyomarsi, K. Role of cell cycle in mediating sensitivity to radiotherapy. *Int. J. Radiat. Oncol. Biol. Phys.* **2004**, *59*, 928–942. [CrossRef] [PubMed]

93. Holahan, E.V.; Stuart, P.K.; Dewey, W.C. Enhancement of survival of CHO cells by acidic pH after x irradiation. *Radiat. Res.* **1982**, *89*, 433–435. [CrossRef] [PubMed]

94. Rottinger, E.M.; Mendonca, M.; Gerweck, L.E. Modification of pH induced cellular inactivation by irradiation-glial cells. *Int. J. Radiat. Oncol. Biol. Phys.* **1980**, *6*, 1659–1662. [CrossRef]

95. Freeman, M.L.; Holahan, E.V.; Highfield, D.P.; Raaphorst, G.P.; Spiro, I.J.; Dewey, W.C. The effect of pH on hyperthermic and X-ray induced cell killing. *Int. J. Radiat. Oncol. Biol. Phys.* **1981**, *7*, 211–216. [CrossRef]

96. Freeman, M.L.; Sierra, E. An acidic extracellular environment reduces the fixation of radiation damage. *Radiat. Res.* **1984**, *97*, 154–161. [CrossRef] [PubMed]

97. Lee, H.S.; Park, H.J.; Lyons, J.C.; Griffin, R.J.; Auger, E.A.; Song, C.W. Radiation-induced apoptosis in different pH environments in vitro. *Int. J. Radiat. Oncol. Biol. Phys.* **1997**, *38*, 1079–1087. [CrossRef]

98. Ojeda, F.; Skardova, I.; Guarda, M.I.; Maldonado, C.; Folch, H. Radiation-induced apoptosis in thymocytes: pH sensitization. *J. Biosci.* **1996**, *51*, 432–434.

99. Park, H.J.; Lee, S.H.; Chung, H.; Rhee, Y.H.; Lim, B.U.; Ha, S.W.; Griffin, R.J.; Lee, H.S.; Song, C.W.; Choi, E.K. Influence of environmental pH on G2-phase arrest caused by ionizing radiation. *Radiat. Res.* **2003**, *159*, 86–93. [CrossRef]

100. Koukourakis, M.I.; Bentzen, S.M.; Giatromanolaki, A.; Wilson, G.D.; Daley, F.M.; Saunders, M.I.; Dische, S.; Sivridis, E.; Harris, A.L. Endogenous markers of two separate hypoxia response pathways (hypoxia inducible factor 2 alpha and carbonic anhydrase 9) are associated with radiotherapy failure in head and neck cancer patients recruited in the CHART randomized trial. *J. Clin. Oncol.* **2006**, *24*, 727–735. [CrossRef] [PubMed]

101. Duivenvoorden, W.C.; Hopmans, S.N.; Gallino, D.; Farrell, T.; Gerdes, C.; Glennie, D.; Lukka, H.; Pinthus, J.H. Inhibition of carbonic anhydrase IX (CA9) sensitizes renal cell carcinoma to ionizing radiation. *Oncol. Rep.* **2015**, *34*, 1968–1976. [CrossRef] [PubMed]

102. Balcer-Kubiczek, E.K. Apoptosis in radiation therapy: A double-edged sword. *Exp. Oncol.* **2012**, *34*, 277–285. [PubMed]

103. Dubois, L.; Peeters, S.; Lieuwes, N.G.; Geusens, N.; Thiry, A.; Wigfield, S.; Carta, F.; McIntyre, A.; Scozzafava, A.; Dogne, J.M.; et al. Specific inhibition of carbonic anhydrase IX activity enhances the in vivo therapeutic effect of tumor irradiation. *Radiother. Oncol.* **2011**, *99*, 424–431. [CrossRef] [PubMed]

104. Dubois, L.; Peeters, S.G.; van Kuijk, S.J.; Yaromina, A.; Lieuwes, N.G.; Sarava, R.; Biemans, R.; Rami, M.; Parvathaneni, N.K.; Vullo, D.; et al. Targeting carbonic anhydrase IX by nitroimidazole based sulfamides enhances the therapeutic effect of tumor irradiation: A new concept of dual targeting drugs. *Radiother. Oncol.* **2013**, *108*, 523–528. [CrossRef] [PubMed]

105. Dent, P.; Yacoub, A.; Contessa, J.; Caron, R.; Amorino, G.; Valerie, K.; Hagan, M.P.; Grant, S.; Schmidt-Ullrich, R. Stress and radiation-induced activation of multiple intracellular signaling pathways. *Radiat. Res.* **2003**, *159*, 283–300. [CrossRef]

106. Brand, T.M.; Iida, M.; Luthar, N.; Starr, M.M.; Huppert, E.J.; Wheeler, D.L. Nuclear EGFR as a molecular target in cancer. *Radiother. Oncol.* **2013**, *108*, 370–377. [CrossRef] [PubMed]

107. Dittmann, K.; Mayer, C.; Kehlbach, R.; Rodemann, H.P. Radiation-induced caveolin-1 associated EGFR internalization is linked with nuclear EGFR transport and activation of DNA-PK. *Mol. Cancer* **2008**, *7*, 69. [CrossRef] [PubMed]

108. Dittmann, K.; Mayer, C.; Fehrenbacher, B.; Schaller, M.; Raju, U.; Milas, L.; Chen, D.J.; Kehlbach, R.; Rodemann, H.P. Radiation-induced epidermal growth factor receptor nuclear import is linked to activation of DNA-dependent protein kinase. *J. Biol. Chem.* **2005**, *280*, 31182–31189. [CrossRef] [PubMed]

109. Dittmann, K.; Mayer, C.; Fehrenbacher, B.; Schaller, M.; Kehlbach, R.; Rodemann, H.P. Nuclear epidermal growth factor receptor modulates cellular radio-sensitivity by regulation of chromatin access. *Radiother. Oncol.* **2011**, *99*, 317–322. [CrossRef] [PubMed]

110. Lee, K.M.; Choi, E.J.; Kim, I.A. microrna-7 increases radiosensitivity of human cancer cells with activated EGFR-associated signaling. *Radiother. Oncol.* **2011**, *101*, 171–176. [CrossRef] [PubMed]

111. Dorai, T.; Sawczuk, I.S.; Pastorek, J.; Wiernik, P.H.; Dutcher, J.P. The role of carbonic anhydrase IX overexpression in kidney cancer. *Eur. J. Cancer* **2005**, *41*, 2935–2947. [CrossRef] [PubMed]

112. Ogawa, K.; Yoshioka, Y.; Isohashi, F.; Seo, Y.; Yoshida, K.; Yamazaki, H. Radiotherapy targeting cancer stem cells: Current views and future perspectives. *Anticancer Res.* **2013**, *33*, 747–754. [PubMed]

113. Pang, L.Y.; Sanders, L.; Argyle, D.J. Epidermal growth factor receptor activity is elevated in glioma cancer stem cells and is required to maintain chemotherapy and radiation resistance. *Oncotarget* **2017**, *8*, 72494–72512. [CrossRef] [PubMed]

114. Higgins, G.S.; Krause, M.; McKenna, W.G.; Baumann, M. Personalized radiation oncology: Epidermal growth factor and other receptor tyrosine kinase inhibitors. *Recent Results Cancer Res.* **2016**, *198*, 107–122. [CrossRef] [PubMed]

115. Baumann, M.; Krause, M.; Dikomey, E.; Dittmann, K.; Dorr, W.; Kasten-Pisula, U.; Rodemann, H.P. EGFR-targeted anti-cancer drugs in radiotherapy: Preclinical evaluation of mechanisms. *Radiother. Oncol.* **2007**, *83*, 238–248. [CrossRef] [PubMed]

116. Skvortsova, I.; Skvortsov, S.; Stasyk, T.; Raju, U.; Popper, B.A.; Schiestl, B.; von Guggenberg, E.; Neher, A.; Bonn, G.K.; Huber, L.A.; et al. Intracellular signaling pathways regulating radioresistance of human prostate carcinoma cells. *Proteomics* **2008**, *8*, 4521–4533. [CrossRef] [PubMed]

117. Miyaguchi, M.; Takeuchi, T.; Morimoto, K.; Kubo, T. Correlation of epidermal growth factor receptor and radiosensitivity in human maxillary carcinoma cell lines. *Acta Otolaryngol.* **1998**, *118*, 428–431. [PubMed]

118. Sheridan, M.T.; O'Dwyer, T.; Seymour, C.B.; Mothersill, C.E. Potential indicators of radiosensitivity in squamous cell carcinoma of the head and neck. *Radiat. Oncol. Investig.* **1997**, *5*, 180–186. [CrossRef]

119. Nyati, M.K.; Morgan, M.A.; Feng, F.Y.; Lawrence, T.S. Integration of EGFR inhibitors with radiochemotherapy. *Nat. Rev. Cancer* **2006**, *6*, 876–885. [CrossRef] [PubMed]

120. Thariat, J.; Milas, L.; Ang, K.K. Integrating radiotherapy with epidermal growth factor receptor antagonists and other molecular therapeutics for the treatment of head and neck cancer. *Int. J. Radiat. Oncol. Biol. Phys.* **2007**, *69*, 974–984. [CrossRef] [PubMed]

121. Perkins, N.D. The diverse and complex roles of NF-B subunits in cancer. *Nat. Rev. Cancer* **2012**, *12*, 121–132. [CrossRef] [PubMed]

122. Xu, L.; Fidler, I.J. Acidic pH-induced elevation in interleukin 8 expression by human ovarian carcinoma cells. *Cancer Res.* **2000**, *60*, 4610–4616. [PubMed]

123. Fukumura, D.; Xu, L.; Chen, Y.; Gohongi, T.; Seed, B.; Jain, R.K. Hypoxia and acidosis independently up-regulate vascular endothelial growth factor transcription in brain tumors in vivo. *Cancer Res.* **2001**, *61*, 6020–6024. [PubMed]

124. Dunn, S.M.; Coles, L.S.; Lang, R.K.; Gerondakis, S.; Vadas, M.A.; Shannon, M.F. Requirement for nuclear factor (NF)-κB p65 and NF-interleukin-6 binding elements in the tumor necrosis factor response region of the granulocyte colony-stimulating factor promoter. *Blood* **1994**, *83*, 2469–2479. [PubMed]

125. Bellocq, A.; Suberville, S.; Philippe, C.; Bertrand, F.; Perez, J.; Fouqueray, B.; Cherqui, G.; Baud, L. Low environmental pH is responsible for the induction of nitric-oxide synthase in macrophages. Evidence for involvement of nuclear factor-κB activation. *J. Biol. Chem.* **1998**, *273*, 5086–5092. [CrossRef] [PubMed]

126. Kim, B.Y.; Kim, K.A.; Kwon, O.; Kim, S.O.; Kim, M.S.; Kim, B.S.; Oh, W.K.; Kim, G.D.; Jung, M.; Ahn, J.S. NF-κB inhibition radiosensitizes Ki-Ras-transformed cells to ionizing radiation. *Carcinogenesis* **2005**, *26*, 1395–1403. [CrossRef] [PubMed]

127. Ahmed, K.M.; Dong, S.; Fan, M.; Li, J.J. Nuclear Factor-κB p65 inhibits mitogen-activated protein kinase signaling pathway in radioresistant breast cancer cells. *Mol. Cancer Res.* **2006**, *4*, 945–955. [CrossRef] [PubMed]

128. Ahmed, K.M.; Li, J.J. NF-κB-mediated adaptive resistance to ionizing radiation. *Free Radic. Biol. Med.* **2008**, *44*, 1–13. [CrossRef] [PubMed]

129. Cataldi, A.; Rapino, M.; Centurione, L.; Sabatini, N.; Grifone, G.; Garaci, F.; Rana, R. NF-κB activation plays an antiapoptotic role in human leukemic K562 cells exposed to ionizing radiation. *J. Cell. Biochem.* **2003**, *89*, 956–963. [CrossRef] [PubMed]

130. Guo, G.; Yan-Sanders, Y.; Lyn-Cook, B.D.; Wang, T.; Tamae, D.; Ogi, J.; Khaletskiy, A.; Li, Z.; Weydert, C.; Longmate, J.A.; et al. Manganese superoxide dismutase-mediated gene expression in radiation-induced adaptive responses. *Mol. Cell. Biol.* **2003**, *23*, 2362–2378. [CrossRef] [PubMed]

131. Shaw, L.M. Integrin function in breast carcinoma progression. *J. Mammary Gland Biol. Neoplasia* **1999**, *4*, 367–376. [CrossRef] [PubMed]

132. Sethi, T.; Rintoul, R.C.; Moore, S.M.; MacKinnon, A.C.; Salter, D.; Choo, C.; Chilvers, E.R.; Dransfield, I.; Donnelly, S.C.; Strieter, R.; et al. Extracellular matrix proteins protect small cell lung cancer cells against apoptosis: A mechanism for small cell lung cancer growth and drug resistance in vivo. *Nat. Med.* **1999**, *5*, 662–668. [CrossRef] [PubMed]

133. Aoudjit, F.; Vuori, K. Integrin signaling inhibits paclitaxel-induced apoptosis in breast cancer cells. *Oncogene* **2001**, *20*, 4995–5004. [CrossRef] [PubMed]

134. Nam, J.M.; Onodera, Y.; Bissell, M.J.; Park, C.C. Breast cancer cells in three dimensional culture display an enhanced radioresponse after coordinate targeting of integrin α5β1 and fibronectin. *Cancer Res.* **2010**, *70*, 5238–5248. [CrossRef] [PubMed]

135. Eke, I.; Deuse, Y.; Hehlgans, S.; Gurtner, K.; Krause, M.; Baumann, M.; Shevchenko, A.; Sandfort, V.; Cordes, N. (1) Integrin/FAK/cortactin signaling is essential for human head and neck cancer resistance to radiotherapy. *J. Clin. Investig.* **2011**, *122*, 1529–1540. [CrossRef] [PubMed]

136. Park, C.C.; Zhang, H.J.; Yao, E.S.; Park, C.J.; Bissell, M.J. β1 integrin inhibition dramatically enhances radiotherapy efficacy in human breast cancer xenografts. *Cancer Res.* **2008**, *68*, 4398–4405. [CrossRef] [PubMed]

137. Cordes, N.; Seidler, J.; Durzok, R.; Geinitz, H.; Brakebusch, C. 1-integrin-mediated signaling essentially contributes to cell survival afterradiation-induced genotoxic injury. *Oncogene* **2006**, *25*, 1378–1390. [CrossRef] [PubMed]

138. Ahmed, K.M.; Zhang, H.; Park, C.C. NF-κB Regulates radioresistance mediated by β1-integrin in three-dimensional culture of breast cancer cells. *Cancer Res.* **2013**, *73*, 3737–3748. [CrossRef] [PubMed]

139. Nam, J.M.; Chung, Y.; Hsu, H.C.; Park, C.C. 1 integrin targeting to enhance radiation therapy. *Int. J. Radiat. Biol.* **2009**, *85*, 923–928. [CrossRef] [PubMed]

140. Chafe, S.C.; Lou, Y.; Sceneay, J.; Vallejo, M.; Hamilton, M.J.; McDonald, P.C.; Bennewith, K.L.; Moller, A.; Dedhar, S. Carbonic anhydrase IX promotes myeloid-derived suppressor cell mobilization and establishment of a metastatic niche by stimulating G-CSF production. *Cancer Res.* **2015**, *75*, 996–1008. [CrossRef] [PubMed]

141. Singh, V.K.; Fatanmi, O.O.; Singh, P.K.; Whitnall, M.H. Role of radiation-induced granulocyte colony-stimulating factor in recovery from whole body gamma-irradiation. *Cytokine* **2012**, *58*, 406–414. [CrossRef] [PubMed]

142. Waddick, K.G.; Song, C.W.; Souza, L.; Uckun, F.M. Comparative analysis of the in vivo radioprotective effects of recombinant granulocyte colony-stimulating factor (G-CSF), recombinant granulocyte-macrophage CSF, and their combination. *Blood* **1991**, *77*, 2364–2371. [PubMed]

143. Yang, X.; Huang, P.; Wang, F.; Xu, Z. Expression of granulocyte colony-stimulating factor receptor in rectal cancer. *World J. Gastroenterol.* **2014**, *20*, 1074–1078. [CrossRef] [PubMed]

144. Studebaker, A.W.; Storci, G.; Werbeck, J.L.; Sansone, P.; Sasser, A.K.; Tavolari, S.; Huang, T.; Chan, M.W.; Marini, F.C.; Rosol, T.J.; et al. Fibroblasts isolated from common sites of breast cancer metastasis enhance cancer cell growth rates and invasiveness in an interleukin-6-dependent manner. *Cancer Res.* **2008**, *68*, 9087–9095. [CrossRef] [PubMed]

145. Schoppmann, S.F.; Jesch, B.; Friedrich, J.; Jomrich, G.; Maroske, F.; Birner, P. Phosphorylation of signal transducer and activator of transcription 3 (STAT3) correlates with Her-2 status, carbonic anhydrase 9 expression and prognosis in esophageal cancer. *Clin. Exp. Metastasis* **2012**, *29*, 615–624. [CrossRef] [PubMed]

146. Bonner, J.A.; Trummell, H.Q.; Willey, C.D.; Plants, B.A.; Raisch, K.P. Inhibition of STAT-3 results in radiosensitization of human squamous cell carcinoma. *Radiother. Oncol.* **2009**, *92*, 339–344. [CrossRef]

147. Lau, J.; Ilkhanizadeh, S.; Wang, S.; Miroshnikova, Y.A.; Salvatierra, N.A.; Wong, R.A.; Schmidt, C.; Weaver, V.M.; Weiss, W.A.; Persson, A.I. See comment in PubMed commons below STAT3 blockade inhibits radiation-induced malignant progression in glioma. *Cancer Res.* **2015**, *75*, 4302–4311. [CrossRef] [PubMed]

148. Wu, X.; Tang, W.; Marquez, R.T.; Li, K.; Highfill, C.A.; He, F.; Lian, J.; Lin, J.; Fuchs, J.R.; Ji, M.; et al. Overcoming chemo/radio-resistance of pancreatic cancer by inhibiting STAT3 signaling. *Oncotarget* **2016**, *7*, 11708–11723. [CrossRef] [PubMed]

149. Wu, C.T.; Chen, M.F.; Chen, W.C.; Hsieh, C.C. The role of IL-6 in the radiation response of prostate cancer. *Radiat. Oncol.* **2013**, *8*, 159. [CrossRef] [PubMed]

150. Matsuoka, Y.; Nakayama, H.; Yoshida, R.; Hirosue, A.; Nagata, M.; Tanaka, T.; Kawahara, K.; Sakata, J.; Arita, H.; Nakashima, H.; et al. IL-6 controls resistance to radiation by suppressing oxidative stress via the Nrf2-antioxidant pathway in oral squamous cell carcinoma. *Br. J. Cancer* **2016**, *115*, 1234–1244. [CrossRef] [PubMed]

151. Karin, M.; Greten, F.R. NF-κB: Linking inflammation and immunity to cancer development and progression. *Nat. Rev. Immunol.* **2005**, *5*, 749–759. [CrossRef] [PubMed]

152. Sattler, U.G.; Meyer, S.S.; Quennet, V.; Hoerner, C.; Knoerzer, H.; Fabian, C.; Yaromina, A.; Zips, D.; Walenta, S.; Baumann, M.; et al. Glycolytic metabolism and tumour response to fractionated irradiation. *Radiother. Oncol.* **2010**, *94*, 102–109. [CrossRef] [PubMed]

153. Hirschhaeuser, F.; Sattler, U.G.; Mueller-Klieser, W. Lactate: A metabolic key player in cancer. *Cancer Res.* **2011**, *71*, 6921–6925. [CrossRef] [PubMed]

154. Groussard, C.; Morel, I.; Chevanne, M.; Monnier, M.; Cillard, J.; Delamarche, A. Free radical scavenging and antioxidant effects of lactate ion: An in vitro study. *J. Appl. Physiol.* **2000**, *89*, 169–175. [CrossRef] [PubMed]

155. Halestrap, A.P. The monocarboxylate transporter family—Structure and functional characterization. *IUBMB Life* **2012**, *64*, 1–9. [CrossRef] [PubMed]

156. Garcia, C.K.; Goldstein, J.L.; Pathak, R.K.; Anderson, R.G.; Brown, M.S. Molecular characterization of a membrane transporter for lactate, pyruvate, and other monocarboxylates: Implications for the Cori cycle. *Cell* **1994**, *76*, 865–873. [CrossRef]

157. Dimmer, K.S.; Friedrich, B.; Lang, F.; Deitmer, J.W.; Bröer, S. The low-affinity monocarboxylate transporter MCT4 is adapted to the export of lactate in highly glycolytic cells. *Biochem. J.* **2000**, *350*, 219–227. [CrossRef] [PubMed]

158. Pertega-Gomes, N.; Vizcain, J.R.; Miranda-Goncalves, V.; Pinheiro, C.; Silva, J.; Pereira, H.; Monteiro, P.; Henrique, R.M.; Reis, R.M.; Lopes, C.; et al. Monocarboxylate transporter 4 (MCT4) and CD147 overexpression is associated with poor prognosis in prostate cancer. *BMC Cancer* **2011**, *11*, 312. [CrossRef] [PubMed]

159. Doyen, J.; Trastour, C.; Ettore, F.; Peyrottes, I.; Toussant, N.; Gal, J.; Ilc, K.; Roux, D.; Parks, S.K.; Ferrero, J.M.; et al. Expression of hypoxia-inducible monocarboxylate transporter MCT4 is increased in triple negative breast cancer and correlates independently with clinical outcome. *Biochem. Biophys. Res. Commun.* **2014**, *451*, 54–61. [CrossRef] [PubMed]

160. Pinheiro, C.; Longatto-Filho, A.; Scapulatempo, C.; Ferreira, L.; Martins, S.; Pellerin, L.; Rodrigues, M.; Alves, V.A.; Schmitt, F.; Baltazar, F. Increased expression of monocarboxylate transporters 1, 2, and 4 in colorectal carcinomas. *Virchows Arch.* **2008**, *452*, 139–146. [CrossRef] [PubMed]

161. Sonveaux, P.; Vegran, F.; Schroeder, T.; Wergin, M.C.; Verrax, J.; Rabbani, Z.N.; De Saedeleer, C.J.; Kennedy, K.M.; Diepart, C.; Jordan, B.F.; et al. Targeting lactate-fueled respiration selectively kills hypoxic tumor cells in mice. *J. Clin. Investig.* **2008**, *118*, 3930–3942. [CrossRef] [PubMed]

162. Bola, B.M.; Chadwick, A.L.; Michopoulos, F.; Blount, K.G.; Telfer, B.A.; Williams, K.J.; Smith, P.D.; Critchlow, S.E.; Stratford, I.J. Inhibition of monocarboxylate transporter-1 (MCT1) by AZD3965 enhances radiosensitivity by reducing lactate transport. *Mol. Cancer Ther.* **2014**, *13*, 2805–2816. [CrossRef] [PubMed]

163. Le Floch, R.; Chiche, J.; Marchiq, I.; Naiken, T.; Ilc, K.; Murray, C.M.; Critchlow, S.E.; Roux, D.; Simon, M.P.; Pouysségur, J. CD147 subunit of lactate/H+ symporters MCT1 and hypoxia-inducible MCT4 is critical for energetics and growth of glycolytic tumors. *Proc. Natl. Acad. Sci. USA* **2011**, *108*, 16663–16668. [CrossRef] [PubMed]

164. Wu, J.; Li, Y.; Dang, Y.Z.; Gao, H.X.; Jiang, J.L.; Chen, Z.N. HAB18G/CD147 promotes radioresistance in hepatocellular carcinoma cells: A potential role for integrin 1 signaling. *Mol. Cancer Ther.* **2015**, *14*, 553–563. [CrossRef] [PubMed]

165. Ju, X.Z.; Yang, J.M.; Zhou, X.Y.; Li, Z.T.; Wu, X.H. EMMPRIN expression as a prognostic factor in radiotherapy of cervical cancer. *Clin. Cancer Res.* **2008**, *14*, 494–501. [CrossRef] [PubMed]

166. Huang, X.Q.; Chen, X.; Xie, X.X.; Zhou, Q.; Li, K.; Li, S.; Shen, L.F.; Su, J. Co-expression of CD147 and GLUT-1 indicates radiation resistance and poor prognosis in cervical squamous cell carcinoma. *Int. J. Clin. Exp. Pathol.* **2014**, *7*, 1651–1666. [PubMed]

167. Becker, H.M.; Deitmer, J.W. Nonenzymatic proton handling by carbonic anhydrase II during H⁺-lactate cotransport via monocarboxylate transporter 1. *J. Biol. Chem.* **2008**, *283*, 21655–21667. [CrossRef] [PubMed]

168. Stridh, M.; Alt, M.D.; Wittmann, S.; Heidtmann, H.; Aggarwal, M.; Riederer, B.; Seidler, U.; Wennemuth, G.; McKenna, R.; Deitmer, J.W.; et al. Lactate flux in astrocytes is enhanced by a non-catalytic action of carbonic anhydrase II. *J. Physiol.* **2012**, *590*, 2333–2351. [CrossRef] [PubMed]

169. Panisova, E.; Kery, M.; Sedlakova, O.; Brisson, L.; Debreova, M.; Sboarina, M.; Sonveaux, P.; Pastorekova, S.; Svastova, E. Lactate stimulates CA IX expression in normoxic cancer cells. *Oncotarget* **2017**, *8*, 77819–77838. [CrossRef] [PubMed]

170. Axelson, H.; Fredlund, E.; Ovenberger, M.; Landberg, G.; Pahlman, S. Hypoxia-induced dedifferentiation of tumor cells—A mechanism behind heterogeneity and aggressiveness of solid tumors. *Semin. Cell Dev. Biol.* **2005**, *16*, 554–563. [CrossRef] [PubMed]

171. Sansone, P.; Storci, G.; Giovannini, C.; Pandolfi, S.; Pianetti, S.; Taffurelli, M.; Santini, D.; Ceccarelli, C.; Chieco, P.; Bonafé, M. p66Shc/Notch-3 interplay controls self-renewal and hypoxia survival in human stem/progenitor cells of the mammary gland expanded in vitro as mammospheres. *Stem Cells* **2007**, *25*, 807–815. [CrossRef] [PubMed]

172. Storci, G.; Sansone, P.; Trere, D.; Tavolari, S.; Taffurelli, M.; Ceccarelli, C.; Guarnieri, T.; Paterini, P.; Pariali, M.; Montanaro, L.; et al. The basal-like breast carcinoma phenotype is regulated by SLUG gene expression. *J. Pathol.* **2008**, *214*, 25–37. [CrossRef] [PubMed]

173. Currie, M.J.; Beardsley, B.E.; Harris, G.C.; Gunningham, S.P.; Dachs, G.U.; Dijkstra, B.; Morrin, H.R.; Wells, J.E.; Robinson, B.A. Immunohistochemical analysis of cancer stem cell markers in invasive breast carcinoma and associated ductal carcinoma in situ: Relationships with markers of tumor hypoxia and microvascularity. *Hum. Pathol.* **2013**, *44*, 402–411. [CrossRef] [PubMed]

174. Phillips, T.M.; McBride, W.H.; Pajonk, F. The response of CD24⁻/low/CD44⁺ breast cancer-initiating cells to radiation. *J. Natl. Cancer Inst.* **2006**, *8*, 1777–1785. [CrossRef] [PubMed]

175. Chang, L.; Graham, P.; Hao, J.; Ni, J.; Deng, J.; Bucci, J.; Malouf, D.; Gillatt, D.; Li, Y. Cancer stem cells and signaling pathways in radioresistance. *Oncotarget* **2016**, *7*, 11002–11017. [CrossRef] [PubMed]

176. Ogawa, K.; Murayama, S.; Mori, M. Predicting the tumour response to radiotherapy using microarray analysis. *Oncol. Rep.* **2007**, *18*, 1243–1248. [PubMed]

177. Woodward, W.A.; Chen, M.S.; Behbod, F.; Alfaro, M.P.; Buchholz, T.A.; Rosen, J.M. WNT/catenin mediates radiation resistance of mouse mammary progenitor cells. *Proc. Natl. Acad. Sci. USA* **2007**, *104*, 618–623. [CrossRef] [PubMed]

178. Zhang, M.; Behbod, F.; Atkinson, R.L.; Landis, M.D.; Kittrell, F.; Edwards, D.; Medina, D.; Tsimelzon, A.; Hilsenbeck, S.; Green, J.E.; et al. Identification of tumor-initiating cells in a p53-null mouse model of breast cancer. *Cancer Res.* **2008**, *68*, 4674–4682. [CrossRef] [PubMed]

179. Desai, A.; Webb, B.; Gerson, S.L. CD133+ cells contribute to radioresistance via altered regulation of DNA repair genes in human lung cancer cells. *Radiother. Oncol.* **2014**, *110*, 538–545. [CrossRef] [PubMed]

180. Chen, Y.; Zhang, F.; Tsai, Y.; Yang, X.; Yang, L.; Duan, S.; Wang, X.; Keng, P.; Lee, S.O. IL-6 signaling promotes DNA repair and prevents apoptosis in CD133+ stem-like cells of lung cancer after radiation. *Radiat. Oncol.* **2015**, *10*, 227. [CrossRef] [PubMed]

181. Hjelmeland, A.B.; Wu, Q.; Heddleston, J.M.; Choudhary, G.S.; MacSwords, J.; Lathia, J.D.; McLendon, R.; Lindner, D.; Sloan, A.; Rich, J.N. Acidic stress promotes a glioma stem cell phenotype. *Cell Death Differ.* **2011**, *18*, 829–840. [CrossRef] [PubMed]

182. Martinez-Outschoorn, U.E.; Prisco, M.; Ertel, A.; Tsirigos, A.; Lin, Z.; Pavlides, S.; Wang, C.; Flomenberg, N.; Knudsen, E.S.; Howell, A.; et al. Ketones and lactate increase cancer cell "stemness," driving recurrence, metastasis and poor clinical outcome in breast cancer: Achieving personalized medicine via Metabolo-Genomics. *Cell Cycle* **2011**, *10*, 1271–1286. [CrossRef] [PubMed]

metabolites

MDPI

Review

Rethinking the Combination of Proton Exchanger Inhibitors in Cancer Therapy

Elisabetta Iessi [1], Mariantonia Logozzi [1], Davide Mizzoni [1], Rossella Di Raimo [1], Claudiu T. Supuran [2] and Stefano Fais [1,*]

[1] Department of Oncology and Molecular Medicine, National Institute of Health, Viale Regina Elena 299, 00161 Rome, Italy; elisabetta.iessi@iss.it (E.I.); mariantonia.logozzi@iss.it (M.L.); davide.mizzoni@iss.it (D.M.); rosella.diraimo@iss.it (R.D.R.)
[2] Dipartimento Neurofarba, Sezione di Scienze Farmaceutiche, Laboratorio di Chimica Bioinorganica, Università degli Studi di Firenze, Polo Scientifico, Via U. Schiff 6, Sesto Fiorentino, 50019 Florence, Italy; claudiu.supuran@unifi.it
* Correspondence: stefano.fais@iss.it; Tel.: +39-06-4990-3195; Fax: +39-06-4990-2436

Received: 21 November 2017; Accepted: 21 December 2017; Published: 23 December 2017

Abstract: Microenvironmental acidity is becoming a key target for the new age of cancer treatment. In fact, while cancer is characterized by genetic heterogeneity, extracellular acidity is a common phenotype of almost all cancers. To survive and proliferate under acidic conditions, tumor cells up-regulate proton exchangers and transporters (mainly V-ATPase, Na^+/H^+ exchanger (NHE), monocarboxylate transporters (MCTs), and carbonic anhydrases (CAs)), that actively extrude excess protons, avoiding intracellular accumulation of toxic molecules, thus becoming a sort of survival option with many similarities compared with unicellular microorganisms. These systems are also involved in the unresponsiveness or resistance to chemotherapy, leading to the protection of cancer cells from the vast majority of drugs, that when protonated in the acidic tumor microenvironment, do not enter into cancer cells. Indeed, as usually occurs in the progression versus malignancy, resistant tumor clones emerge and proliferate, following a transient initial response to a therapy, thus giving rise to more malignant behavior and rapid tumor progression. Recent studies are supporting the use of a cocktail of proton exchanger inhibitors as a new strategy against cancer.

Keywords: acidity; hypoxia; pH; carbonic anhydrases; V-ATPases; proton pump inhibitors; carbonic anhydrase inhibitors

1. Introduction

Tumor cells often grow in a hypoxic microenvironment where there is low nutrient supply, and they upregulate glycolysis to sustain their high proliferation rate [1,2]. Growing evidence suggests that cancer cells take up much more glucose than normal cells, and mainly process it through aerobic glycolysis, the so-called "Warburg effect" [3,4]. This phenomenon leads to the conversion of one molecule of glucose into two molecules of lactic acid and 2 H^+ to produce 2 ATP, compared to the 36 ATP produced by oxidative metabolism [1,2]. Thus, tumor cells implement glycolysis, promoting an abnormally high rate of glucose utilization, which in turn, leads to the accumulation of lactic acid and to the production of a large amount of H^+ associated with proton efflux and extracellular pH reduction [5]. Associated with high glycolysis rate, high levels of carbon dioxide produced during mitochondrial respiration of oxygenated cancer cells may also contribute to a substantial release of H^+ into the tumor environment [6–10]. The complete oxidation of one glucose to carbonic dioxide yields 6 HCO_3^- and 6 H^+, leading to three times greater production of H^+ than when glucose is converted to lactate, significantly accounting for tumor extracellular acidosis [6–10]. Uncontrolled growth, lactic and carbonic acid production, low blood and nutrient supply, contribute to the generation of a tumor microenvironment that is extremely toxic for either normal or more differentiated cells, and therefore, progressively selects cells able to survive in these adverse conditions. It is therefore conceivable that malignant cancer cells survive in this hostile microenvironment, thanks to the upregulation of the expression and activity of several proton extrusion mechanisms [11], which release protons and lactate into extracellular environment, avoiding the acidification of the cytosol. Among proton flux regulators are vacuolar H^+-ATPases (V-ATPases), Na^+/H^+ exchanger (NHE), monocarboxylate transporters (MCTs), carbonic anhydrase IX (CA-IX) [11–14], and Na^+/HCO_3 co-transporters (NBC) [15] (Figure 1).

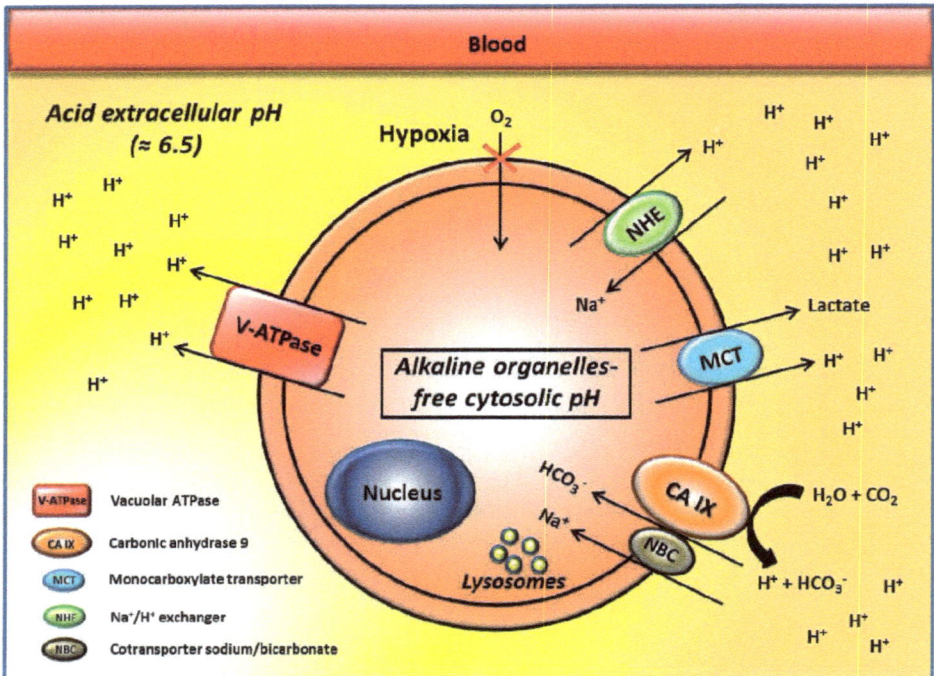

Figure 1. Proton flux regulators and their role in cancer. The aberrant expression and activity of proton exchangers leads to acidification of the tumor microenvironment and creates a reversed pH gradient across the plasma membrane leading to extracellular acidity and an alkaline, organelle-free cytosol.

Two of the most studied proton flux regulators are vacuolar H^+-ATPases (V-ATPase) [16–18] and carbonic anhydrase (CA) IX/XII [19–22]. The aberrant activity of these proton extruders creates a reversed pH gradient across the plasma membrane that is considered a hallmark of malignancy: extracellular acidity and alkaline conditions in the organelle-free cytosol, while normal cells show a neutral pH extracellularly, with a weakly acidic organelle-free cytosolic pH [23,24]. For this reason, the tumor pH gradient is called "reversed pH gradient" [5,16,25–28]. More precisely, the pH of tumor microenvironments have been shown to range between 6.0 and 6.8, with median values around 6.5, and the level of acidity was related with tumor malignancy [29–32]. Direct measurements of both intratumoral pO_2 and pH revealed either a spatial heterogeneity between hypoxia and acidosis gradients [10,33,34], meaning that the areas of hypoxia and acidosis in tumors may not overlap in mouse tumor models, and a lack of correlation between CA-IX expression and various hypoxia markers [35–37]. In many tumors, chronic exposure to acidic pH has been reported to promote invasiveness, metastatic behavior, and resistance to cytotoxic agents [38–42]. All in all, tumor extracellular acidity is considered a crucial phenotype of malignant tumors, subjecting cancer cells to a sort of selective pressure, that independently from the tumor histotype, leads to the development of cells able to survive in such a hostile microenvironment. Notably, normal cells at pH ranging from very acidic to weakly acidic die or are entirely blocked in their functions [43,44]. For this reason, the acidic pH of solid tumors has been proposed as a therapeutic target and a drug delivery system for selective anticancer treatments [25,32,45]. Indeed, inhibition of these pH regulation systems has been reported to lead to potent antitumor effects [13]. Therefore, approaches aimed at inhibiting the proteins involved in pH regulation are now under exploitation for the design of novel promising alternative therapeutic anticancer strategies. NHE inhibitors were the first studied as interesting pharmacological agents for interfering with tumor hypoxia/acidosis [12]. Considering their significant toxicity and the lack of isoform-selectivity for other proteins involved in these processes (such as the MCTs) [46,47], the attention has been focused on the V-ATPases and CAs. Indeed, we have already reported that a specific inhibition of H^+ release through proton pump inhibitors (PFIs) was able to induce acidification of the tumor cell cytosol [48], and acidic vesicle retention within tumor cells, with consequent increase in the antitumor activity of chemotherapeutic drugs [49], and significant antitumor effects [31,32,48,50]. Moreover, growing evidence in the literature is supporting the evidence that inhibition of CAs has potent antiproliferative and antimetastatic action [22,51]. A recent study has shown for the first time that the combination of proton pump and CA inhibitors (PPIs and CAIs) leads to a more efficient antitumor effect as compared to single treatments, representing the first attempt aimed at targeting, in a unique treatment, two important mechanisms involved in tumor acidification (i.e., tumor acidity and hypoxia) [52]. Herein, we will review the fields of the proton pump and CA inhibitors as promising agents in the management of solid tumors. This review will also attempt to emphasize the importance of using a cocktail of proton exchangers inhibitors as a new and innovative therapeutic strategy against cancer.

2. The pH Regulators Vacuolar H^+-ATPases and Carbonic Anhydrase IX/XII

Tumor microenvironments are characterized by hypoxia, low blood supply, and acidity, which favor the generation of a microenvironment that is hostile and highly toxic for normal or more differentiated cells, inducing progressively, the selection against a more aggressive phenotype. Therefore, to avoid intracellular accumulation of toxic molecules, tumor cells upregulate the expression and activity of proton exchangers that maintain a relative neutral or even alkaline intracellular pH, through pumping protons into the extracellular environment or within the lumen of some membrane-bound organelles. Proton pump exchangers and CAs have been reported to be largely responsible for this hypoxic and acidic microenvironment [17,53,54].

Vacuolar-type ATPase (V-ATPase) is a ubiquitous proton pump shuttling protons from the cytoplasm towards intracellular organelles, and from inside to outside the cell plasma membrane [17,27,55,56]. It is a complex multi-subunit protein, composed of a transmembrane subunit, named V0 complex, devoted to proton transfer and a cytoplasmic portion, named V1 complex, that provides the necessary energy for proton translocation [55]. Its expression and activity are upregulated in many cancer cells. Augmented expression of V-ATPase is considered to be a well-designed compensatory mechanism that confers survival and growth advantages to cancer cells [57–61]. In tumor cells, the extrusion of protons by V-ATPases causes intracellular alkalinization and extracellular acidification, which are important mechanisms favoring the increased activation of extracellular metalloproteinases, thus contributing to tumor cell survival and growth, motility, invasion, metastasis, resistance to apoptosis, and multidrug resistance [62–68]. The V-ATPases are also expressed in vacuolar membranes (i.e., lysosomes), where they are involved in the transport of H^+ into the lumen of intracellular organelles. Lysosomal acidification by V-ATPases has been reported to be under the control of lactate dehydrogenase B (LDHB), and to be facilitated by a physical interaction between LDHB and V-ATPase at the lysosomal surface [69]. Data obtained by Lu and collaborators [70] strongly demonstrated that inhibition of the V-ATPase has an antineoplastic activity [45]. In fact, the inhibition of V-ATPase function via knockdown of ATP6L expression using siRNA suppresses cancer metastasis by decreased proton extrusion, and downregulated protease activity [70]. Thus, this proton pump may be considered a suitable target for the development of novel anticancer strategy.

Carbonic anhydrases (CAs) are cellular pH regulators with a key role in the maintenance of pH homeostasis in cancer cells, thus representing suitable targets for anticancer therapies. CAs are a family of metalloenzymes that catalyze the reversible hydration of carbonic dioxide to bicarbonate and protons [22,71,72]. They are present in several tissues such us the gastrointestinal tract, the reproductive tract, the nervous system, kidney, lungs, skin, and eyes [20,22,73–75]. CAs are involved in respiration and acid–base equilibrium, electrolyte secretion, bone resorption, calcification, ureagenesis, gluconeogenesis, and lipogenesis [20,72]. Several isoforms are known, which are subdivided according to their location: membrane-bound/transmembrane, cytosolic, mitochondrial and secreted [21,22,72]. CA-IX and CA-XII are the two major tumor-related CA isoforms [76,77]. The two enzymes are transmembrane, multi-domain proteins formed of a short intra-cytosolic tail, a transmembrane short domain, an extracellular catalytic domain and a proteoglycan (PG)-like domain. CA-IX and CA-XII contribute significantly to the acidification of tumor microenvironment together with lactic acid production. Indeed, their inhibition has been reported to revert this phenomenon [19]. These two CA isoforms are predominantly found in hypoxic tumors with restricted expression in normal tissue, where they seem to be in their catalytically inactive state. Thus, their inhibition showed an anticancer effect with less side effects compared to other anticancer drugs [72,78]. For this reason, CA-IX and CA-XII have been considered attractive targets for cancer therapy, being validated recently as antitumor/antimetastatic targets [22,79].

Taking into account all these considerations, V-ATPases and tumor-associated CAs have been considered good candidates and ideal targets for the design of novel and innovative anticancer therapy, which interfere with tumor microenvironmental acidification, thus gaining a renewal interest in the last decade for oncologists.

3. V-ATPase Inhibitors

A bulk of evidence correlated hyperfunction of V-ATPases with tumor migration and invasion, multidrug resistance and metastatic processes. Consequently, inhibition of V-ATPase has become an attractive and promising strategy to counteract the tumor hypoxic and acidic microenvironment and to develop novel drugs for the benefit of cancer patients. Different anti-VATPases compounds have then been assessed as potential anticancer agents. Several studies are currently ongoing in order to better investigate in vitro and in vivo, in both preclinical and clinical settings, the binding properties and the mode of inhibition of V-ATPase inhibitors.

Among the V-ATPase inhibitors, the first identified and most frequently used are the natural compounds of microbial origin, bafilomycins, and concanamycins (belonging to plecomacrolide antibiotics) [80,81]. They are lipophilic compounds that have been reported to inhibit growth and to induce apoptosis in human cancer cells [59,82–87]. Other molecules capable of inhibiting V-ATPases via different mechanisms of action, such as benzolactone enamides, archazolid, and indolyls, were later discovered [88–92]. These compound, like bafilomycin A1 or concanamycin, have also investigated as anticancer agents [93,94]. Unfortunately, many reports evidenced high cytotoxicity of these ATPase inhibitors for normal cells, probably because V-ATPase is ubiquitously expressed and active in all types of cells, strongly hampering their potential clinical applications [16,45,55].

Our group focused the attention on another class of promising V-ATPase inhibitors, the family of proton pump inhibitors (PPIs), that include omeprazole, esomeprazole, lansoprazole, pantoprazole, and rabeprazole [95]. They are currently used as antiacid drugs against peptic diseases, including gastroesophageal reflux disease, peptic ulcers or functional dyspepsia [96]. These compounds are weak bases that need an acidic milieu in order to be transformed into the active molecule. Indeed, after protonation in the acidic spaces of the stomach, PPIs irreversibly bind to the cysteine residues of proton pump, dramatically inhibiting proton translocation and acidification of the extracellular environment [97] (Figure 2).

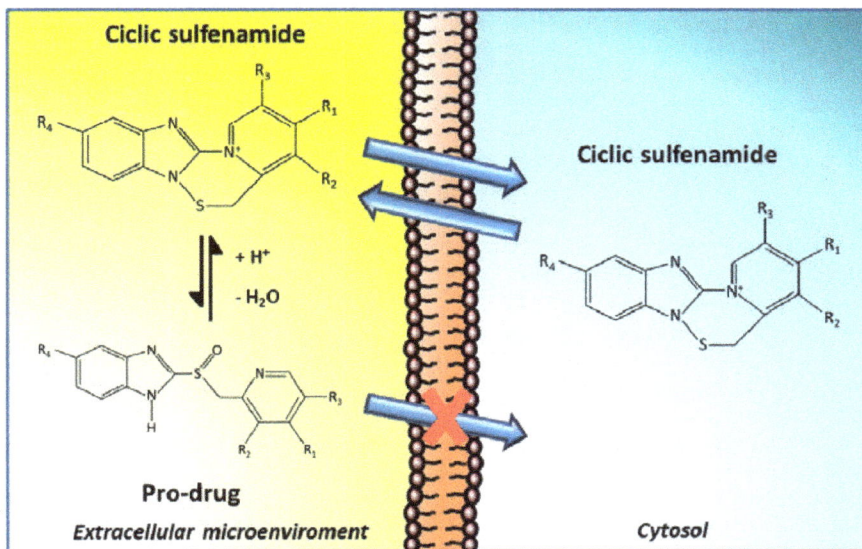

Figure 2. Proton pump inhibitors (PPIs) mechanism of action. PPIs are weak base pro-drugs, that once in the acidic extracellular environment surrounding tumors, can be protonated, thereby reducing their ability to cross the membrane of cells. PPIs then bind irreversibly to proton pumps, dramatically inhibiting their activity, leading to inhibition of proton translocation across the plasma membrane, which in turn, induces alkalization of the tumor microenvironment.

The specific targets of PPIs are H-ATPases contained within the lumen of gastric parietal cells, and through a lesser activity, they inhibit the activity of V-ATPases, thus blocking proton transport across membranes [12,13,45,95]. Some reports also refer to that PPIs could be useful in blocking ATPase activity in tumor cells. These agents did not show relevant systemic toxicity for normal cells, even in prolonged treatments and at very high dosages [98]. Therefore, PPIs have represented an attractive possibility as V-ATPases inhibitors, as they require an acidic environment to be activated, such as that found in the tumor microenvironment, which provides the possibility of tumor specific selectivity, and thus, of targeted anticancer strategy [56]. Therefore, several in vitro and in vivo studies have been carried out on PPIs, in both preclinical and clinical settings, for the design of novel anticancer therapy which target specifically tumor acidity.

3.1. PPIs as Therapeutic Agents

The first evidence supporting the use of PPIs in cancer was provided in 2004 when our group demonstrated that pre-treatment with PPIs (such as omeprazole, esomeprazole, or pantoprazole) in vitro and in vivo (i) reverted chemoresistance of different human tumor cell lines to cisplatin, 5-fluorouracile and doxorubicin; and (ii) increased the sensitivity of drug-sensitive cells to anticancer agents [49]. These effects were mediated by the intracellular retention of chemotherapeutic agents into the lysosomal organelles, associated with a "normalization" of the pH gradients of the tumor cells [49]. Additional studies investigated the response of B-cell lymphoma cell lines and acute lymphoblastic leukemia (ALL) bone marrow blasts, to omeprazole and esomeprazole. PPIs were able to induce a cytotoxic effect against both B-cell tumors and pre-B acute lymphoblastic leukemia cells obtained from patients with acute lymphoblastic leukemia (ALL). The cytotoxic effect was exerted through (i) activation of reactive oxygen species (ROS); (ii) perturbation of lysosomal membranes; (iii) alkalinization of acidic vesicles; (iv) acidification of the cytosol; (v) caspase-independent cell death [48]. PPIs were also able to significantly delay human tumor growth in SCID mice engrafted with B cell lymphomas [48]. Similarly, PPIs exerted a strong tumor inhibitory action against different human melanoma cell lines as well as in tumor-bearing mice [31]. Treatments with PPIs in different pH conditions, induced a marked cytotoxicity that was exerted through a caspase-dependent mechanism strongly influenced by low pH. In vivo studies in SCID mice engrafted with human melanoma pointed out that the tumor growth delay induced by PPIs was consistent with a clear reduction of pH gradients in tumor tissue [31]. We further demonstrated either in vitro and in vivo that pre-treatment with PPIs were able to improve the cytotoxic activity of suboptimal doses of the chemotherapeutic agent paclitaxel against human melanoma cells [32]. Then, we focused on identifying the most effective and promising agent within the PPI family, and discovered that lansoprazole was the most effective single agent against tumor cells [50]. Based on this finding, we investigated the efficacy of lansoprazole against multiple myeloma cell lines, showing that this drug was capable of inducing a strong tumor inhibitory action against this troublesome neoplasm, leading to a caspase-independent cell death [99]. Lastly, we showed that acidity represents a potent mechanism of tumor immune escape, and that PPIs increase the immune reaction against tumors [25,44,100]. Beside the direct toxic effects, PPIs were also able to inhibit mTOR signaling and other metabolic pathways in gastric carcinoma cell line, and to potentiate the antitumor effectiveness of Adriamycin in mice harboring gastric carcinoma xenografts [101–106]. In gastric cancer models, PPIs induced cancer stem cell depletion which passed through inhibition of proliferation, sphere formation, and 5-fluorouracil chemoresistance [107]. PPIs were also tested on breast cancers. Different studies demonstrated that pre-treatments with lansoprazole were able to potentiate the cytotoxic effect of doxorubicin, with no significant effect on non-neoplastic breast epithelial cells [108–114]. In esophageal cancer cell lines, esomeprazole significantly reduced cell viability, adhesion, and migration, as well as enhanced the cytotoxic effects of cisplatin and 5-fluorouracil [115]. Moreover, in ovarian carcinomas, omeprazole pre-treatments have been reported to induce a synergistic effect with paclitaxel on tumor growth in orthotopic and patient-derived xenograft mouse models [116]. Comparable results have been obtained by other groups that have been focused their investigations on hepatocellular, pancreatic, prostatic, and brain tumors [85,117–122].

Therefore, these results provided the proof of concept that PPIs may be considered, not only chemosensitizer agents, but also a new class of antineoplastic drugs, and described the background for a series of clinical studies aimed at supporting the use of PPIs as chemosensitizers.

3.2. PPIs in Clinical Trials

Up to now, two main clinical trials in humans investigated the effect of combined application of PPIs and chemotherapy treatment in cancer. The studies were performed in either osteosarcomas or metastatic breast cancer patient (MBC) [123,124]. The results showed that pre-treatment with PPIs increased the effectiveness of neoadjuvant chemotherapy in osteosarcomas patients, increasing the overall response rate [123]. Confirmation came with the second trial that enrolled women with metastatic breast cancer [124]. This phase II clinical trial aimed at evaluating whether esomeprazole improves efficacy of docetaxel and cisplatin treatment in metastatic breast cancer. The results showed that high dose PPIs proved beneficial and improved chemotherapy efficacy [124]. Indeed, patients that underwent esomeprazole treatment experienced the highest response rates and the longest survivals.

Another phase I clinical trial assessed the effectiveness and safety of the combined use of pantoprazole and doxorubicin in patients with advanced solid tumors, identifying the dose of 240 mg as the baseline for future phase II studies in patients with castration-resistant prostate cancer [125]. Moreover, a study on head and neck tumor patients confirmed that treatments with PPIs increase the overall survival of patients [126].

A more recent study on three patients affected with gastrointestinal cancer, refractory to standard chemotherapy, reported that high-dose of the PPI rabeprazole were able to sensitize human metastatic colorectal cancer to metronomic capecitabine, leading to a good quality of life with acceptable side effects [127]. This combined approach (rabeprazole with capecitabine) is currently under investigation in an approved clinical II trial, proving to be beneficial for the patients [128].

Lastly, two studies in companion animals with spontaneous tumors have been performed in order to explore the effect of PPIs treatment in combination with chemotherapy. Interestingly, PPIs were able to reverse chemoresistance in refractory tumors, both hematopoietic (lymphoma) or solid (melanoma and squamous cell carcinoma) [129]. Moreover, PPIs increased the efficacy of metronomic chemotherapy, independently from the tumor histotype or the animal species affected by cancer [130].

All these studies provided the first clinical evidence that PPIs pretreatment could be easily included into the standard protocols in clinical oncology with a clear benefit for patients having the less favorable prognostic factors. Indeed, pretreatment with PPIs, by inhibiting proton pumps, induced a decrease of the protonation of extracellular tumor environment, in turn allowing the chemotherapeutics to be fully effective, improving the effectiveness of either chemical and biological drugs against cancer. Thus, tumor alkalinization could improve the outcome of patients by counteracting tumor chemoresistance.

4. Inhibitors of Carbonic Anhydrase IX/XII

In recent years, the involvement of CA-IX/XII in generating the peculiar pH gradients of tumor cells is becoming more and more evident [12]. In fact, inhibition of the enzyme catalytic activity with fluorescent sulfonamides has been reported to be able to revert this phenomenon [19]. Moreover, accumulating experimental evidence recognizes that disruption of CA-IX by gene knockdown or inhibition of its catalytic activity with small molecules and/or antibodies strongly correlates with both extracellular and intracellular pH, tumor growth [78,131–133], tumor cell migration/invasion [134–136], chemo- or radiotherapy resistance [137,138].

As a consequence, different pharmacological inhibitors that specifically target the tumor-associated isoforms CA-IX and -XII were developed and tested for their antitumor activity during the last years [20].

4.1. CA-IX/XII Inhibitors as Therapeutic Agents

Many sulfonamide, sulfamate, sulfamide, and coumarine CA inhibitors were reported to efficiently target CA-IX [139–141]. The compounds specifically designed for targeting CA-IX were: (i) fluorescent sulfonamides, used for imaging purposes [19,142]; (ii) positively or negatively-charged compounds that inhibit selectively extracellular CAs [74]; (iii) nanoparticles coated with CAs [143]; (iv) monoclonal antibodies, among which M75 is a highly specific anti-CAIX mAb targeting the PG domain of CA-IX, discovered by Pastorekova's group [144–146].

The most studied and important CA-IX/XII inhibitors are the sulfonamides, that exert their potent CA inhibitory properties by binding to the catalytic site of Cas, blocking then, its function [147,148]. Several inhibitors of the sulfonamide type have been identified for both CA-IX [76] and CA-XII [77]. Unfortunately, most of these compounds did not show specificity for the inhibition of the tumor-associated isoforms IX and XII versus the remaining CAs. Therefore, different sulfonamide derivates, more selective inhibitors of the tumor-associated CAs (CA-IX and XII) were designed and developed for targeting these agents, mainly through structure-based drug design approaches [149]. With the discovery of the X-ray crystal structure of CA-IX by the De Simone's group in 2009 [150], the drug-design studies of CA inhibitors targeting isoform IX/XII were highly favored, and highly isoform-selective inhibitors were identified [139–141]. These compounds have been shown, over the past years, to be very promising in anticancer therapies. For instance, a fluorescent sulfonamide with high affinity for CA-IX was generated. This compound was used to determine the role of CA-IX in tumor acidification and for imaging purposes [19]. Indeed, Dubois et al. showed that fluorescent sulfonamides accumulate selectively in the hypoxic regions of xenograft animals with transplanted hypoxic colorectal cancers [142,151]. The in vivo proof of concept that sulfonamide CA-IX inhibitors may show antitumor effects has been first published by Neri's group [152]. Similar studies from different laboratories on diverse models and cancer types demonstrated that sulfonamide CA-IX/XII inhibitors have a profound effect in inhibiting the growth of primary tumors, and great in vivo anti-metastatic effects [79,152–156]. These inhibitors were also able to induce depletion of cancer stem cells in models of breast cancer metastasis, leading, in turn, to diminished tumor growth and metastasis, and showing a promising efficacy for the recurrence of some cancers [157]. Supuran and coworkers developed a sulfonamide-derived compound, SLC-0111 (also known as WBI-5111), that strongly and specifically inhibits CA-IX, in vitro and in vivo [79,154]. This compound inhibits tumor growth and metastasis formation alone or in combination with antineoplastic drug, and decreased the cancer stem cell population in breast cancer models [79,154,158]. It has now completed the phase I clinical trials and is scheduled for phase II trials later this year [22,156].

Coumarins are another class of CA-IX and XII inhibitors. These natural compounds, together with the highly isoforms-selective CA-Is, are selective against the tumor-associated isoforms IX and XII at nanomolar concentrations, being ineffective against the broadly expressed isoforms CA-I and CA-II [153,159]. One of these derivatives, a glycosyl coumarin, was recently shown to inhibit the growth of primary tumors and the formation of metastasis in the highly aggressive 4T1 syngeneic mouse metastatic breast cancer [153]. These inhibitors are still under investigation in pre-clinical studies [160].

Monoclonal antibodies may represent another venue for the selective CA-IX and XII inhibition [144,161]. M75 is a highly specific anti-CAIX monoclonal antibody targeting the proteoglycan domain of CA-IX, discovered by Pastorekova's group [78] and widely used in immunohistochemical and Western blot studies. WX-G250 (known also as girentuximab), the first CA-IX inhibitor to enter clinical trials [162], is another chimeric monoclonal antibody that is actually in phase III clinical trials, as an adjuvant therapy for the treatment of non-metastasized renal cell carcinoma. CA-XII specific monoclonal antibodies have also been generated and characterized. The first and most significant anti-CAXII monoclonal antibody (6A10) was created by Battke and coworkers [144]. The antibody binds to the catalytic domain of CA-XII, inhibiting its activity at nanomolar concentrations. 6A10 was shown to successfully inhibiting the growth of tumor spheroids in vitro and efficiently decreased tumor growth in vivo [144,163].

4.2. CA-IX Inhibitors in Clinical Trials

In the past few years, the tumor-associated cell surface CA isoform, CA-IX, has been validated as new anticancer drug target for the treatment and imaging of cancers expressing this enzyme, and some therapeutic strategies against CA-IX have already entered in clinical studies.

Serum levels of CA-IX have been explored as a potential biomarker for the treatment response in patients with metastatic renal cell cancer (mRCC) in a pilot human trial [164]. The study collected blood samples from 91 patients with mRCC and 32 healthy individuals, and associated serum CA-IX levels with the occurrence of tumor progression (stage, tumor grade, tumor size, recurrence, and metastasis), suggesting that CA-IX may be a valuable diagnostic and prognostic tool in RCC [164]. Serum CA-IX levels have also been suggested as a potential biomarker to predict the outcome for patients with mRCC [165]. Correlation between CA-IX level of expression and recurrence, survival, and clinical response to therapy was also observed in patients with high-risk, nonmetastatic renal cell carcinoma [166] and laryngeal cancer [167].

A selective small molecule, an ureido-substituted benzenesulfonamide derivative, SLC-0111, entered, in 2014, a phase I clinical trial for the treatment of advanced, metastatic solid tumors to evaluate the safety, tolerability, and pharmacokinetics of this compound (NCT02215850). Although the study was completed in March 2016, the results have not yet been published.

A novel small molecule radiotracer, 18F-VM4-037, that binds to CA-IX in clinically localized kidney tumors, has been submitted to a phase II clinical trial sponsored by National Cancer Institute

(NCI) (NCT01712685). The main objective of the study was to test the safety and effectiveness of 18F-VM4-037 during imaging studies of kidney cancer. The study enrolled 12 patients with renal cell carcinoma, and demonstrated that it is a well-tolerated agent which allows imaging of CA-IX expression both in primary tumors and metastases, highlighting 18F-VM4-037 as a useful drug in the evaluation of metastatic ccRCC lesions [168].

Regarding antibodies, the monoclonal antibody G250 was the first monoclonal antibody anti-CAIX introduced in a phase I/II clinical trials in combination with interferon-alpha-2a in metastatic renal cell carcinoma patients (mRCC), who are at a high risk of recurrence after resection of the primary tumor [162]. The study on 31 patients with mRCC treated with both drugs demonstrated that the treatment was safe, well tolerated, and led to clinical disease stabilization [162]. This antibody also progressed to a phase III clinical trial as an adjuvant therapy for the treatment of patients with metastatic renal cell carcinoma (NCT00087022). This study had proposed to investigate the efficacy and safety of adjuvant G250. The adjuvant treatment demonstrated a safe and well tolerated profile. However, G250 had no clinical benefit for the patients [169]. Antibodies were also proposed as imaging agents for CA-IX positive tumors, originally by Neri's group [152]. Therefore, radiolabelled chimeric G250 was also tested as a valuable imaging agent for the diagnosis of patients with renal cell carcinoma. Zirconium-89- labeled girentuximab has been developed and entered in a phase II/III clinical trial (NCT02883153) to study the impact of the Zirconium-89-girentuximab in clinical management. Unfortunately, even if the study is completed, the results have not been published yet. Lastly, radioimmunotherapy with Lutetium 177–labeled girentuximab are also well tolerated in metastatic ccRCC patients [170].

5. Cocktail of Proton Exchanger Inhibitors as a Novel Therapeutic Approach against Cancer

Triggered from such encouraging results, we explored the hypothesis that PPIs could increase the effectiveness of CA-IX inhibitors against very malignant human melanoma cells, fully expressing the enzyme [52]. To this purpose, human melanoma cells have been treated with potent CA-IX inhibitors, the sulfamates S4, and *p*-nitrophenyl derivative FC9-399A (selective ureido-sulfamate derivatives), in combination with lansoprazole [52]. First of all, we observed that treatment with these CA-IX inhibitors induced a slight but significant inhibition of melanoma cell growth, which was strongly impaired under acidic conditions, typical of malignant cell growth [52]. We postulated that the impairment of CA-IX inhibitor activity was probably due to their neutralization by protonation outside the cells induced by acidity. When we compared the combination treatment at suboptimal doses (lansoprazole followed by one of the two CA-IX inhibitors), we observed a more efficient and significantly increased tumor cell growth inhibition, and a straightforward cytotoxic effect against metastatic melanoma cells, compared to each single agent [52]. We also demonstrated that the effect of combined treatments was not due to changes in the CA-IX protein expression, but rather probably to induction of an inhibition in CA-IX activity [52]. The pre-treatment of human melanoma cells with lansoprazole significantly improved the antitumor effect of CA-IX inhibitors, probably because lansoprazole, fully active in acidic conditions, induced an alkalinization of the tumor extracellular environment, in turn leading to stronger and the greatest activation and effectiveness of CA-IX inhibitors. Therefore, our results provided the first evidence that combinations of lansoprazole with two different CA-IX inhibitors were more effective than single treatments in inhibiting cell proliferation and inducing cell death in human melanoma cells. These results were supported by another recent study where we combined the proton pump inhibitor lansoprazole with the inhibitor of the reverse transcriptase, Efavirenz, in order to target in a unique treatment two new oncotypes, i.e., proton pumps and reverse transcriptase [171]. Again, the results clearly showed that pre-treatment of human melanoma cells with the proton pump inhibitor lansoprazole significantly improved the Efavirenz antitumor effect [171]. These two recent works of ours highlighted, for the first time, that the combination of proton pump and CA-IX/reverse transcriptase inhibitors possess a more efficient antitumor action compared to single treatments. Actually, they represent the first attempt aimed at combining either different proton exchangers inhibitors or a proton exchanger with the reverse transcriptase inhibitor in a unique antitumor approach, with conceivably more effectiveness and less toxicity.

Altogether, the results of these two studies clearly supported the hypothesis that treatment aimed at buffering the acidic tumor microenvironment, through the use of PPIs, that contrary to other drugs are activated in acidic conditions, may improve the effectiveness of CA-IX and reverse transcriptase inhibitors, significantly improving their antitumor effects. Thus, they open the way to novel and alternative antitumor strategies that are more specific, and probably less toxic for tumor patients. Of course, clinical trials obtained with lansoprazole followed by CA-IX or reverse transcriptase inhibitors are needed in order to further support the use of combination therapies that these studies were proposing.

6. Conclusions

Despite the great efforts of the scientific community in finding proper treatments for cancer, the responsiveness of human tumors to chemotherapy has not changed in the last decades, and resistance or refractoriness to chemotherapeutic drugs has become a key problem in the therapy of tumor patients and still remains unsolved. Therefore, novel antitumor strategies, which are more specific and probably less toxic, have become an urgent medical need. During the last decades, tumor metabolism and microenvironmental acidity are increasingly considered important determinants of tumor progression and drug resistance. Therefore, the mechanisms controlling the acidic pH of solid tumors have been proposed as selective and specific therapeutic targets in setting up novel antitumor strategies. Proton exchangers, whose expression and activity are upregulated by hypoxia and acidity, have gained attention in the last years, due to their crucial function in determining the acidification of the tumor microenvironment. In particular, V-ATPases and CA-IX represented interesting targets for the development of novel approaches in anticancer therapy. For these reasons, several V-ATPases and CA-IX inhibitors have been tested for their antitumor activity. Investigations reported that several inhibitors of the tumor-associated carbonic anhydrase isoform IX, and of V-ATPases, had a clear antineoplastic action, and may be useful to be combined and tested for developing novel antitumor therapies. We tested combinations of lansoprazole, targeting proton pumps, followed by CA-IX inhibitors, and observed that combined treatments, while inducing the alkalinization of tumor microenvironment, led to an increased effectiveness of CA-IX inhibitors against very malignant human melanoma cells [52]. The same effect was observed combining lansoprazole with the inhibitor of the reverse transcriptase, another important hallmark of cancers [171]. These results clearly supported the hypothesis that an approach aimed at targeting tumor extracellular acidity, combining two or more proton exchanger inhibitors with different antitumor actions, may open the way to the development of innovative and alternative antitumor strategies that are more specific, effective, and hopefully less toxic for tumor patients. Of course, data from clinical trials obtained through combination of PPIs with CA-IX inhibitors are needed, in order to further support the use of proton pump and CA-IX inhibitor combination therapies, and to translate these results to the patients. Finally, the results of these studies give support to new investigations aimed at the setting up of either hybrids or combined molecules containing both proton pump and CA-IX inhibitors.

Acknowledgments: This study was supported by a grant from the Italian Ministry of Health.

Author Contributions: Elisabetta Iessi conceptualized and designed the structure of the article, and wrote the article. Mariantonia Logozzi and Rossella Di Raimo assisted in literature search. Davide Mizzoni prepared figures presented in this review. Claudiu T. Supuran edited and supervised the manuscript. Stefano Fais edited and supervised the manuscript draft and figures. All authors were involved in the critical reading and editing of the review. All authors have read and approved the final version of the manuscript.

Conflicts of Interest: The authors declare no conflict of interest.

References

1. Gatenby, R.A.; Gillies, R.J. Why do cancers have high aerobic glycolysis? *Nat. Rev. Cancer* **2004**, *4*, 891–899. [CrossRef] [PubMed]
2. Vander Heiden, M.G.; Cantley, L.C.; Thompson, C.B. Understanding the Warburg effect: The metabolic requirements of cell proliferation. *Science* **2009**, *324*, 1029–1033. [CrossRef] [PubMed]
3. Warburg, O. On the origin of cancer cells. *Science* **1956**, *123*, 309–314. [CrossRef] [PubMed]
4. Cairns, R.A.; Harris, I.S.; Mak, T.W. Regulation of cancer cell metabolism. *Nat. Rev. Cancer* **2011**, *11*, 85–95. [CrossRef] [PubMed]
5. Tredan, O.; Galmarini, C.M.; Patel, K.; Tannock, I.F. Drug resistance and the solid tumor microenvironment. *J. Natl. Cancer Inst.* **2007**, *99*, 1441–1454. [CrossRef] [PubMed]
6. Helmlinger, G.; Sckell, A.; Dellian, M.; Forbes, N.S.; Jain, R.K. Acid production in glycolysis-impaired tumors provides new insights into tumor metabolism. *Clin. Cancer Res.* **2002**, *8*, 1284–1291. [PubMed]
7. Newell, K.; Franchi, A.; Pouyssegur, J.; Tannock, I. Studies with glycolysis-deficient cells suggest that production of lactic acid is not the only cause of tumor acidity. *Proc. Natl. Acad. Sci. USA* **1993**, *90*, 1127–1131. [CrossRef] [PubMed]
8. Yamagata, M.; Hasuda, K.; Stamato, T.; Tannock, I.F. The contribution of lactic acid to acidification of tumours: Studies of variant cells lacking lactate dehydrogenase. *Br. J. Cancer* **1998**, *77*, 1726–1731. [CrossRef] [PubMed]
9. Mookerjee, S.A.; Goncalves, R.L.; Gerencser, A.A.; Nicholls, D.G.; Brand, M.D. The contributions of respiration and glycolysis to extracellular acid production. *Biochim. Biophys. Acta* **2015**, *1847*, 171–181. [CrossRef] [PubMed]
10. Corbet, C.; Feron, O. Tumour acidosis: From the passenger to the driver's seat. *Nat. Rev. Cancer* **2017**, *17*, 577–593. [CrossRef] [PubMed]
11. Izumi, H.; Torigoe, T.; Ishiguchu, H.; Uramoto, H.; Yoshida, Y.; Tanabe, M.; Ise, T.; Murakami, T.; Yoshida, T.; Nomoto, M.; et al. Cellular pH regulators: Potentially promising molecular targets for cancer chemotherapy. *Cancer Treat. Rev.* **2003**, *29*, 541–549. [CrossRef]
12. Parks, S.K.; Chiche, J.; Pouysségur, J. Disrupting proton dynamics and energy metabolism for cancer therapy. *Nat. Rev. Cancer* **2013**, *13*, 611–623. [CrossRef] [PubMed]
13. Spugnini, E.P.; Sonveaux, P.; Stock, C.; Perez-Sayans, M.; De Milito, A.; Avnet, S.; Garcìa, A.G.; Harguindey, S.; Fais, S. Proton channels and exchangers in cancer. *Biochim. Biophys. Acta* **2015**, *1848*, 2715–2726. [CrossRef] [PubMed]
14. Spugnini, E.; Fais, S. Proton pump inhibition and cancer therapeutics: A specific tumor targeting or it is a phenomenon secondary to a systemic buffering? *Semin. Cancer Biol.* **2017**, *43*, 111–118. [CrossRef] [PubMed]
15. Gorbatenko, A.; Olesen, C.W.; Boedtkjer, E.; Pedersen, S.F. Regulation and roles of bicarbonate transporters in cancer. *Front. Physiol.* **2014**, *5*, 130. [CrossRef] [PubMed]
16. De Milito, A.; Fais, S. Proton pump inhibitors may reduce tumour resistance. *Expert Opin. Pharmacother.* **2005**, *6*, 1049–1054. [CrossRef] [PubMed]
17. Nishi, T.; Forgac, M. The vacuolar (H^+)-ATPases-nature's most versatile proton pumps. *Nat. Rev. Mol. Cell Biol.* **2002**, *3*, 94–103. [CrossRef] [PubMed]
18. Martinez-Zaguilan, R.; Lynch, R.M.; Martinez, G.M.; Gillies, R.J. Vacuolar-type H(+)-ATPases are functionally expressed in plasma membranes of human tumor cells. *Am. J. Physiol.* **1993**, *265*, C1015–C1029. [PubMed]
19. Svastová, E.; Hulíková, A.; Rafajová, M.; Zat'ovicová, M.; Gibadulinová, A.; Casini, A.; Cecchi, A.; Scozzafava, A.; Supuran, C.T.; Pastorek, J.; et al. Hypoxia activates the capacity of tumor-associated carbonic anhydrase IX to acidify extracellular pH. *FEBS Lett.* **2004**, *577*, 439–445. [CrossRef] [PubMed]
20. Supuran, C.T. Carbonic anhydrases: Novel therapeutic applications for inhibitors and activators. *Nat. Rev. Drug Discov.* **2008**, *7*, 168–181. [CrossRef] [PubMed]
21. Supuran, C.T. Carbonic anhydrase inhibition/activation: Trip of a scientist around the world in the search of novel chemotypes and drug targets. *Curr. Pharm. Des.* **2010**, *16*, 3233–3245. [CrossRef] [PubMed]
22. Supuran, C.T. Carbonic Anhydrase Inhibition and the Management of Hypoxic Tumors. *Metabolites* **2017**, *7*, 48. [CrossRef] [PubMed]
23. Huber, V.; De Milito, A.; Harguindey, S.; Reshkin, S.J.; Wahl, M.L.; Rauch, C.; Chiesi, A.; Pouysségur, J.; Gatenby, R.A.; Rivoltini, L.; et al. Proton dynamics in cancer. *J. Transl. Med.* **2010**, *8*, 57. [CrossRef] [PubMed]

24. Webb, B.A.; Chimenti, M.; Jacobson, M.P.; Barber, D.L. Dysregulated pH: A perfect storm for cancer progression. *Nat. Rev. Cancer* **2011**, *11*, 671–677. [CrossRef] [PubMed]

25. Fais, S.; Venturi, G.; Gatenby, B. Microenvironmental acidosis in carcinogenesis and metastases: New strategies in prevention and therapy. *Cancer Metastasis Rev.* **2014**, *33*, 1095–1108. [CrossRef] [PubMed]

26. Barar, J.; Omidi, Y. Dysregulated pH in tumor microenvironment checkmatescancer therapy. *Bioimpacts* **2013**, *3*, 149–162. [PubMed]

27. Daniel, C.; Bell, C.; Burton, C.; Harguindey, S.; Reshkin, S.J.; Rauch, C. The role of proton dynamics in the development and maintenance of multidrug resistance in cancer. *Biochim. Biophys. Acta* **2013**, *1832*, 606–617. [CrossRef] [PubMed]

28. Gillies, R.J.; Raghunand, N.; Karczmar, G.S.; Bhujwalla, Z.M. MRI of the tumor microenvironment. *J. Magn. Reson. Imaging* **2002**, *16*, 430–450. [CrossRef] [PubMed]

29. Gallagher, F.A.; Kettunen, M.I.; Day, S.E.; Hu, D.E.; Ardenkjaer-Larsen, J.H.; Zandt, Ri.; Jensen, P.R.; Karlsson, M.; Golman, K.; Lerche, M.H.; et al. Magnetic resonance imaging of pH in vivo using hyperpolarized ^{13}C-labelled bicarbonate. *Nature* **2008**, *453*, 940–943. [CrossRef] [PubMed]

30. Van Sluis, R.; Bhujwalla, Z.M.; Raghunand, N.; Ballesteros, P.; Alvarez, J.; Cerdán, S.; Galons, J.P.; Gillies, R.J. In vivo imaging of extracellular pH using 1H MRSI. *Magn. Reson. Med.* **1999**, *41*, 743–750. [CrossRef]

31. De Milito, A.; Canese, R.; Marino, M.L.; Borghi, M.; Iero, M.; Villa, A.; Venturi, G.; Lozupone, F.; Iessi, E.; Logozzi, M.; et al. pH-dependent antitumor activity of proton pump inhibitors against human melanoma is mediated by inhibition of tumor acidity. *Int. J. Cancer* **2010**, *127*, 207–219. [CrossRef] [PubMed]

32. Azzarito, T.; Venturi, G.; Cesolini, A.; Fais, S. Lansoprazole induces sensitivity to suboptimal doses of paclitaxel in human melanoma. *Cancer Lett.* **2015**, *356*, 697–703. [CrossRef] [PubMed]

33. Helmlinger, G.; Yuan, F.; Dellian, M.; Jain, R.K. Interstitial pH and pO$_2$ gradients in solid tumors in vivo: High-resolution measurements reveal a lack of correlation. *Nat. Med.* **1997**, *3*, 177–182. [CrossRef] [PubMed]

34. Vaupel, P.W.; Frinak, S.; Bicher, H.I. Heterogeneous oxygen partial pressure and pH distribution in C3H mouse mammary adenocarcinoma. *Cancer Res.* **1981**, *41*, 2008–2013. [PubMed]

35. Bittner, M.I.; Wiedenmann, N.; Bucher, S.; Hentschel, M.; Mix, M.; Rücker, G.; Weber, W.A.; Meyer, P.T.; Werner, M.; Grosu, A.L.; et al. Analysis of relation between hypoxia PET imaging and tissue-based biomarkers during head and neck radiochemotherapy. *Acta Oncol.* **2016**, *55*, 1299–1304. [CrossRef] [PubMed]

36. Le, Q.T.; Kong, C.; Lavori, P.W.; O'byrne, K.; Erler, J.T.; Huang, X.; Chen, Y.; Cao, H.; Tibshirani, R.; Denko, N.; et al. Expression and prognostic significance of a panel of tissue hypoxia markers in head-and-neck squamous cell carcinomas. *Int. J. Radiat. Oncol. Biol. Phys.* **2007**, *69*, 167–175. [CrossRef] [PubMed]

37. Rademakers, S.E.; Lok, J.; van der Kogel, A.J.; Bussink, J.; Kaanders, J.H. Metabolic markers in relation to hypoxia; staining patterns and colocalization of pimonidazole, HIF-1α, CAIX, LDH-5, GLUT-1, MCT1 and MCT4. *BMC Cancer* **2011**, *11*, 167. [CrossRef] [PubMed]

38. Lee, A.H.; Tannock, I.F. Heterogeneity of intracellular pH and of mechanisms that regulate intracellular pH in populations of cultured cells. *Cancer Res.* **1998**, *58*, 1901–1908. [PubMed]

39. Martinez-Zaguilan, R.; Seftor, E.A.; Seftor, R.E.; Chu, Y.W.; Gillies, R.J.; Hendrix, M.J. Acidic pH enhances the invasive behavior of human melanoma cells. *Clin. Exp. Metastasis* **1996**, *14*, 176–186. [CrossRef] [PubMed]

40. Rofstad, E.K.; Mathiesen, B.; Kindem, K.; Galappathi, K. Acidic extracellular pH promotes experimental metastasis of human melanoma cells in athymic nude mice. *Cancer Res.* **2006**, *66*, 6699–6707. [PubMed]

41. Wachsberger, P.R.; Landry, J.; Storck, C.; Davis, K.; O'Hara, M.D.; Owen, C.S.; Leeper, D.B.; Coss, R.A. Mammalian cells adapted to growth at pH 6.7 have elevated HSP27 levels and are resistant to cisplatin. *Int. J. Hyperth.* **1997**, *13*, 251–255. [CrossRef]

42. Raghunand, N.; Mahoney, B.; van Sluis, R.; Baggett, B.; Gillies, R.J. Acute metabolic alkalosis enhances response of C3H mouse mammary tumors to the weak base mitoxantrone. *Neoplasia* **2001**, *3*, 227–235. [CrossRef] [PubMed]

43. Lugini, L.; Matarrese, P.; Tinari, A.; Lozupone, F.; Federici, C.; Iessi, E.; Gentile, M.; Luciani, F.; Parmiani, G.; Rivoltini, L.; et al. Cannibalism of live lymphocytes by human metastatic but not primary melanoma cells. *Cancer Res.* **2006**, *66*, 3629–3638. [CrossRef] [PubMed]

44. Calcinotto, A.; Filipazzi, P.; Grioni, M.; Iero, M.; De Milito, A.; Ricupito, A.; Cova, A.; Canese, R.; Jachetti, E.; Rossetti, M.; et al. Modulation of microenvironment acidity reverses anergy in human and murine tumor-infiltrating T lymphocytes. *Cancer Res.* **2012**, *72*, 2746–2756. [CrossRef] [PubMed]

45. Fais, S.; De Milito, A.; You, H.; Qin, W. Targeting vacuolar H$^+$-ATPases as a new strategy against cancer. *Cancer Res.* **2007**, *67*, 10627–10630. [CrossRef] [PubMed]

46. Pettersen, E.O.; Ebbesen, P.; Gieling, R.G.; Williams, K.J.; Dubois, L.; Lambin, P.; Ward, C.; Meehan, J.; Kunkler, I.H.; Langdon, S.P.; et al. Targeting tumour hypoxia to prevent cancer metastasis. From biology, biosensing and technology to drug development: The METOXIA consortium. *J. Enzyme Inhib. Med. Chem.* **2015**, *30*, 689–721. [CrossRef] [PubMed]

47. Perez-Sayans, M.; Garcia-Garcia, A.; Scozzafava, A.; Supuran, C.T. Inhibition of V-ATPase and carbonic anhydrases as interference strategy with tumor acidification processes. *Curr. Pharm. Des.* **2012**, *18*, 1407–1413. [CrossRef] [PubMed]

48. De Milito, A.; Iessi, E.; Logozzi, M.; Lozupone, F.; Spada, M.; Marino, M.L.; Federici, C.; Perdicchio, M.; Matarrese, P.; Lugini, L.; et al. Proton pump inhibitors induce apoptosis of human B-cell tumors through a caspase-independent mechanism involving reactive oxygen species. *Cancer. Res.* **2007**, *67*, 5408–5417. [CrossRef] [PubMed]

49. Luciani, F.; Spada, M.; De Milito, A.; Molinari, A.; Rivoltini, L.; Montinaro, A.; Marra, M.; Lugini, L.; Logozzi, M.; Lozupone, F.; et al. Effect of proton pump inhibitor pretreatment on resistance of solid tumors to cytotoxic drugs. *J. Natl. Cancer Inst.* **2004**, *96*, 1702–1713. [CrossRef] [PubMed]

50. Lugini, L.; Federici, C.; Borghi, M.; Azzarito, T.; Marino, M.L.; Cesolini, A.; Spugnini, E.P.; Fais, S. Proton pump inhibitors while belonging to the same family of generic drugs show different anti-tumor effect. *J. Enzyme Inhib. Med. Chem.* **2016**, *31*, 538–545. [CrossRef] [PubMed]

51. Thiry, A.; Supuran, C.T.; Masereel, J.M.; Dogne´, J.M. Recent developments of carbonic anhydrase inhibitors as potential anticancer drugs. *J. Med. Chem.* **2008**, *51*, 3051–3056. [CrossRef] [PubMed]

52. Federici, C.; Lugini, L.; Marino, M.L.; Carta, F.; Iessi, E.; Azzarito, T.; Supuran, C.T.; Fais, S. Lansoprazole and carbonic anhydrase IX inhibitors sinergize against human melanoma cells. *J. Enzyme Inhib. Med. Chem.* **2016**, *31*, 119–125. [CrossRef] [PubMed]

53. Reshkin, S.J.; Cardone, R.A.; Harguindey, S. Na$^+$–H$^+$ exchanger, pH regulation and cancer. *Recent Pat Anticancer Drug Discov.* **2013**, *8*, 85–99. [CrossRef] [PubMed]

54. Swayampakula, M.; McDonald, P.C.; Vallejo, M.; Coyaud, E.; Chafe, S.C.; Westerback, A.; Venkateswaran, G.; Shankar, J.; Gao, G.; Laurent, E.M.N.; et al. The interactome of metabolic enzyme carbonic anhydrase IX reveals novel roles in tumor cell migration and invadopodia/MMP14-mediated invasion. *Oncogene* **2017**, *36*, 6244–6261. [CrossRef] [PubMed]

55. Forgac, M. Vacuolar ATPases: Rotary proton pumps in physiology and pathophysiology. *Nat. Rev. Mol. Cell Biol.* **2007**, *8*, 917–929. [CrossRef] [PubMed]

56. Spugnini, E.P.; Citro, G.; Fais, S. Proton pump inhibitors as anti vacuolar-ATPases drugs: A novel anticancer strategy. *J. Exp. Clin. Cancer Res.* **2010**, *29*, 44. [CrossRef] [PubMed]

57. Von Schwarzenberg, K.; Wiedmann, R.M.; Oak, P.; Schulz, S.; Zischka, H.; Wanner, G.; Efferth, T.; Trauner, D.; Vollmar, A.M. Mode of cell death induction by pharmacological vacuolar H$^+$-ATPase (V-ATPase) inhibition. *J. Biol. Chem.* **2013**, *288*, 1385–1396. [CrossRef] [PubMed]

58. Schempp, C.M.; von Schwarzenberg, K.; Schreiner, L.; Kubisch, R.; Muller, R.; Wagner, E.; Vollmar, A.M. V-ATPase inhibition regulates anoikis resistance and metastasis of cancer cells. *Mol. Cancer Ther.* **2014**, *13*, 926–937. [CrossRef] [PubMed]

59. Sennoune, S.R.; Bakunts, K.; Martinez, G.M.; Chua-Tuan, J.L.; Kebir, Y.; Attaya, M.N.; Martínez-Zaguilán, R. Vacuolar H$^+$-ATPase in human breast cancer cells with distinct metastatic potential: Distribution and functional activity. *Am. J. Physiol. Cell Physiol.* **2004**, *286*, C1443–C1452. [CrossRef] [PubMed]

60. Martinez-Zaguilan, R.; Raghunand, N.; Lynch, R.M.; Bellamy, W.; Martinez, G.M.; Rojas, B.; Smith, D.; Dalton, W.S.; Gillies, R.J. pH and drug resistance. I. functional expression of plasmalemmal V-type H$^+$-ATPase in drug-resistant human breast carcinoma cell lines. *Biochem. Pharmacol.* **1999**, *57*, 1037–1046. [CrossRef]

61. Raghunand, N.; Martinez-Zaguilan, R.; Wright, S.H.; Gillies, R.J. pH and drug resistance. II. Turnover of acidic vesicles and resistance to weakly basic chemotherapeutic drugs. *Biochem. Pharmacol.* **1999**, *57*, 1047–1058. [CrossRef]

62. Von Schwarzenberg, K.; Lajtos, T.; Simon, L.; Muller, R.; Vereb, G.; Vollmar, A.M. V-ATPase inhibition overcomes trastuzumab resistance in breast cancer. *Mol. Oncol.* **2014**, *8*, 9–19. [CrossRef] [PubMed]

63. Perez-Sayans, M.; Somoza-Martin, J.M.; Barros-Angueira, F.; Diz, P.G.; Rey, J.M.; Garcia-Garcia, A. Multidrug resistance in oral squamous cell carcinoma: The role of vacuolar ATPases. *Cancer Lett.* **2010**, *295*, 135–143. [CrossRef] [PubMed]

64. Huang, L.; Lu, Q.; Han, Y.; Li, Z.; Zhang, Z.; Li, X. ABCG2/V-ATPase was associated with the drug resistance and tumor metastasis of esophageal squamous cancer cells. *Diagn. Pathol.* **2012**, *7*, 180. [CrossRef] [PubMed]

65. Xu, J.; Xie, R.; Liu, X.; Wen, G.; Jin, H.; Yu, Z.; Jiang, Y.; Zhao, Z.; Yang, Y.; Ji, B.; et al. Expression and functional role of vacuolar H$^+$-ATPase in human hepatocellular carcinoma. *Carcinogenesis* **2012**, *33*, 2432–2440. [CrossRef] [PubMed]

66. Chung, C.; Mader, C.C.; Schmitz, J.C.; Atladottir, J.; Fitchev, P.; Cornwell, M.L.; Koleske, A.J.; Crawford, S.E.; Gorelick, F. The vacuolar-ATPase modulates matrix metalloproteinase isoforms in human pancreatic cancer. *Lab. Investig.* **2011**, *91*, 732–743. [CrossRef] [PubMed]

67. Lu, Q.; Lu, S.; Huang, L.; Wang, T.; Wan, Y.; Zhou, C.X.; Zhang, C.; Zhang, Z.; Li, X. The expression of V-ATPase is associated with drug resistance and pathology of non-small-cell lung cancer. *Diagn. Pathol.* **2013**, *8*, 145. [CrossRef] [PubMed]

68. Perut, F.; Avnet, S.; Fotia, C.; Baglio, S.R.; Salerno, M.; Hosogi, S.; Kusuzaki, K.; Baldini, N. V-ATPase as an effective therapeutic target for sarcomas. *Exp. Cell Res.* **2014**, *320*, 21–32. [CrossRef] [PubMed]

69. Brisson, L.; Bański, P.; Sboarina, M.; Dethier, C.; Danhier, P.; Fontenille, M.J.; Van Hée, V.F.; Vazeille, T.; Tardy, M.; Falces, J.; et al. Lactate Dehydrogenase B Controls Lysosome Activity and Autophagy in Cancer. *Cancer Cell* **2016**, *30*, 418–431. [CrossRef] [PubMed]

70. Lu, X.; Qin, W.; Li, J.; Tan, N.; Pan, D.; Zhang, H.; Xie, L.; Yao, G.; Shu, H.; Yao, M.; et al. The growth and metastasis of human hepatocellular carcinoma xenografts are inhibited by small interfering RNA targeting to the subunit ATP6L of proton pump. *Cancer Res.* **2005**, *65*, 6843–6849. [CrossRef] [PubMed]

71. Ekinci, D.; Cavdar, H.; Durdagi, S.; Oktay, T.; Murat, Ş.; Claudiu, T.S. Structure-activity relationships for the interaction of 5,10-dihydroindeno[1,2-b]indole derivatives with human and bovine carbonic anhydrase isoforms I, II, III, IV and VI. *Eur. J. Med. Chem.* **2012**, *49*, 68–73. [CrossRef] [PubMed]

72. Supuran, C.T. Inhibition of carbonic anhydrase IX as a novel anticancer mechanism. *World J. Clin. Oncol.* **2012**, *3*, 98–103. [CrossRef] [PubMed]

73. Said, H.M.; Supuran, C.T.; Hageman, C.; Staab, A.; Polat, B.; Katzer, A.; Scozzafava, A.; Anacker, J.; Flentje, M.; Vordermark, D. Modulation of carbonic anhydrase 9 (CA9) in human brain cancer. *Curr. Pharm. Des.* **2010**, *16*, 3288–3299. [CrossRef] [PubMed]

74. Neri, D.; Supuran, C.T. Interfering with pH regulation in tumours as a therapeutic strategy. *Nat. Rev. Drug Discov.* **2011**, *10*, 767–777. [CrossRef] [PubMed]

75. Pastorekova, S.; Parkkila, S.; Pastorek, J.; Supuran, C.T. Carbonic anhydrases: Current state of the art, therapeutic applications and future prospects. *J. Enzyme Inhib. Med. Chem.* **2004**, *19*, 199–229. [CrossRef] [PubMed]

76. Vullo, D.; Franchi, M.; Gallori, E.; Pastorek, J.; Scozzafava, A.; Pastorekova, S.; Supuran, C.T. Carbonic anhydrase inhibitors: Inhibition of the tumor-associated isozyme IX with aromatic and heterocyclic sulfonamides. *Bioorg. Med. Chem. Lett.* **2003**, *13*, 1005–1009. [CrossRef]

77. Vullo, D.; Innocenti, A.; Nishimori, I.; Pastorek, J.; Scozzafava, A.; Pastorekova, S.; Supuran, C.T. Carbonic anhydrase inhibitors. Inhibition of the transmembrane isozyme XII with sulfonamides—A new target for the design of antitumor and antiglaucoma drugs? *Bioorg. Med. Chem. Lett.* **2005**, *15*, 963–969. [CrossRef] [PubMed]

78. Zatovicova, M.; Jelenska, L.; Hulikova, A.; Csaderova, L.; Ditte, Z.; Ditte, P.; Goliasova, T.; Pastorek, J.; Pastorekova, S. Carbonic anhydrase IX as an anticancer therapy target: Preclinical evaluation of internalizing monoclonal antibody directed to catalytic domain. *Curr. Pharm. Des.* **2010**, *16*, 3255–3263. [CrossRef] [PubMed]

79. Lou, Y.; McDonald, P.C.; Oloumi, A.; Chia, S.; Ostlund, C.; Ahmadi, A.; Kyle, A.; Auf dem Keller, U.; Leung, S.; Huntsman, D.; et al. Targeting tumor hypoxia: Suppression of breast tumor growth and metastasis by novel carbonic anhydrase IX inhibitors. *Cancer Res.* **2011**, *71*, 3364–3376. [CrossRef] [PubMed]

80. Bowman, E.J.; Siebers, A.; Altendorf, K. Bafilomycins: A class of inhibitors of membrane ATPases from microorganisms, animal cells, and plant cells. *Proc. Natl. Acad. Sci. USA* **1988**, *85*, 7972–7976. [CrossRef] [PubMed]

81. Dröse, S.; Bindseil, K.U.; Bowman, E.J.; Siebers, A.; Zeeck, A.; Altendorf, K. Inhibitory effect of modified bafilomycins and concanamycins on P- and V-type adenosinetriphosphatases. *Biochemistry* **1993**, *32*, 3902–3906. [CrossRef] [PubMed]

82. McSheehy, P.M.; Troy, H.; Kelland, L.R.; Judson, I.R.; Leach, M.O.; Griffiths, J.R. Increased tumour extracellular pH induced by Bafilomycin A1 inhibits tumour growth and mitosis in vivo and alters 5-fluorouracil pharmacokinetics. *Eur. J. Cancer* **2003**, *39*, 532–540. [CrossRef]

83. Lim, J.H.; Park, J.W.; Kim, M.S.; Park, S.K.; Johnson, R.S.; Chun, Y.S. Bafilomycin induces the p21-mediated growth inhibition of cancer cells under hypoxic conditions by expressing hypoxia-inducible factor-1alpha. *Mol. Pharmacol.* **2006**, *70*, 1856–1865. [CrossRef] [PubMed]

84. Ohta, T.; Arakawa, H.; Futagami, F.; Fushida, S.; Kitagawa, H.; Kayahara, M.; Nagakawa, T.; Miwa, K.; Kurashima, K.; Numata, M.; et al. Bafilomycin A1 induces apoptosis in the human pancreatic cancer cell line Capan-1. *J. Pathol.* **1998**, *185*, 324–330. [CrossRef]

85. Morimura, T.; Fujita, K.; Akita, M.; Nagashima, M.; Satomi, A. The proton pump inhibitor inhibits cell growth and induces apoptosis in human hepatoblastoma. *Pediatr. Surg. Int.* **2008**, *24*, 1087–1094. [CrossRef] [PubMed]

86. Nakashima, S.; Hiraku, Y.; Tada-Oikawa, S.; Hishita, T.; Gabazza, E.C.; Tamaki, S.; Imoto, I.; Adachi, Y.; Kawanishi, S. Vacuolar H$^+$-ATPase inhibitor induces apoptosis via lysosomal dysfunction in the human gastric cancer cell line MKN-1. *J. Biochem.* **2003**, *134*, 359–364. [CrossRef] [PubMed]

87. Hishita, T.; Tada-Oikawa, S.; Tohyama, K.; Miura, Y.; Nishihara, T.; Tohyama, Y.; Yoshida, Y.; Uchiyama, T.; Kawanishi, S. Caspase-3 activation by lysosomal enzymes in cytochrome *c*-independent apoptosis in myelodysplastic syndrome-derived cell line P39. *Cancer Res.* **2001**, *61*, 2878–2884. [PubMed]

88. Erickson, K.L.; Beutler, J.A.; Cardellina, J.H., II; Boyd, M.R. Salicylihalamides A and B, novel cytotoxic macrolides from the marine sponge Haliclona sp. *J. Org. Chem.* **1997**, *62*, 8188–8192. [CrossRef] [PubMed]

89. Sasse, F.; Steinmetz, H.; Hofle, G.; Reichenbach, H. Archazolids, new cytotoxic macrolactones from *Archangium gephyra* (Myxobacteria). Production, isolation, physico-chemical and biological properties. *J. Antibiot. Tokyo* **2003**, *56*, 520–525. [CrossRef] [PubMed]

90. Gagliardi, S.; Nadler, G.; Consolandi, E.; Parini, C.; Morvan, M.; Legave, M.N.; Belfiore, P.; Zocchetti, A.; Clarke, G.D.; James, I.; et al. 4-Pentadienamides: Novel and selective inhibitors of the vacuolar H$^+$-ATPase of osteoclasts with bone antiresorptive activity. *J. Med. Chem.* **1998**, *41*, 1568–1573. [CrossRef] [PubMed]

91. Nadler, G.; Morvan, M.; Delimoge, I.; Belfiore, P.; Zocchetti, A.; James, I.; Zembryki, D.; Lee-Rycakzewski, E.; Parini, C.; Consolandi, E.; et al. (2Z,4E)-5-(5,6-dichloro-2-indolyl)-2-methoxy-N-(1,2,2,6,6-pentamethylpiperidin-4-yl)-2,4-pentadienamide, a novel, potent and selective inhibitor of the osteoclast V-ATPase. *Bioorg. Med. Chem. Lett.* **1998**, *8*, 3621–3626. [CrossRef]

92. Huss, M.; Wieczorek, H. Inhibitors of V-ATPases: Old and new players. *J. Exp. Biol.* **2009**, *212*, 341–346. [CrossRef] [PubMed]

93. Lebreton, S.; Jaunbergs, J.; Roth, M.G.; Ferguson, D.A.; De Brabander, J.K. Evaluating the potential of Vacuolar ATPase inhibitors as anticancer agents and multigramsynthesis of the potent salicylihalamide analog saliphenylhalamide. *Bioorg. Med. Chem. Lett.* **2008**, *18*, 5879–5883. [CrossRef] [PubMed]

94. Wiedmann, R.M.; von Schwarzenberg, K.; Palamidessi, A.; Schreiner, L.; Kubisch, R.; Liebl, J.; Schempp, C.; Trauner, D.; Vereb, G.; Zahler, S.; et al. The V-ATPase-inhibitor archazolid abrogates tumor metastasis via inhibition of endocytic activation of the rho-GTPase Rac1. *Cancer Res.* **2012**, *72*, 5976–5987. [PubMed]

95. Mullin, J.M.; Gabello, M.; Murray, L.J.; Farrell, C.P.; Bellows, J.; Wolov, K.R.; Kearney, K.R.; Rudolph, D.; Thornton, J.J. Proton pump inhibitors: Actions and reactions. *Drug Discov. Today* **2009**, *14*, 647–660. [CrossRef] [PubMed]

96. Schwartz, M.D. Dyspepsia, peptic ulcer disease, and esophageal reflux disease. *West. J. Med.* **2002**, *176*, 98–103. [PubMed]

97. Ward, R.M.; Kearns, G.L. Proton pump inhibitors in pediatrics: Mechanism of action, pharmacokinetics, pharmacogenetics, and pharmacodynamics. *Paediatr. Drugs* **2013**, *15*, 119–131. [CrossRef] [PubMed]

98. Han, Y.M.; Hahm, K.B.; Park, J.M.; Hong, S.P.; Kim, E.H. Paradoxically augmented anti-tumorigenic action of proton pump inhibitor and Gastrinin APCMin/+ intestinal polyposis model. *Neoplasia* **2014**, *16*, 73–83. [CrossRef] [PubMed]

99. Canitano, A.; Iessi, E.; Spugnini, E.P.; Federici, C.; Fais, S. Proton pump inhibitors induce a caspase-independent antitumor effect against human multiple myeloma. *Cancer Lett.* **2016**, *376*, 278–283. [CrossRef] [PubMed]

100. Bellone, M.; Calcinotto, A.; Filipazzi, P.; De Milito, A.; Fais, S.; Rivoltini, L. The acidity of the tumor microenvironment is a mechanism of immune escape that can be overcome by proton pump inhibitors. *Oncoimmunology* **2013**, *2*, e22058. [CrossRef] [PubMed]

101. Chen, M.; Huang, S.L.; Zhang, X.Q.; Zhang, B.; Zhu, H.; Yang, V.W.; Zou, X.P. Reversal effects of pantoprazole on multidrug resistance in human gastric adenocarcinoma cells by down-regulating the V-ATPases/mTOR/HIF-1α/P-gp and MRP1 signaling pathway in vitro and in vivo. *J. Cell. Biochem.* **2012**, *113*, 2474–2487. [CrossRef] [PubMed]

102. Yeo, M.; Kim, D.K.; Kim, Y.B.; Oh, T.Y.; Lee, J.E.; Cho, S.W.; Kim, H.C.; Hahm, K.B. Selective induction of apoptosis with proton pump inhibitor in gastric cancer cells. *Clin. Cancer Res.* **2004**, *10*, 8687–8696. [CrossRef] [PubMed]

103. Yeo, M.; Kim, D.K.; Park, H.J.; Cho, S.W.; Cheong, J.Y.; Lee, K.J. Blockage of intracellular proton extrusion with proton extrusions with proton pump inhibitor induces apoptosis in gastric cancer. *Cancer Sci.* **2008**, *99*, 185. [CrossRef] [PubMed]

104. Gu, M.; Zhang, Y.; Zhou, X.; Ma, H.; Yao, H.; Ji, F. Rabeprazole exhibits antiproliferative effects on human gastric cancer cell lines. *Oncol. Lett.* **2014**, *8*, 1739–1744. [CrossRef] [PubMed]

105. Huang, S.; Chen, M.; Ding, X.; Zhang, X.; Zou, X. Proton pump inhibitor selectively suppresses proliferation and restores the chemosensitivity of gastric cancer cells by inhibiting STAT3 signaling pathway. *Int. Immunopharmacol.* **2013**, *17*, 585–592. [CrossRef] [PubMed]

106. Shen, Y.; Chen, M.; Huang, S.; Zou, X. Pantoprazole inhibits human gastric adenocarcinoma SGC-7901 cells by downregulating the expression of pyruvate kinase M2. *Oncol. Lett.* **2016**, *11*, 717–722. [CrossRef] [PubMed]

107. Feng, S.; Zheng, Z.; Feng, L.; Yang, L.; Chen, Z.; Lin, Y.; Gao, Y.; Chen, Y. Proton pump inhibitor pantoprazole inhibits the proliferation, self-renewal and chemoresistance of gastric cancer stem cells via the EMT/β-catenin pathways. *Oncol. Rep.* **2016**, *36*, 3207–3214. [CrossRef] [PubMed]

108. Patel, K.J.; Lee, C.; Tan, Q.; Tannock, I.F. Use of the proton pump inhibitor pantoprazole to modify the distribution and activity of doxorubicin: A potential strategy to improve the therapy of solid tumors. *Clin. Cancer Res.* **2013**, *19*, 6766–6776. [CrossRef] [PubMed]

109. Yu, M.; Lee, C.; Wang, M.; Tannock, I.F. Influence of the proton pump inhibitor lansoprazole on distribution and activity of doxorubicin in solid tumors. *Cancer Sci.* **2015**, *106*, 1438–1447. [CrossRef] [PubMed]

110. Fan, S.; Niu, Y.; Tan, N.; Wu, Z.; Wang, Y.; You, H.; Ke, R.; Song, J.; Shen, Q.; Wang, W.; et al. LASS2 enhances chemosensitivity of breast cancer by counteracting acidic tumor microenvironment through inhibiting activity of V-ATPase proton pump. *Oncogene* **2013**, *32*, 1682–1690. [CrossRef] [PubMed]

111. Glunde, K.; Guggino, S.E.; Solaiyappan, M.; Pathak, A.P.; Ichikawa, Y.; Bhujwalla, Z.M. Extracellular acidification alters lysosomal trafficking in human breast cancer cells. *Neoplasia* **2003**, *5*, 533–545. [CrossRef]

112. Jin, U.H.; Lee, S.O.; Pfent, C.; Safe, S. The aryl hydrocarbon receptor ligand omeprazole inhibits breast cancer cell invasion and metastasis. *BMC Cancer* **2014**, *14*, 498. [CrossRef] [PubMed]

113. Robey, I.F.; Martin, N.K. Bicarbonate and dichloroacetate: Evaluating pH altering therapies in a mouse model for metastatic breast cancer. *BMC Cancer* **2011**, *11*, 235. [CrossRef] [PubMed]

114. Udelnow, A.; Kreyes, A.; Ellinger, S.; Landfester, K.; Walther, P.; Klapperstueck, T.; Wohlrab, J.; Goh, W.; Sleptsova-Freidrich, I.; Petrovic, N. Use of proton pump inhibitors as adjunct treatment for triple-negative breast cancers. An introductory study. *J. Pharm. Pharm. Sci.* **2014**, *17*, 439–446.

115. Lindner, K.; Borchardt, C.; Schopp, M.; Burgers, A.; Stock, C.; Hussey, D.J.; Haier, J.; Hummel, R. Proton pump inhibitors (PPIs) impact on tumour cell survival, metastatic potential and chemotherapy resistance, and affect expression of resistance-relevant miRNAs in esophageal cancer. *J. Exp. Clin. Cancer Res.* **2014**, *33*, 73. [CrossRef] [PubMed]

116. Lee, Y.Y.; Jeon, H.K.; Hong, J.E.; Cho, Y.J.; Ryu, J.Y.; Choi, J.J.; Lee, S.H.; Yoon, G.; Kim, W.Y.; Do, I.G. Proton pump inhibitors enhance the effects of cytotoxic agentsin chemoresistant epithelial ovarian carcinoma. *Oncotarget* **2015**, *6*, 35040–35050. [PubMed]

117. Song, J.; Ge, Z.; Yang, X.; Luo, Q.; Wang, C.; You, H.; Ge, T.; Deng, Y.; Lin, H.; Cui, Y.; et al. Hepatic stellate cells activated by acidic tumor microenvironment promote the metastasis of hepatocellular carcinoma via osteopontin. *Cancer Lett.* **2015**, *356*, 713–720. [CrossRef] [PubMed]

118. Henne-Bruns, D.; Knippschild, U.; Würl, P. Omeprazole inhibits proliferation and modulates autophagy in pancreatic cancer cells. *PLoS ONE* **2011**, *6*, e20143.

119. Jin, U.H.; Kim, S.B.; Safe, S. Omeprazole inhibits pancreatic cancer cell invasion through a nongenomic aryl hydrocarbon receptor pathway. *Chem. Res. Toxicol.* **2015**, *28*, 907–918. [CrossRef] [PubMed]

120. Ihling, A.; Ihling, C.H.; Sinz, A.; Gekle, M. Acidosis-induced changes in proteome patterns of the prostate cancer-derived tumor cell line AT-1. *J. Proteome Res.* **2015**, *14*, 3996–4004. [CrossRef] [PubMed]

121. Harris, R.J.; Cloughesy, T.F.; Liau, L.M.; Prins, R.M.; Antonios, J.P.; Li, D.; Yong, W.H.; Pope, W.B.; Lai, A.; Nghiemphu, P.L.; et al. pH-weighted molecular imaging of gliomas using amine chemical exchange saturation transfer MRI. *Neuro Oncol.* **2015**, *17*, 1514–1524. [CrossRef] [PubMed]

122. Coman, D.; Huang, Y.; Rao, J.U.; De Feyter, H.M.; Rothman, D.L.; Juchem, C.; Hyder, F. Imaging the intratumoral-peritumoral extracellular pH gradient of gliomas. *NMR Biomed.* **2016**, *29*, 309–319. [CrossRef] [PubMed]

123. Ferrari, S.; Perut, F.; Fagioli, F.; Brach Del Prever, A.; Meazza, C.; Parafioriti, A.; Picci, P.; Gambarotti, M.; Avnet, S.; Baldini, N.; et al. Proton pump inhibitor chemosensitization in human osteosarcoma: From the bench to the patients' bed. *J. Transl. Med.* **2013**, *11*, 268. [CrossRef] [PubMed]

124. Wang, B.Y.; Zhang, J.; Wang, J.L.; Sun, S.; Wang, Z.H.; Wang, L.P.; Zhang, Q.L.; Lv, F.F.; Cao, E.Y.; Shao, Z.M.; et al. Intermittent high dose proton pump inhibitor enhances the antitumor effects of chemotherapy in metastatic breast cancer. *J. Exp. Clin. Cancer Res.* **2015**, *34*, 85. [CrossRef] [PubMed]

125. Brana, I.; Ocana, A.; Chen, E.X.; Razak, A.R.; Haines, C.; Lee, C.; Douglas, S.; Wang, L.; Siu, L.L.; Tannock, I.F.; et al. A phase I trial of pantoprazole in combination with doxorubicin in patients with advanced solid tumors: Evaluation of pharmacokinetics of both drugs and tissue penetration of doxorubicin. *Investig. New Drugs* **2014**, *32*, 1269–1277. [CrossRef] [PubMed]

126. Papagerakis, S.; Bellile, E.; Peterson, L.A.; Pliakas, M.; Balaskas, K.; Selman, S.; Hanauer, D.; Taylor, J.M.; Duffy, S.; Wolf, G. Proton pump inhibitors and histamine 2 blockers are associated with improved overall survival in patients with head and neck squamous carcinoma. *Cancer Prev. Res.* **2014**, *7*, 1258–1269. [CrossRef] [PubMed]

127. Falcone, R.; Roberto, M.; D'Antonio, C.; Romiti, A.; Milano, A.; Onesti, C.E.; Marchetti, P.; Fais, S. High-doses of proton pumps inhibitors in refractory gastro-intestinal cancer: A case series and the state of art. *Dig. Liver Dis.* **2016**, *48*, 1503–1505. [CrossRef] [PubMed]

128. Marchetti, P.; Milano, A.; D'Antonio, C.; Romiti, A.; Falcone, R.; Roberto, M.; Fais, S. Association between proton pump inhibitors and metronomic capecitabine as salvage treatment for patients with advanced gastrointestinal tumors: A randomized phase II trial. *Clin. Colorectal Cancer* **2016**, *15*, 377–380. [CrossRef] [PubMed]

129. Spugnini, E.P.; Baldi, A.; Buglioni, S.; Carocci, F.; de Bazzichini, G.M.; Betti, G.; Pantaleo, I.; Menicagli, F.; Citro, G.; Fais, S. Lansoprazole as a rescue agent in chemoresistant tumors: A phase I/II study in companion animals with spontaneously occurring tumors. *J. Transl. Med.* **2011**, *9*, 221. [CrossRef] [PubMed]

130. Spugnini, E.P.; Buglioni, S.; Carocci, F.; Francesco, M.; Vincenzi, B.; Fanciulli, M.; Fais, S. High dose lansoprazole combined with metronomic chemotherapy: A phase I/II study in companion animals with spontaneously occurring tumors. *J. Transl. Med.* **2014**, *12*, 225. [CrossRef] [PubMed]

131. Chiche, J.; Ilc, K.; Laferriere, J.; Trottier, E.; Dayan, F.; Mazure, N.M.; Brahimi-Horn, M.C.; Pouyssegur, J. Hypoxia-inducible carbonic anhydrase IX and XII promote tumor cell growth by counteracting acidosis through the regulation of the intracellular pH. *Cancer Res.* **2009**, *69*, 358–368. [CrossRef] [PubMed]

132. Gondi, G.; Mysliwietz, J.; Hulikova, A.; Jen, J.P.; Swietach, P.; Kremmer, E.; Zeidler, R. Antitumor efficacy of a monoclonal antibody that inhibits the activity of cancer-associated carbonic anhydrase XII. *Cancer Res.* **2013**, *73*, 6494–6503. [CrossRef] [PubMed]

133. Chiche, J.; Brahimi-Horn, M.C.; Pouysségur, J. Tumour hypoxia induces a metabolic shift causing acidosis: A common feature in cancer. *J. Cell. Mol. Med.* **2010**, *14*, 771–794. [CrossRef] [PubMed]

134. Pastorek, J.; Pastorekova, S. Hypoxia-induced carbonic anhydrase IX as a target for cancer therapy: From biology to clinical use. *Semin. Cancer Biol.* **2015**, *31*, 52–64. [CrossRef] [PubMed]

135. Svastova, E.; Witarski, W.; Csaderova, L.; Kosik, I.; Skvarkova, L.; Hulikova, A.; Zatovicova, M.; Barathova, M.; Kopacek, J.; Pastorek, J. Carbonic anhydrase IX interacts with bicarbonate transporters in lamellipodia and increases cell migration via its catalytic domain. *J. Biol. Chem.* **2012**, *287*, 3392–3402. [CrossRef] [PubMed]

136. Parks, S.K.; Cormerais, Y.; Marchiq, I.; Pouysségur, J. Hypoxia optimises tumour growth by controlling nutrient import and acidic metabolite export. *Mol. Asp. Med.* **2016**, *47*, 3–14. [CrossRef] [PubMed]

137. Doyen, J.; Parks, S.K.; Marcie, S.; Pouysségur, J.; Chiche, J. Knock-down of hypoxia-induced carbonic anhydrases IX and XII radiosensitizes tumor cells by increasing intracellular acidosis. *Front. Oncol.* **2012**, *2*, 199. [CrossRef] [PubMed]

138. Dubois, L.; Peeters, S.; Lieuwes, N.G.; Geusens, N.; Thiry, A.; Wigfield, S.; Carta, F.; McIntyre, A.; Scozzafava, A.; Dogné, J.M.; et al. Specific inhibition of carbonic anhydrase IX activity enhances the in vivo therapeutic effect of tumor irradiation. *Radiother. Oncol.* **2011**, *99*, 424–431. [CrossRef] [PubMed]

139. Supuran, C.T. How many carbonic anhydrase inhibition mechanisms exist? *J. Enzyme Inhib. Med. Chem.* **2016**, *31*, 345–360. [CrossRef] [PubMed]

140. Supuran, C.T. Advances in structure-based drug discovery of carbonic anhydrase inhibitors. *Expert Opin. Drug Discov.* **2017**, *12*, 61–88. [CrossRef] [PubMed]

141. Supuran, C.T. Structure-based drug discovery of carbonic anhydrase inhibitors. *J. Enzyme Inhib. Med. Chem.* **2012**, *27*, 759–772. [CrossRef] [PubMed]

142. Dubois, L.; Lieuwes, N.G.; Maresca, A.; Thiry, A.; Supuran, C.T.; Scozzafava, A.; Wouters, B.G.; Lambin, P. Imaging of CA IX with fluorescent labelled sulfonamides distinguishes hypoxic and (re)-oxygenated cells in a xenograft tumour model. *Radiother. Oncol.* **2009**, *92*, 423–428. [CrossRef] [PubMed]

143. Stiti, M.; Cecchi, A.; Rami, M.; Abdaoui, M.; Barragan-Montero, V.; Scozzafava, A.; Guari, Y.; Winum, J.Y.; Supuran, C.T. Carbonic anhydrase inhibitor coated gold nanoparticles selectively inhibit the tumor-associated isoform IX over the cytosolic isozymes I and II. *J. Am. Chem. Soc.* **2008**, *130*, 16130–16131. [CrossRef] [PubMed]

144. Battke, C.; Kremmer, E.; Mysliwietz, J.; Gondi, G.; Dumitru, C.; Brandau, S.; Lang, S.; Vullo, D.; Supuran, C.; Zeidler, R. Generation and characterization of the first inhibitory antibody targeting tumour-associated carbonic anhydrase XII. *Cancer Immunol. Immunother.* **2011**, *60*, 649–658. [CrossRef] [PubMed]

145. Hoeben, B.A.; Kaanders, J.H.; Franssen, G.M.; Troost, E.G.; Rijken, P.F.; Oosterwijk, E.; van Dongen, G.A.; Oyen, W.J.; Boerman, O.C.; Bussink, J. PET of hypoxia with ^{89}Zr-labeled cG250-F(ab')$_2$ in head and neck tumors. *J. Nucl. Med.* **2010**, *51*, 1076–1083. [CrossRef] [PubMed]

146. Perez-Sayans, M.; Suarez-Penaranda, J.M.; Pilar, G.D.; Supuran, C.T.; Pastorekova, S.; Barros-Angueira, F.; Gándara-Rey, J.M.; García-García, A. Expression of CA-IX is associated with advanced stage tumors and poor survival in oral squamous cell carcinoma patients. *J. Oral Pathol. Med.* **2012**, *41*, 667–674. [CrossRef] [PubMed]

147. Winum, J.Y.; Vullo, D.; Casini, A.; Montero, J.L.; Scozzafava, A.; Supuran, C.T. Carbonic anhydrase inhibitors: Inhibition of transmembrane, tumor-associated isozyme IX, and cytosolic isozymes I and II with aliphatic sulfamates. *J. Med. Chem.* **2003**, *46*, 5471–5477. [CrossRef] [PubMed]

148. Winum, J.Y.; Vullo, D.; Casini, A.; Montero, J.L.; Scozzafava, A.; Supuran, C.T. Carbonic anhydrase inhibitors. Inhibition of cytosolic isozymes I and II and transmembrane, tumor-associated isozyme IX with sulfamates including EMATE also acting as steroid sulfatase inhibitors. *J. Med. Chem.* **2003**, *46*, 2197–2204. [CrossRef] [PubMed]

149. Guler, O.O.; De Simone, G.; Supuran, C.T. Drug design studies of the novel antitumor targets carbonic anhydrase IX and XII. *Curr. Med. Chem.* **2010**, *17*, 1516–1526. [CrossRef] [PubMed]

150. Alterio, V.; Hilvo, M.; Di Fiore, A.; Supuran, C.T.; Pan, P.; Parkkila, S.; Scaloni, A.; Pastorek, J.; Pastorekova, S.; Pedone, C.; et al. Crystal structure of the catalytic domain of the tumor-associated human carbonic anhydrase IX. *Proc. Natl. Acad. Sci. USA* **2009**, *106*, 16233–16238. [CrossRef] [PubMed]

151. Dubois, L.; Douma, K.; Supuran, C.T.; Chiu, R.K.; van Zandvoort, M.A.; Pastoreková, S.; Scozzafava, A.; Wouters, B.G.; Lambin, P. Imaging the hypoxia surrogate marker CA IX requires expression and catalytic activity for binding fluorescent sulfonamide inhibitors. *Radiother. Oncol.* **2007**, *83*, 367–373. [CrossRef] [PubMed]

152. Ahlskog, J.K.; Dumelin, C.E.; Trüssel, S.; Mårlind, J.; Neri, D. In vivo targeting of tumor-associated carbonic anhydrases using acetazolamide derivatives. *Bioorg. Med. Chem. Lett.* **2009**, *19*, 4851–4856. [CrossRef] [PubMed]

153. McDonald, P.C.; Winum, J.Y.; Supuran, C.T.; Dedhar, S. Recent developments in targeting carbonic anhydrase IX for cancer therapeutics. *Oncotarget* **2012**, *3*, 84–97. [CrossRef] [PubMed]

154. Pacchiano, F.; Carta, F.; McDonald, P.C.; Lou, Y.; Vullo, D.; Scozzafava, A.; Dedhar, S.; Supuran, C.T. Ureido-substituted benzene sulfonamides potently inhibit carbonic anhydrase IX and show antimetastatic activity in a model of breast cancer metastasis. *J. Med. Chem.* **2011**, *54*, 1896–1902. [CrossRef] [PubMed]

155. Gieling, R.G.; Babur, M.; Mamnani, L.; Burrows, N.; Telfer, B.A.; Carta, F.; Winum, J.Y.; Scozzafava, A.; Supuran, C.T.; Williams, K.J. Antimetastatic effect of sulfamate carbonic anhydrase IX inhibitors in breast carcinoma xenografts. *J. Med. Chem.* **2012**, *55*, 5591–5600. [CrossRef] [PubMed]

156. Supuran, C.T.; Winum, J.Y. Carbonic anhydrase IX inhibitors in cancer therapy: An update. *Future Med. Chem.* **2015**, *7*, 1407–1414. [CrossRef] [PubMed]

157. Lock, F.E.; McDonald, P.C.; Lou, Y.; Serrano, I.; Chafe, S.C.; Ostlund, C.; Aparicio, S.; Winum, J.Y.; Supuran, C.T.; Dedhar, S. Targeting carbonic anhydrase IX depletes breast cancer stem cells within the hypoxic niche. *Oncogene* **2013**, *32*, 5210–5219. [CrossRef] [PubMed]

158. Boyd, N.H.; Walker, K.; Fried, J.; Hackney, J.R.; McDonald, P.C.; Benavides, G.A.; Spina, R.; Audia, A.; Scott, S.E.; Libby, C.J.; et al. Addition of carbonic anhydrase 9 inhibitor SLC-0111 to temozolomide treatment delays glioblastoma growth in vivo. *JCI Insight* **2017**, *2*. [CrossRef] [PubMed]

159. Maresca, A.; Scozzafava, A.; Supuran, C.T. 7,8-disubstituted- but not 6,7-disubstituted coumarins selectively inhibit the transmembrane, tumor-associated carbonic anhydrase isoforms IX and XII over the cytosolic ones I and II in the low nanomolar/subnanomolar range. *Bioorg. Med. Chem. Lett.* **2010**, *20*, 7255–7258. [CrossRef] [PubMed]

160. Touisni, N.; Maresca, A.; McDonald, P.C.; Lou, Y.; Scozzafava, A.; Dedhar, S.; Winum, J.Y.; Supuran, C.T. Glycosyl coumarin carbonic anhydrase IX and XII inhibitors strongly attenuate the growth of primary breast tumors. *J. Med. Chem.* **2011**, *54*, 8271–8277. [CrossRef] [PubMed]

161. Stillebroer, A.B.; Mulders, P.F.; Boerman, O.C.; Oyen, W.J.; Oosterwijk, E. Carbonic anhydrase IX in renal cell carcinoma: Implications for prognosis, diagnosis, and therapy. *Eur. Urol.* **2010**, *58*, 75–83. [CrossRef] [PubMed]

162. Siebels, M.; Rohrmann, K.; Oberneder, R.; Stahler, M.; Haseke, N.; Beck, J.; Hofmann, R.; Kindler, M.; Kloepfer, P.; Stief, C. A clinical phase I/II trial with the monoclonal antibody cG250 (RENCAREX®) and interferon-alpha-2a in metastatic renal cell carcinoma patients. *World J. Urol.* **2011**, *29*, 121–126. [CrossRef] [PubMed]

163. Mishra, C.B.; Kumari, S.; Angeli, A.; Monti, S.M.; Buonanno, M.; Prakash, A.; Tiwari, M.; Supuran, C.T. Design, synthesis and biological evaluation of N-(5-methyl-isoxazol-3-yl/1,3,4-thiadiazol-2-yl)-4-(3-substitutedphenylureido) benzenesulfonamides as human carbonic anhydrase isoenzymes I, II, VII and XII inhibitors. *J. Enzyme Inhib. Med. Chem.* **2016**, *31*, 174–179. [CrossRef] [PubMed]

164. Li, G.; Feng, G.; Gentil-Perret, A.; Genin, C.; Tostain, J. Serum carbonic anhydrase 9 level is associated with postoperative recurrence of conventional renal cell cancer. *J. Urol.* **2008**, *180*, 510–514. [CrossRef] [PubMed]

165. Gigante, M.; Li, G.; Ferlay, C.; Perol, D.; Blanc, E.; Paul, S.; Zhao, A.; Tostain, J.; Escudier, B.; Negrier, S.; et al. Prognostic value of serum CA9 in patients with metastatic clear cell renal cell carcinoma under targeted therapy. *Anticancer Res.* **2012**, *32*, 5447–5451. [PubMed]

166. Chamie, K.; Klöpfer, P.; Bevan, P.; Störkel, S.; Said, J.; Fall, B.; Belldegrun, A.S.; Pantuck, A.J. Carbonic anhydrase-IX score is a novel biomarker that predicts recurrence and survival for high-risk, nonmetastatic renal cell carcinoma: Data from the phase III ARISER clinical trial. *Urol. Oncol.* **2015**, *33*, e25–e33. [CrossRef] [PubMed]

167. Rademakers, S.E.; Hoogsteen, I.J.; Rijken, P.F.; Oosterwijk, E.; Terhaard, C.H.; Doornaert, P.A.; Langendijk, J.A.; van den Ende, P.; Takes, R.; De Bree, R.; et al. Pattern of CAIX expression is prognostic for outcome and predicts response to ARCON in patients with laryngeal cancer treated in a phase III randomized trial. *Radiother. Oncol.* **2013**, *108*, 517–522. [CrossRef] [PubMed]

168. Turkbey, B.; Lindenberg, M.L.; Adler, S.; Kurdziel, K.A.; McKinney, Y.L.; Weaver, J.; Vocke, C.D.; Anver, M.; Bratslavsky, G.; Eclarinal, P.; et al. PET/CT imaging of renal cell carcinoma with ^{18}F-VM4-037: A phase II pilot study. *Abdom. Radiol.* **2016**, *41*, 109–118. [CrossRef] [PubMed]

169. Belldegrun, A.S.; Chamie, K.; Kloepfer, P.; Fall, B.; Bevan, P.; Störkel, S. ARISER: A randomized double blind phase III study to evaluate adjuvant cG250 treatment versus placebo in patients with high-risk ccRCC—Results and implications for adjuvant clinical trials. *J. Clin. Oncol.* **2013**, *31*, 4507.

170. Stillebroer, A.B.; Boerman, O.C.; Desar, I.M.; Boers-Sonderen, M.J.; van Herpen, C.M.; Langenhuijsen, J.F.; Smith-Jones, P.M.; Oosterwijk, E.; Oyen, W.J.; Mulders, P.F. Phase 1 radioimmunotherapy study with lutetium 177-labeled anti-carbonic anhydrase IX monoclonal antibody girentuximab in patients with advanced renal cell carcinoma. *Eur. Urol.* **2013**, *64*, 478–485. [CrossRef] [PubMed]

171. Lugini, L.; Sciamanna, I.; Federici, C.; Iessi, E.; Spugnini, E.P.; Fais, S. Antitumor effect of combination of the inhibitors of two new oncotargets: Proton pumps and reverse transcriptase. *Oncotarget* **2017**, *8*, 4147–4155. [CrossRef] [PubMed]

metabolites

MDPI

Review

Amino Acids as Building Blocks for Carbonic Anhydrase Inhibitors

Niccolò Chiaramonte *, Maria Novella Romanelli [iD], Elisabetta Teodori and Claudiu T. Supuran [iD]

Department of Neuroscience, Psychology, Drug Research and Child's Health, Section of Pharmaceutical and Nutraceutical Sciences, University of Florence, Via Ugo Schiff 6, 50019 Sesto Fiorentino, Italy; novella.romanelli@unifi.it (M.N.R.); elisabetta.teodori@unifi.it (E.T.); claudiu.supuran@unifi.it (C.T.S.)
* Correspondence: niccolo.chiaramonte@unifi.it, Tel.: +39-055-457-3700

Received: 9 May 2018; Accepted: 23 May 2018; Published: 24 May 2018

Abstract: Carbonic anhydrases (CAs) are a superfamily of metalloenzymes widespread in all life, classified into seven genetically different families (α–θ). These enzymes catalyse the reversible hydration of carbonic anhydride (CO_2), generating bicarbonate (HCO_3^-) and protons (H^+). Fifteen isoforms of human CA (hCA I–XV) have been isolated, their presence being fundamental for the regulation of many physiological processes. In addition, overexpression of some isoforms has been associated with the outbreak or progression of several diseases. For this reason, for a long time CA inhibitors (CAIs) have been used in the control of glaucoma and as diuretics. Furthermore, the search for new potential CAIs for other pharmacological applications is a very active field. Amino acids constitute the smallest fundamental monomers of protein and, due to their useful bivalent chemical properties, are widely used in organic chemistry. Both proteinogenic and non-proteinogenic amino acids have been extensively used to synthesize CAIs. This article provides an overview of the different strategies that have been used to design new CAIs containing amino acids, and how these bivalent molecules influence the properties of the inhibitors.

Keywords: carbonic anhydrase; enzyme inhibition; metalloenzymes; amino acid; glaucoma; tumors

1. Introduction

The interconversion between CO_2 and HCO_3^- is fundamental for the successful flow of biochemical processes in all living cells [1]. Metabolic conversion is ensured by the hydration of carbonic anhydride which leads to the corresponding soluble ion, bicarbonate, and the release of a proton [2,3]. Due to the high number of physiological process in which these chemical species are essential, the optimal balance between them is vital for all life forms [2,4–7]. This equilibrium is regulated by a superfamily of metalloenzyme, the carbonic anhydrases (CAs, EC 4.2.1.1) [1]. By accelerating this normally slow reaction [8], CAs fulfil the metabolic needs connected to CO_2/bicarbonate and protons [2,4,6]. In fact, CAs are among the most effective catalysts known in nature [3] as their turnover number (k_{cat}) can reach the value of 10^6 s^{-1} [9]. Seven genetically different families of these ubiquitous enzymes were identified (α–η) [2]; all of them possess a bivalent metal ion fundamental for catalysis, as the apoenzyme is devoid of activity [3]. Mammalian CAs belong to the α-family and they are characterized by the presence of a bivalent zinc ion within the active site [1]. This Zn^{2+} coordinates a water molecule, or a hydroxide ion in the activated form of the enzyme, making the hydration of CO_2 a fast process [1,2,10].

Human carbonic anhydrases (hCAs) are widespread in the organism, varying for tissue distribution and subcellular localization. Fifteen α-isozymes (hCA I–XV) have been isolated and characterized so far, but only 12 of them are catalytically active [4,5,11]. hCAs are involved in pH and CO_2 homeostasis but also in the regulation of many crucial physiological processes, such as gluconeogenesis, lipogenesis or electrolyte secretion in a variety of tissues and organs [2,4–6]. hCA I and II are the widest expressed isoforms and, together with hCA IV, were identified in the anterior chamber of the eye, being responsible in this organ for the production of bicarbonate, the main constituent of aqueous humor [5,12]. hCA IX and XII can be defined as tumor-associated proteins, due to their massive expression in many hypoxic cancers [4,5,13,14]. These isozymes are transmembrane proteins, with an extracellular catalytic domain; they are fundamental for the survival of tumor cells under stressful conditions, due to their ability to generate a differential pH microenvironment, resulting in increased tumor growth [5,13,15,16]. While CA activators do not have for the moment any approved pharmacological application [17], CA inhibitors (CAIs) are commonly used in therapy, mainly as anti-glaucoma agents, anti-epileptics and diuretics, while other therapeutic applications are still under investigation (anti-cancer, anti-obesity agents, and others) [5,12,18–21].

2. Carbonic Anhydrase Inhibitors (CAIs)

In the CAs active site the bivalent metal ion normally possesses a tetrahedral geometry, since it interacts with three amino acids moieties and coordinates the water molecule/hydroxide ion. In α-CAs, the Zn^{2+} ligands are three histidine residues (Figure 1) and, together with other key amino acids, they form a very particular and strictly conserved catalytic site within each family of carbonic anhydrase [1–3]. It is, therefore, possible to recognize two different regions in the cavity of the active site, a first half composed exclusively of hydrophobic amino acids and a second one where only hydrophilic residues are present [22], leading to an amphiphilic site architecture [2].

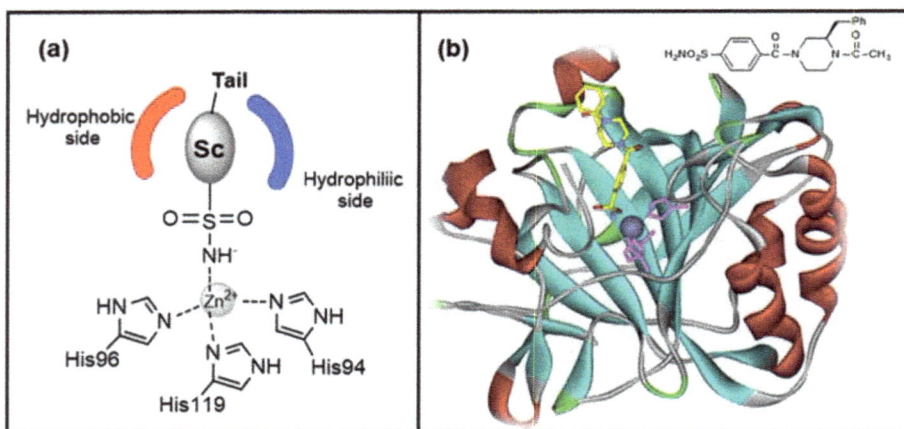

Figure 1. (a) Schematization of the interaction between a generic sulfonamide and the Zn^{2+} in the carbonic anhydrase (CA) active site. Sc = Scaffold (b) X-ray structure of (S)-4-(4-acetyl-3-benzylpiperazine-1-carbonyl)-benzenesulfonamide (**49b**, discussed in Section 7, shown with carbon atoms in yellow,) bound in the active site of hCA I (Protein Data Bank (PDB) entry 6EVR) [23]. The sulfonamide group interacts with the Zn^{2+} (dark grey sphere) which is coordinated by three His residues (in purple).

Depending on the mechanism of action, five different classes of CAIs are known (Figure 2) [21,24]:

1. Zinc binders, i.e., compounds that chelates the bivalent metal ion of the active site. This interaction interrupts the coordination between the Zn^{2+} atom and the water molecule/hydroxide ion and consequently blocks the enzymatic activity [21,24,25]. The mechanism is schematized in Figure 1a: the scaffold of these molecules (reported as "Sc") may interact with one or both the halves of the active site, stabilizing the interaction with the ion in a tetrahedral geometry. This is the most

important class of inhibitors, to which belong sulfonamides and their isosteres (sulfamates or sulfamides), dithiocarbamates, hydroxamate, etc. [3,24]. Sulfonamides are the most widely studied CAIs with at least 20 compounds in clinical use for decades [24]. Some examples (acetazolamide, brinzolamide and dorzolamide) are shown in Figure 2.

2. Compounds that anchor to the-zinc coordinated water molecule/hydroxide ion, such as phenols and polyamines [3,24,26].
3. Compounds occluding the entrance of the active site (coumarins and their isosters) [3,24,27].
4. Compounds that bind out of the active site, such as 2-(benzylsulfonyl)benzoic acid [24,28].
5. CAIs acting without a known mechanism, such as secondary/tertiary sulfonamides, imatinib, etc. [3,24].

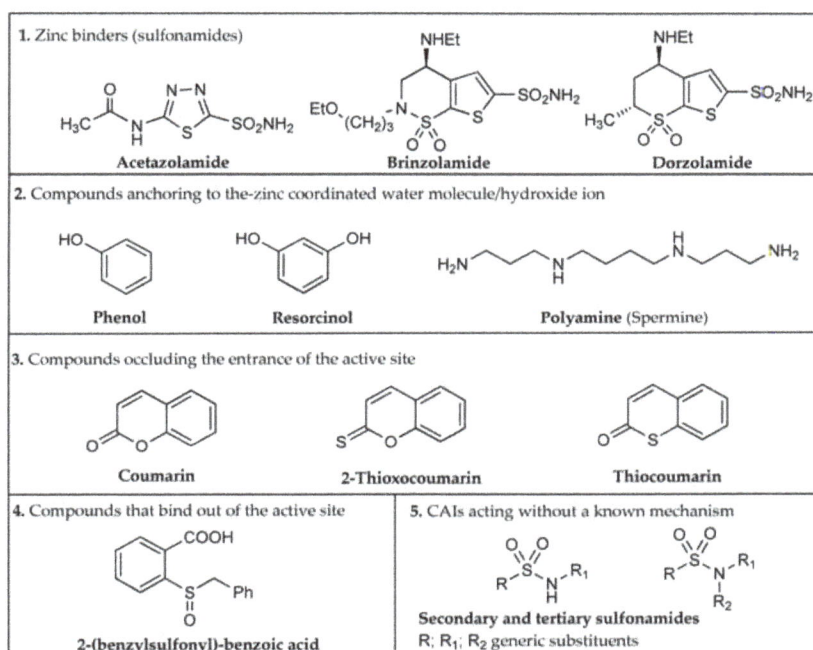

Figure 2. The five classes of carbonic anhydrases inhibitors (CAIs).

As we can note in Figure 2, the five types of inhibitors possess various structures; chemical differences in their pharmacophoric moieties result in different structure activity relationships. Despite this premise, a common problem of CAIs, and in particular of zinc binders, is the lack of selectivity for a specific isoform or the absence of a good water solubility. Many efforts have been made in the last few decades to explore chemical modifications of the CAI structures that could have a positive effect on these properties. In particular, the aim of this review is to provide an overview of the investigated applications of amino acid moieties in the resolution of these problems. This paper is focused on CAIs that carry amino acids or their derivatives in their structures, and how these molecules influence the properties of these inhibitors. Both proteinogenic and non-proteinogenic amino acids have been considered in this analysis and, depending on their synthetic use, five different strategies have been identified.

Amino acids are the primary building blocks of proteins and, of the over 300 naturally occurring, 22 constitute the monomer units of these biological molecules [29]. Amino acids are widely used in organic chemistry; they constitute the smallest fundamental pieces in solid phase peptide synthesis but, due to their characteristics, they are also commonly used in solution synthesis [30]. Their properties range from acidic to basic due to the contemporary presence of a carboxyl and an amino group that, in addition, allow the possibility to easily generate different kind of connections to link the amino acid to other molecules. Beside this, a number of amino acids have reactive groups in their side chains that can be chemically modified [29]. Furthermore, the use of natural α-amino acids allows a broad range of derivatives to be obtained, with the possibility, starting from enantiopure amino acids, to further define the stereochemistry of the side chain. Thanks to ionizing properties that could be exploited to form salts, the carboxy and the amino moieties also ensure good water solubility to these derivatives.

3. Amino Acyl as a Water-Solubilizing Tail

Glaucoma is a group of optical neuropathies associated with progressive loss of visual field, leading to visual impairment and blindness. There is a general agreement that increased intraocular pressure (IOP) is the most important risk factor for the outbreak and progression of this disease [12,31]. It was demonstrated [32] that in glaucomatous patients the inhibition of CAs leads to a reduction of IOP and an amelioration of the symptoms. Due to the wide distribution of the different CA isozymes in the body, the administration of systemic inhibitors usually elicits undesired side effects [12,31,33]. For this reason, the treatment of glaucoma is mainly performed with topically administered CAIs [12,34]; two drugs, brinzolamide and dorzolamide (Figure 2), are available in the clinic. These two inhibitors often produce local side effects, such as ocular burning, a stinging sensation, superficial punctuate keratitis, blurred vision and reddening of the eye. These problems are mainly due to the strong acidic pH of their solution, as they are both administered as hydrochloride salts [12,35].

In order to reduce these inconveniences, many efforts have been made to develop new topically administered CAIs. Among others, a method investigated in depth has been the attachment of water-solubilising tails to known effective CAIs. This method is particularly important for sulfonamides, very potent anti-glaucoma agents often endowed with low water-solubility.

The use of amino acids for this purpose was investigated by many research groups, with a broad variety of approaches. The bivalent chemical nature of these synthons allows two possible series of products to be obtained, the amino-substituted compounds **1** and the carboxy derivatives **2** (Figure 3). In **1**, the free carboxyl group can be treated with strong bases to give salts or can be further functionalized. Also compounds **2** can be endowed with good hydrophilic properties thanks to the free basic amino group: such compounds are ideal for ophthalmologic applications, since the water solutions of their salts with strong acids have a weakly acidic pH, preferred over those of alkaline pH [36].

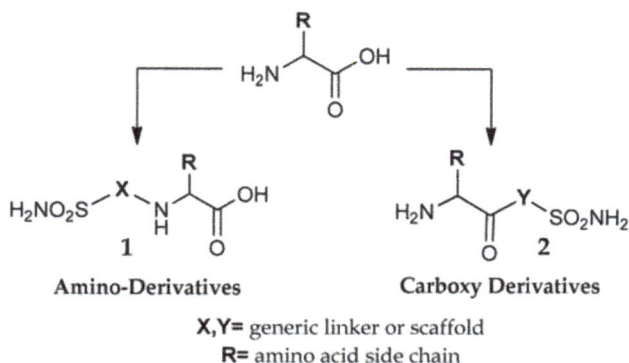

X,Y= generic linker or scaffold
R= amino acid side chain

Figure 3. General structure of amino acids functionalised on the amino group (**1**) or on the carboxy moiety (**2**).

Antonaroli et al. [37], Blackburn's [38] group, and later Barboiu et al. [39] reported some structural manipulations of acetazolamide, where the N-acetyl residue was substituted with different aminoacids, giving compounds with general formula **3** and **4** (Figure 4). The structural modification of this first-generation CA inhibitor led to interesting products: the salts of compounds **3** and **4** showed a very good water solubility, associated with effective inhibitory activities against the enzyme. As an example, the β-alanyl derivatives **3a** was three times more potent on hCA II and over 100 times on hCA I than dorzolamide. Only against the isozymes hCA IV, this compound showed an activity three times lower than the reference drug.

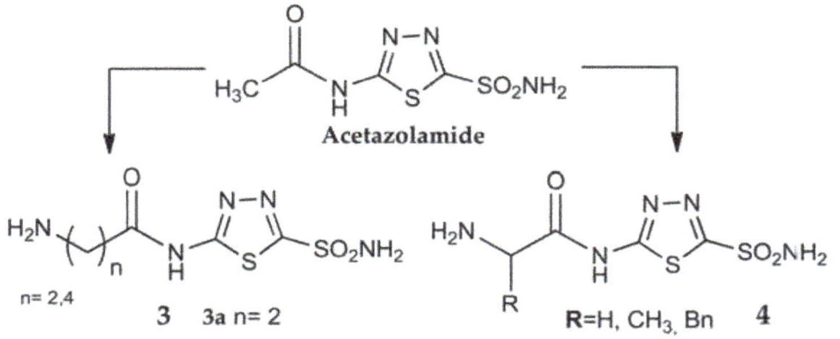

Figure 4. Structure of acetazolamide analogues **3** and **4**.

A similar approach was also investigated by Scozzafava et al. in two separate works [36,40]. Twenty six different sulfonamides containing amino, imino, hydrazino or hydroxyl groups (general formula **5**, Figure 5) reacted with the carboxy moiety of five different glycine derivatives (glycine, sarcosine, creatine, gly-gly and β-alanine). The extremely versatile nature of the carboxyl group allowed a large library of new compounds (general formula **6**) to be generated, differing not only in the structure of the two reagents, but also for the characteristics of the new formed carboxy-derived bonds (i.e., amidic, esteric and hydrazidic bonds).

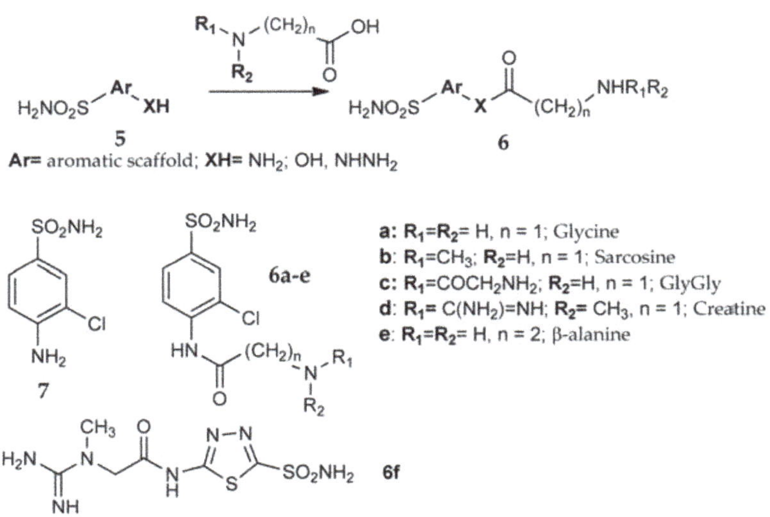

Figure 5. General structure and some examples of the compounds investigated by Scozzafava et al. [36,40].

All the 130 obtained compounds **6** possess very good water solubility as salts of strong acids. They showed a wide range of inhibitory activities against three CA isozymes, hCA I, II and bCA IV. The addition of the aminoacyl/dipeptidyl moiety generally led to an increase of CA inhibitory properties with respect to the corresponding parent sulfonamide. The five amino acid derivatives **6a–f** can be taken as an example: they were from 3 to 18 times more potent than the precursor **7**. In addition, the most active compounds of the series were selected for in vivo studies. The topical application directly into the eye showed IOP-lowering effects both in normotensive and glaucomatous rabbits, a frequently used animal model of glaucoma [41,42]. As an example, the acetazolamide derivative **6f** was a more efficient and long-lasting IOP-lowering agent compared to the reference drug dorzolamide.

It is also possible to enhance the basic nature of the free NH_2 group by conversion into a guanidine moiety, as reported by Ceruso et al. [43]. The carboxyl group of two amino acids, Nα-acetyllysine and γ-aminobutyric acid, was connected to a benzyl or phenetylamine carrying a sulfonamide group, obtaining a first series of basic compounds (**8a** and **9a**, Figure 6); subsequently the amino groups were treated with N,N'-di-Boc-N''-trifluoromethane-sulfonylguanidine, obtaining the guanidine derivatives **8b** and **9b**. This transformation enhanced the solubility of the hydrochloride salts but not their potency, since the guanidine derivatives showed K_i values almost in the same range of the amino-precursor on hCA I, hCA II and *Porphyromonas gingivalis* γ-CA. The same products were also tested against the two tumor-associated isoforms hCA IX and hCA XII [44]. Due to their transmembrane localization and their extracellular catalytic domain, the use of hydrophilic CAIs, endowed with poor cellular permeability, could in principle selectively target these isoforms. The effect of the terminal guanidine moiety on activity was relevant only on hCA XII for some GABA-derivatives: as an example, **9c** showed a K_i value almost nine times lower than the corresponding amino precursor.

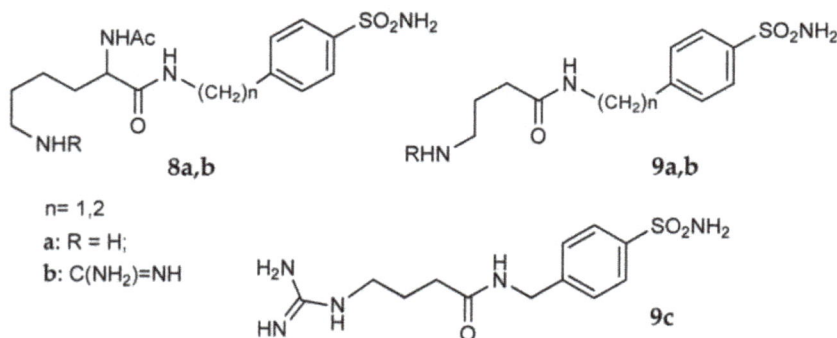

Figure 6. Structure of benzenesulfonamide derivatives **8** and **9**.

The addition of the amino acid by means of the amino group, instead of the carboxy one, was investigated by Casini et al. in two separate, but closely related papers [35,45]. Two series of sulfonamides were synthesized by the reaction of 4-isothiocyanatobenzenesulfonamide **10a** or 4-isothiocyanatomethyl-benzenesulfonamide **10b** with 33 different amino acids and oligopeptides, forming a thioureido linkage between these two portions (general formula **11**, Figure 7). Many newly obtained compounds possess very good water-solubility properties; the presence of at least one free carboxyl group allows the easy formation of sodium salts. Their solutions showed pH values in the range of 6.5–7.0 and, due to these optimal values, no or modest eye irritation effects were observed. Strong inhibitory properties were detected for many derivatives against hCA I, hCA II and bCA IV and, thanks to their strong tendency to concentrate in ocular fluids and tissues, almost 20 compounds showed effective IOP-lowering properties after topical administration.

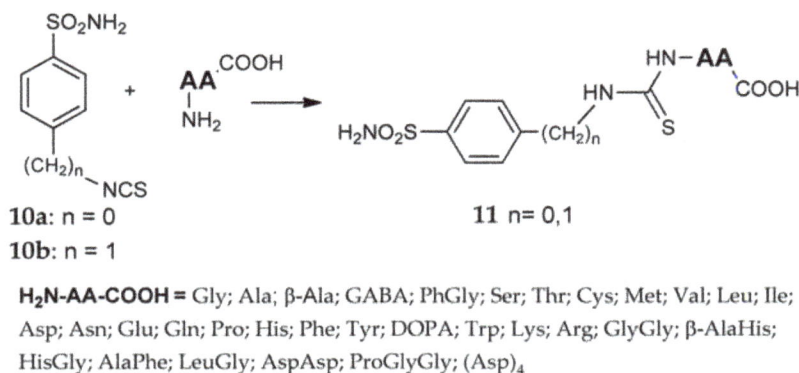

H$_2$N-AA-COOH = Gly; Ala; β-Ala; GABA; PhGly; Ser; Thr; Cys; Met; Val; Leu; Ile; Asp; Asn; Glu; Gln; Pro; His; Phe; Tyr; DOPA; Trp; Lys; Arg; GlyGly; β-AlaHis; HisGly; AlaPhe; LeuGly; AspAsp; ProGlyGly; (Asp)$_4$

Figure 7. Thiourea derivatives studied by Casini et al. [35,45].

Using 4-carboxybenzenesulfonamide **12** or 4-chloro-3-sulfamoylbenzoic acid **13** (Figure 8) Mincione et al. [46] synthesized a new series of inhibitors, where different amino acids or dipeptides were directly linked to the sulfonamide through an amidic bond (general formula **14** and **15**, Figure 8). As previously mentioned for the thioureido derivatives **11**, many newly obtained compounds were able to inhibit three CA isozymes (hCA I, hCA II and bCA IV) showing K$_1$ values in the nanomolar range. The corresponding carboxylate salts, with their good water-solubility, were investigated in IOP-lowering in vivo experiments. The topical administration of these solutions, which possessed pH values in the neutrality range, showed very important and long lasting IOP-lowering effects in rabbits, stronger and longer-lasting on the glaucomatous animals than on the normotensive ones.

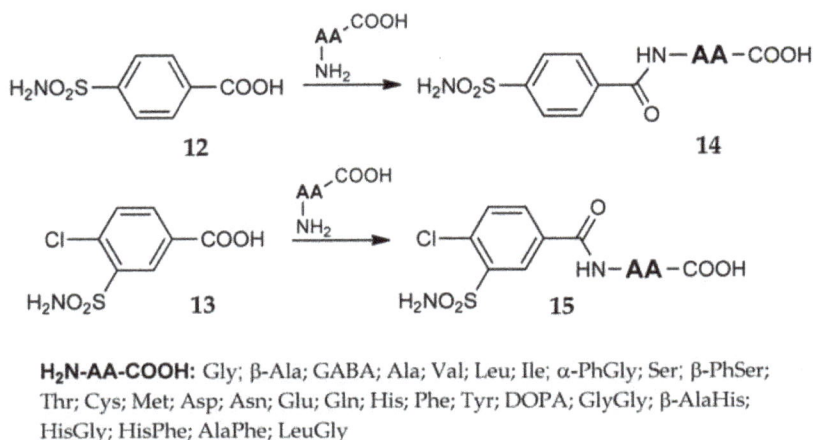

H$_2$N-AA-COOH: Gly; β-Ala; GABA; Ala; Val; Leu; Ile; α-PhGly; Ser; β-PhSer; Thr; Cys; Met; Asp; Asn; Glu; Gln; His; Phe; Tyr; DOPA; GlyGly; β-AlaHis; HisGly; HisPhe; AlaPhe; LeuGly

Figure 8. Sulfamoylbenzamides of amino acids and dipeptides.

4. The Tail Approach to Carbonic Anhydrases Inhibitors Using Amino Acids

As reported in Section 2, CAIs fall into five categories, according to their mechanism of action [21]. Probably, the most important CAIs problem is the lack of selectivity for a specific isoform, due to the well conserved catalytic site among the isozymes. A possible method to gain selectivity is the so called "Tail Approach" [21,24,25,33]. This strategy is based on the structural modification of a portion on the inhibitor (Tail in Figure 1a), usually in a position not fundamental for the Zn-chelating activity, that could potentially interact with the regions surrounding the active site, where the amino acids variability is higher. On this basis, the chemical differences in the amino acids' side chains and the possibility to use also "non-α" amino acids make these molecules optimal candidates for the "tail" role (Figure 9). As is simple to imagine, there is an extremely high variety of possible synthetic pathways that can be used to obtain different compounds by using the carboxy (as in **16**) or the amino moiety (as in **17**).

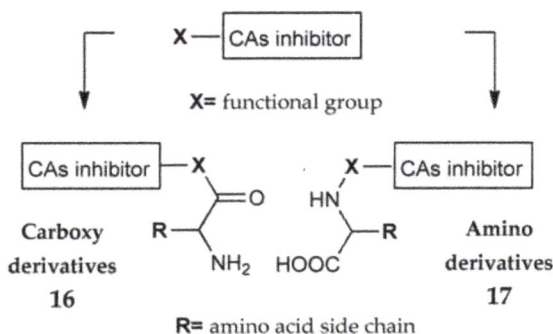

Figure 9. General structure of CAIs decorated with amino acids through the carboxy group (**16**) or the amino moiety (**17**).

An interesting work, focused on the investigation of enzyme-inhibitor interactions, was performed by the groups of Whitesides and Christianson [47–49]: in order to explore if secondary interactions away from the active site could influence inhibitory activity, they prepared three series of derivatives (general formula **18–20**, Figure 10). At first these researchers synthesized a series of oligolglycine- and oligo(ethyleneglycol)-linked benzenesulfonamides (**18**), whose dissociation constant on bCA II (K_d, measured by means of a competitive fluorescence-based assay) was found in the high micromolar range and independent of the polymer length [48]. Later, they condensed several tripeptides to 4-sulfamoylbenzoic acid, obtaining compounds with general formula **19**, endowed with K_d values in the nanomolar range on hCA II [49]. The complex with **19a** was analyzed by means of X-ray crystallography: the phenylglycine moiety was placed near Phe131 and Pro202, establishing hydrophobic interactions and validating the initial design.

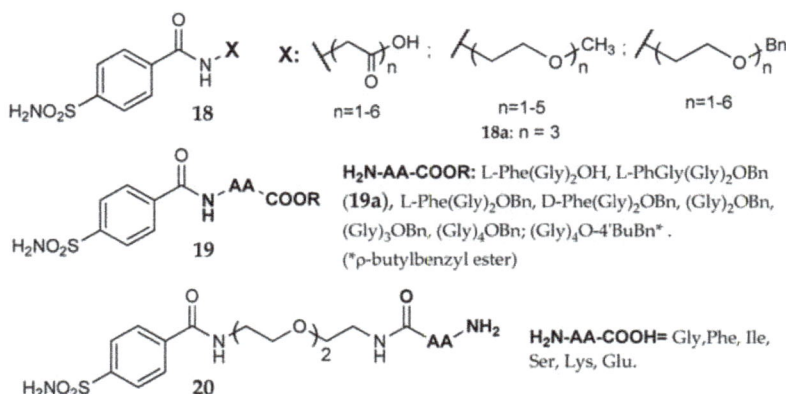

Figure 10. General structures of the compounds investigated in [47–49].

The compound carrying three ethylene glycol moieties (**18a**) was, therefore, chosen by Boriack et al. as primary scaffold for their final investigation [47]. Different amino acids were added to the terminal portion of the poly-ethylene glycol chain; with this design compounds **20** were obtained, carrying a free amino group, and endowed with increased activity (K_d values from 2 to 10 times lower than the parent compound **18a**). This flexible and quite long linker was able to direct the amino acid toward enzyme surfaces far from the zinc ion, in a region where it can stabilize the binding through side interactions. Since in this series of compounds the best terminal pendants were lipophilic amino acids, these researchers suggested that the linker chain delivered the terminal group close to a hydrophobic region composed by Pro201, Pro202 and Leu198, even if the X-ray structures (Protein Data Bank (PDB) codes 1CNW, 1CNX and 1CNY) did not reveal ordered folds of the ethylene glycol linker.

Garaj et al. [50] used the tail approach on a series of novel sulfonamides incorporating the 1,3,5-triazine moiety, and obtained a small set of compounds with affinity from nanomolar to micromolar on hCA I, hCA II and hCa IX. In this series, glycine (m = 1) and β-alanine (m = 2) derivatives **21** (Figure 10) showed good potency on hCA IX, with Ki values 20–30 times lower than on hCA I and II, but independently of m. The library of these derivatives was later expanded by Carta et al. [51], who investigated, among several substituents, a wider number of amino acids in position 2 or 2,4 of the triazinyl ring (general formula **22**, Figure 11). Many compounds showed nanomolar K_i values against the transmembrane tumor-associated CA IX and XII, in addition to CA XIV; the latter isoform is not associated with cancers but is widespread in many tissues such as the kidney, liver and brain among others [3]. As for **21**, higher K_i values were found against the off-target isoforms hCA I and II. In some cases some selectivity was obtained: as an example, compound **22a** showed a Ki 0.96 nM on hCA IX, being on this isoform 6, 10, >550 and >1300 times more potent than on hCA XII, XIV, II and I, respectively. Two derivatives (**22b** and **22c**) displayed significant activity also on hCA VII. The binding of the analogue **22d** was analyzed by means of X-ray crystallography (PDB 2ILI) on hCA II: the chlorine atom was engaged in contacts with the side-chain atoms of Ile91 and Gln92 while the triazinyl ring made π-stacking interactions with the phenyl ring of Phe131. The glycine moiety was directed toward the rim of the active site, apparently not engaged in positive interactions.

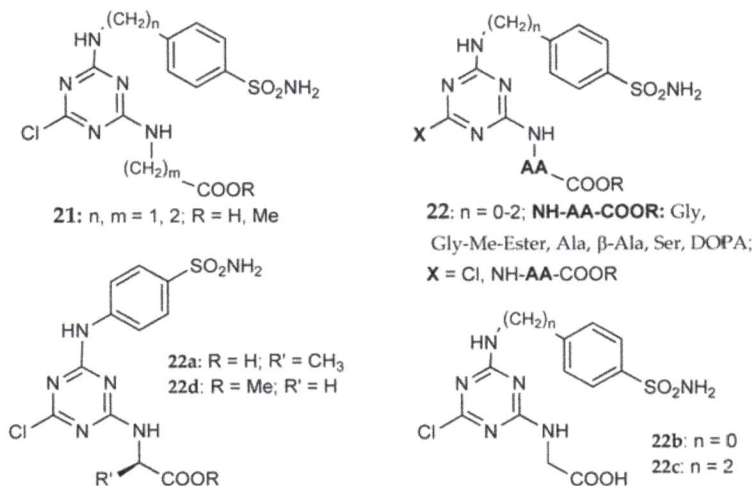

21: n, m = 1, 2; R = H, Me

22: n = 0-2; **NH-AA-COOR**: Gly, Gly-Me-Ester, Ala, β-Ala, Ser, DOPA; X = Cl, **NH-AA-COOR**

22a: R = H; R' = CH₃
22d: R = Me; R' = H

22b: n = 0
22c: n = 2

Figure 11. General structure of triazinyl derivatives **21-22**.

Amino acids were the ideal candidates for the side decoration of fullerene (C$_{60}$), an atypical potential CA inhibitor [52]. Fullerene derivatives possess some interesting properties: as an example, fullevir [53], the sodium salt of fullerene-polyhydro-polyaminocaproic acid, displays significant antiviral, antibacterial and anticancer activity [54–57]. In addition, fullerene possesses a diameter of about 1 nm, a size similar to the width of the active site entrance of most CA isozymes [58]. In order to improve the very low water solubility and to insert groups that could provide selectivity by interacting with residues of the entrance of the catalytic site, some phenylalanine derivatives were investigated as fullerene pendants (general formula **23**, Figure 12). Inhibitory activity was measured on a panel of hCA isozymes (I, II, III, IV, VA, VB, VI, VII, IX, XII, XIII, XIV, XV), finding K_i values in the micromolar range and no selectivity. Computational studies suggested that the inhibition mechanism is the occlusion of the active site through the fullerene cage; it is, therefore, possible that this bulky and extremely rigid scaffold prevents pendants from interacting with crucial amino acid residues, located in a deeper area of the enzyme. This hypothesis seems confirmed by the very low difference in activity measured for the synthesized fullerenes, including those not carrying amino acids pendants.

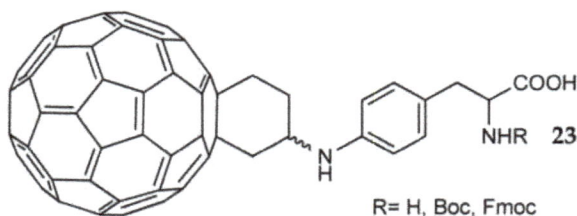

Figure 12. Fullerene-amino acid derivatives studied in [52].

A wide investigation of the "amino acid-tail approach" was performed by Küçükbay et al. They synthesised carboxy-derivatives linking a N-protected amino acid to three different types of CA inhibitors, obtaining amino-sulfonamides (**24–26**) [59], coumarins (**27, 28**), tetrahydroquinolinones **29** [60] and benzothiazoles **30, 31** [61] (Figure 13).

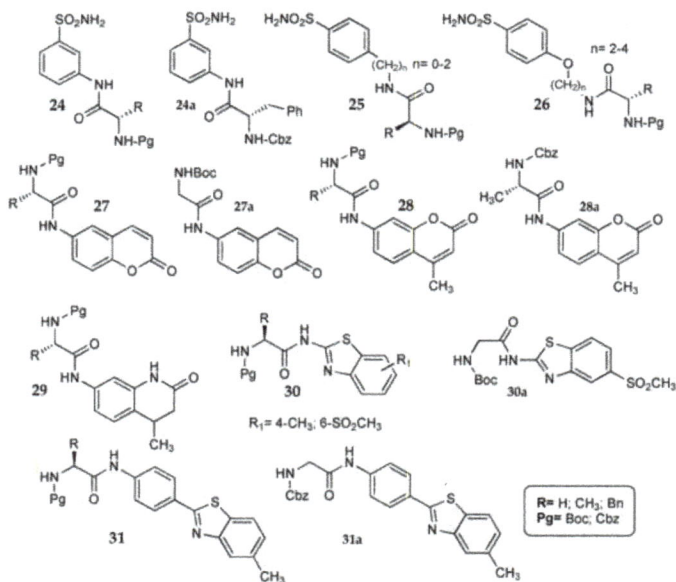

Figure 13. Structure of the compounds investigated by Küçükbay et al. [59–61].

Compounds **24–31** were tested against four CA isozymes (hCA I, II, IV and XII). The results were extremely different between the three series. Sulfonamides **24–26** were the most active and sensitive to structural variation, with K_i values mainly in the nanomolar range. By analysing the data, it is indeed possible to note appreciable differences in activity when the N-protected amino acid was varied, even if it is not simple to infer clear structure activity relationships. Few compounds were endowed with some selectivity toward CA XII: on this isoform, for instance, compound **24a** (K_i 9.5 nM) is 18, 47 and 201 times more potent than on CA IV, II and I, respectively.

Dihydroquinolinones **29** were inactive on the four isoforms, while coumarins **27** and **28** showed micromolar K_i values only against hCA IV and XII [60]. As a matter of fact, this scaffold is already known as preferential inhibitor of the tumor-associated hCA IX and XII [24,62,63]: the addition of an amino acid tail abolished activity against the off-target isozymes hCA I and II. Moreover, two derivatives (**27a** and **28a**, Figure 13) were completely inactive also against hCA IV, thus showing selectivity for hCA XII.

Benzoathiazoles **30** and **31** showed micromolar K_i values almost exclusively on the ubiquitous isoform hCA I, with only three compounds being active also on hCA II. Surprisingly, one N-Boc-glycine derivative (**30a**, Figure 13) showed activity also on hCA XII, with a K_i value nine time lower than that registered against hCA I. Another compound, **31a**, showed activity only on CA II, albeit with low potency (K_i 10 µM).

It is interesting to note that compounds **24–31** inhibit CA by different mechanisms, owing to the diverse scaffolds carrying the amino acid tail. The analysis of this research suggests that the "amino acid tail approach" could be interesting for the derivatization of different kind of CA inhibitors. The pretty simple and quick synthetic methods available to link an amino acid moiety to a generic group allow a large library of structurally-related compounds to be obtained, from which it is easy to study the effects of the substitution on activity. In addition, the hydrophilic nature of these molecules could be useful for enhancing their water-solubility.

5. Amino Acids as Linkers

The bifunctional chemical nature of the amino acids can be used to link together two different substituents, generating molecules with new characteristics. A generic sulfonamide could indeed be connected to another chemical entity through these bivalent linkers (generic formula **32** and **33**, Figure 14). The possibility to protect the amino or the carboxyl moieties, in addition to their different reactivity, allows the easy and selective functionalization of these chemical groups. This strategy could be considered an upgrade of the "amino acid tail approach", on these compounds, beyond the amino acid chain, the side group (**SG** in Figure 14) represents an additional point of variation.

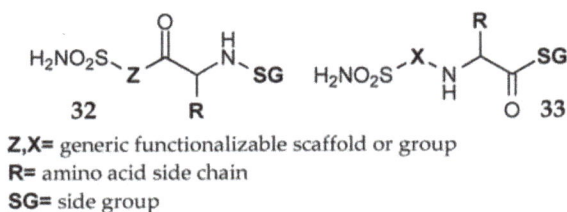

Z,X= generic functionalizable scaffold or group
R= amino acid side chain
SG= side group

Figure 14. General structure of sulfonamide CAIs having an amino acid as linker.

A first example is the previously cited work of Barboiu et al. [39], where compound **3a** (Figure 4) was further functionalized on the free NH_2 with various pendants, affording a wider series of 5-substituted thiadiazole-2-sulfonamides (general formula **34**, Figure 15). In order to obtain derivatives with different physico-chemical properties, compound **3a** (Figure 4) was reacted with 30 different electrophilic reagents (i.e., alkyl/aryl sulfonyl chlorides or fluorides, sulfonic acid cyclic anhydrides or acyl chlorides) and the corresponding products were then tested against isozymes hCA I, II and bCA IV. This series is characterized by a potent inhibitory activity on hCA II, most of the measured K_i values being in the low nanomolar range on this isoform, but in the micromolar range against hCA I and bCA IV. We can cite derivative **34a** as an example: the Barboiu group determined a K_i of 0.75 nM on hCA II while the values measured against hCA I and bCA were, respectively, 125 and 100 nM. There is indeed a strong selectivity for the human isoform II confirming the evidence that this scaffold could be a promising candidate for the development of selective hCA II inhibitors.

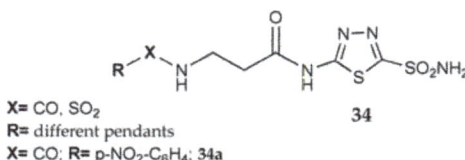

X= CO, SO$_2$
R= different pendants
X= CO; R= p-NO$_2$-C$_6$H$_4$: **34a**

34

Figure 15. Acetazolamide derivatives studied by Barboiu et al.

Kolb et al. patented a series of radioactively labelled sulfonamides [64], with high affinity for hCA IX, that could be potentially used as positron emission tomography (PET) tracers for imaging the hCA IX expression on the surface of cancer cells. The group identified a benzothiazole or *p*-substituted phenyl ring as ideal aromatic scaffolds to carry the sulfonamidic moiety and used click-chemistry to functionalize the inhibitor portion with different groups; a ^{18}F atom was inserted on this tail as radionuclide. They designed four series of compounds and two of them (general formula **35** and **36**, Figure 16) were prepared using an amino acid derivative (azido acid) as building block for the construction of the triazole central ring. In addition, the use of a second amino acid as radionuclide-carrying portion was investigated. With such a strategy, these researchers were able to obtain a large library of compounds, with different chemical characteristics and many possible points of structural variation. Furthermore, the use of L or D amino and azido acids allowed the stereochemical control of the side chain and, consequently, its orientation toward the regions of the enzyme. One of the most promising compound, VM-4037A (**35a**), was tested in tumor xenograft models expressing hCA IX [65]. In addition, its biodistribution and radiation dose were determined in healthy human volunteers [66]; the high uptake of **35a** in liver and kidney indicated that the compound was not suitable for the detection of hCA IX in these tissues.

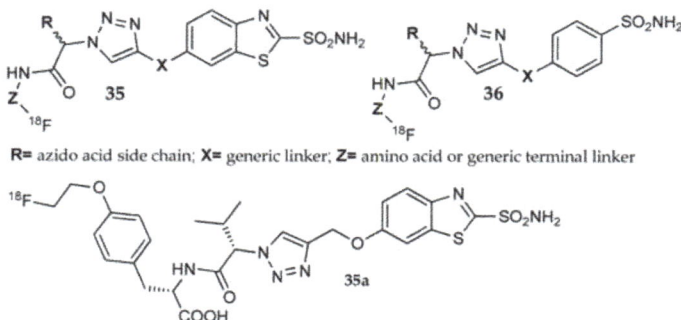

R= azido acid side chain; X= generic linker; Z= amino acid or generic terminal linker

Figure 16. Sulfonamide derivatives carrying a triazole-amino acid as linker.

A quite different approach was used by the Matulis group [67,68] to derive *p*-amino benzenesulfonamide. By reacting the *p*-amino group with acrylic acid, Rutkauskas et al. obtained a series of compounds where a β-alanyl moiety is directly linked to the aromatic scaffold (general formula **37** and **38**, Figure 17). In this instance, the amino acid linker was generated in situ and then functionalized on the free β-carboxy moiety. In addition, the versatile nature of this group allowed its transformation into heterocycles, such as triazoles, oxadiazoles or thiadiazcles [67]. Only in a few instances the activity of this first series of compounds (K_d values determined by a fluorescent thermal shift assay on CA I, II, VI, VII, XII and XIII) was below 1 μM. The functiorualization of the alanyl nitrogen and substitution on the benzenesulfonamide ring were also investigated, adding other possible points of variation [68]. As one can imagine, this large number of different products showed a wide range of activities, with K_d values on CA I, II, XII and XIII from nanomolar to micromolar. The selectivity of some compounds was further tested on a larger panel of hCAs, with some interesting findings. For instance, **39** (Figure 17) showed good potency on CA VB and CA IX (K_d = 5 and 43 nM, respectively) while being almost inactive on the other isoforms. The crystallographic studies performed on **37a** and hCA II allowed the way the flexible β-alanyl linker was folded in the enzyme cavity (PDB code 4Q6E) to be understood.

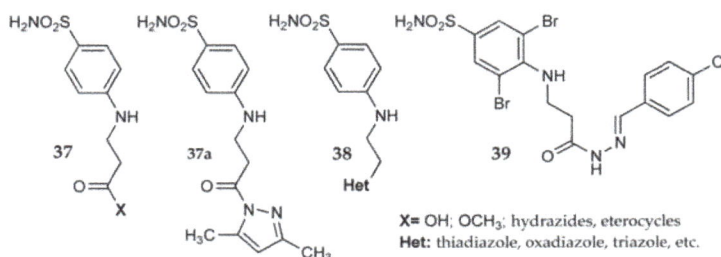

Figure 17. Structure of the β-alanine derivatives studied by the Matulis group.

Ceruso et al. [69] extended the chain using a γ-aminobutiric acid moiety as linker between two different benzenesulfonamides and various aromatic pendants. Compared to **37**, the amino acid orientation was inverted, leaving a free amino group that was reacted with aryl isocyanates. The molecules obtained (general formula **40**, Figure 18) were tested against 13 hCA isoforms, showing good inhibition potency against hCA I, II, VII and XII. This serie lacks isoform selectivity, but a deeper investigation on the effects of the chain length and of the chemical nature of the terminal aromatic pendants could potentially lead to more interesting derivatives.

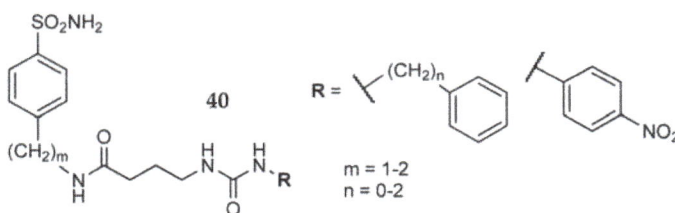

Figure 18. Benzenesulfonamide derivatives incorporating a GABA moiety.

Moeker et al. [70] studied an innovative approach using a sulfamate as zinc binding group (ZBG). These researchers designed a series of glycosylated compounds with poor membrane permeability that could potentially lead to hCA IX selective inhibitors, owing to the extracellular localization of this enzyme. The general structure of these derivatives (**41**, Figure 19) is characterized by a free or acetylated glucose moiety, linked to a panel of primary and secondary amino alcohols through a sulfonamide bridge. Two hydroxylated amino acid (L-serine methyl ester and L-4-(R)-hydroxyproline methyl ester) were investigated as hydroxylated linkers and transformed into the corresponding sulfamate (**41a** and **41b**). In such a way, the ZBG is directly placed on the chiral amino acid side chain. All the synthesized molecules showed nanomolar K_1 values against the tumor-associated isoforms hCA IX and XII [5,13], and lower potency against the off-target isozymes hCA I and II. Particularly interesting is the proline-derivative **41a**, with Ki values of 2 and 1 nanomolar respectively on hCA IX and XII and more than 350 times lower potency on hCA I and II. This innovative method is interesting not only from a synthetic point of view, but also because it allows the ZBG to be kept fixed and functionalizes both the amino and carboxy portions of the amino acid with different residues.

These examples demonstrate that a chiral amino acid can serve either as building block either as linker between side pendants.

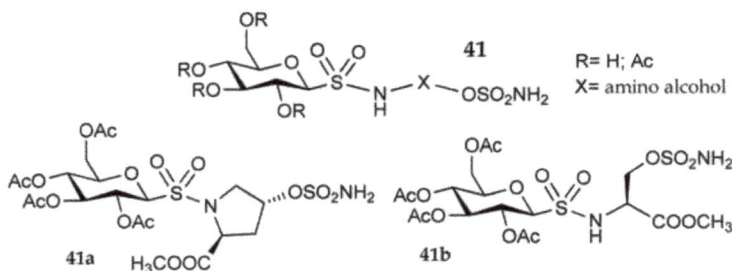

Figure 19. Sulfamate derivatives studied by Moeker et al. [70].

6. Dual Carbonic Anhydrase (CA) and Matrix Metalloproteinase (MMP) Inhibitors

Matrix metalloproteinases (MMPs) are zinc-containing endopeptidases, capable of degrading the extracellular matrix proteins; up-regulation of specific isoforms is associated with various pathologies, including some metastatic cancers [71–73]. These enzymes are potently inhibited by hydroxamates, but also other chemical groups are known to be effective Zn-chelating agents [74]. Since hCA IX and XII are overexpressed in tumors [5,13], the simultaneous inhibition of both enzymes should give a synergic effect.

On these basis, dual inhibitors were designed [75,76], i.e., compounds able to interact with the binding regions of both metalloenzymes. Some amino acids were investigated as the structural backbone of the new dual inhibitors; the derivatization of the carboxy and amino moieties gave sulfonylated amino acid hydroxamates and N-alkyloxy amino acid hydroxamates (general formulas **42–44**, Figure 20). Compounds carrying both a hydroxamic and a protic-sulfonyl groups are, therefore, provided with two different chelating moieties that could potentially interact with both metalloenzymes. Compounds **42** were strong dual inhibitors with Ki values in the low-medium nanomolar range, their inhibitory properties against the two enzymes depending on some structural features [75]. For instance, compounds **42** carrying a NH moiety (**X** = H) were more active on CA (isoform I, II and IV) than on MMPs (isoforms 1, 2, 8 and 9). When X was a benzyl group, a potent activity was found on both enzymes, while a substituent on the benzyl moiety almost abolished activity on CAs. Unfortunately, these substances were not tested on the tumor-associated CA isoforms (IX and XII).

Compounds **43** and **44** were in general less potent on both hCAs and MMPs, probably due to the complete absence of protic sulfonyl moiety that could behave as ZBG [76].

These studies confirm that amino acids, by showing good synthetic properties, are interesting synthons also for the design of hybrid inhibitors, where an important need is the possibility to easily modify many positions with different kind of substituents.

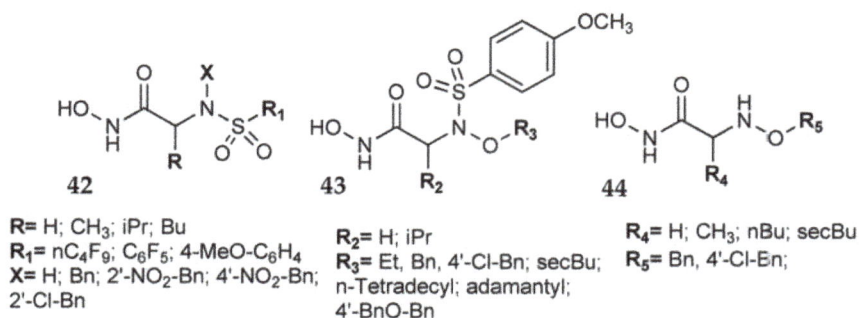

Figure 20. Dual Matrix Metalloproteinase–Carbonic Anhydrase (MMP–CA) inhibitors.

7. Other Approaches

Especially from a synthetic point of view, another interesting approach to the design of novel CAIs is the one reported in a paper by Korkmaz N. et al. [77]. These researchers, aiming to investigate innovative inhibitors, synthesized a series of thiourea derivatives lacking a classic Z3G. As we can note from general structures **45** and **46** in Figure 21, there are not pharmacophoric moieties that could be immediately associated with an inhibitory activity on the enzyme. Nevertheless, micromolar K_i values were measured for these series of compounds against hCA I and II. By analysing the structures, we can observe that lipophilic amino acids were used as linker between two aromatic portions. For **46**, the amino acid carboxy group reacted with *o*-phenylendiamine giving a benzimidazole ring. Therefore, compounds **46** can be considered as amino acid derivatives only from the synthetic point of view, since the carboxy group is lost with the formation of the heterocyclic ring. Compounds **45** and **46** were not extremely potent, but these synthetic pathways could be interesting for the side derivatization of sulfonamidic molecules.

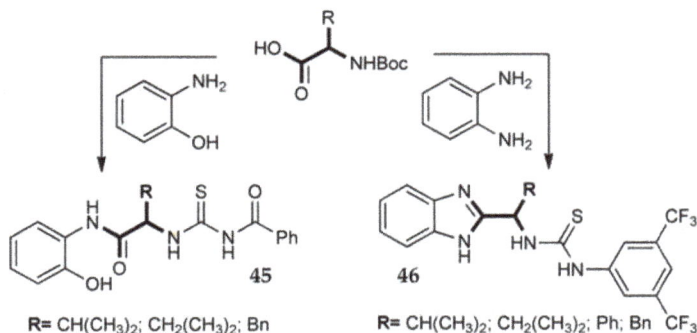

R= CH(CH3)2; CH2(CH3)2; Bn   R= CH(CH3)2; CH2(CH3)2; Ph; Bn CF3

Figure 21. Thiourea CAIs built from amino acids.

Another scaffold lacking a classic chelating moiety is Probenecid, a drug used to treat gout and hyperuricemia. Probenecid, probably thanks to the free carboxy group that coordinates the zinc ion, possesses a weak inhibitory activity against CAs. Mollica et al. [78] investigated the effects of adding different L-amino acid, obtaining probenecid-based amide derivatives (general formula **47**, Figure 22). This modification shifted the carboxy group away from the aromatic ring and introduced a chiral linker. With the aim of studying the impact of the chemical modification of this terminal COOH, these researchers also synthesized the corresponding primary amides. All the compounds showed an interesting profile, their inhibitory activity against hCA IX and XII being higher than against the off-target isozymes hCA I and II. The transformation of the COOH group into $CONH_2$ was also interesting, resulting in a remarkably lower affinity against hCA II. As an example, both the methionine derivatives **47a** and **47b** were very active against the two tumor-associated isoforms, but while the carboxy derivatives possess a low residual activity against hCA I and II, the corresponding amide is completely inactive toward these isoforms. This paper provides a useful overview of the application of the "amino acid tail approach" to an atypical zinc-chelating molecule, confirming the synthetic and pharmacological benefits that could be obtained from this method.

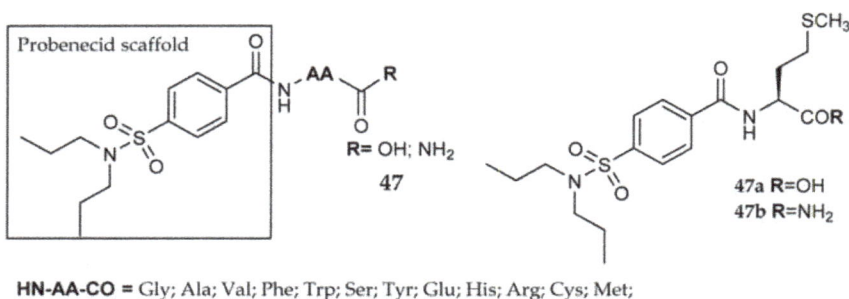

HN-AA-CO = Gly; Ala; Val; Phe; Trp; Ser; Tyr; Glu; His; Arg; Cys; Met;

Figure 22. Probenecid amide derivatives.

In this section we report also the paper published by Fidan et al. [79], in which four aromatic sulfonamides were directly functionalized on the amidic NH_2 group with glycine and phenylalanine (general formula **48**, Figure 23). The derivatization of the zinc binding group with substituents able to interact with the amino acids of the catalytic site was aimed at obtaining an enhancement in the activity of these secondary sulfonamides. However, it must be recalled that the architecture of the active site is very well conserved among the different CA isoforms, so this modification may not confer selectivity to such derivatives. In general, the new compounds did not show a better inhibition profile against hCA I and II as compared to their primary sulfonamide precursors. A substitution directly on the ZBG, probably due to steric hindrance, seems not to be a convenient way to enhance the potency of this kind of sulfonamides. Despite these negative results, this approach should be further investigated, maybe with longer linkers between the sulfonamidic nitrogen and the pendants.

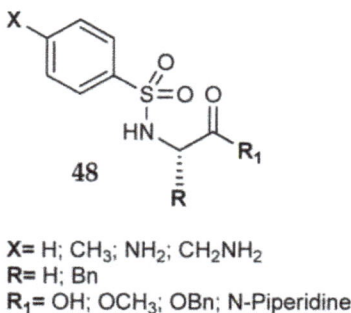

X= H; CH_3; NH_2; CH_2NH_2
R= H; Bn
R_1= OH; OCH_3; OBn; N-Piperidine

Figure 23. Glycine and phenylalanine N-sulfonamides studied by Fidan et al. [79].

Chiral pool synthesis was used by Chiaramonte et al. in order to prepare enantiopure CA inhibitors [23]. Starting from L- or D-phenylalanine, both enantiomers of two series of 2-benzylpiperazines (**49** and **50**, Figure 24), carrying a sulfamoyl-benzoic moiety as ZBG, were obtained.

The amino acid was directly used to build the piperazine scaffold, allowing the stereochemical control of the side benzyl group. With this method, it was possible to take advantage of the versatile synthetic properties of the synthon: the reduction of the intermediate diketopiperazine **51** converted the amino acid into an alkyl derivative without affecting the absolute configuration. The piperazines were further decorated on one nitrogen atom with the ZBG and with different alkyl/acyl/sulfonyl groups on the other one, with the aim to look for selectivity toward the CA isoforms. All the synthesized products were able to inhibit four pharmacological-relevant CA isozymes (hCA I, II, IV and IX), showing K_i values ranging from low nanomolar to micromolar; hCA IX was the least sensitive isoform. The hydrochloric salts of two basic piperazines (**49a** and **50a**, Figure 24), showing very good water solubility, were then selected for in-vivo tests in a rabbit model of glaucoma. Both derivatives were able to reduce IOP with a potency and efficacy similar to the reference drug dorzolamide, confirming that this series is promising for the development of new anti-glaucoma agents. The binding mode was further investigated with X-ray analysis and the structure of the complex of compound **49b**, bound in the hCA I active site, was solved (PDB code 6EVR). In the image, reported in Figure 1b (Section 1), it is possible to appreciate the interaction between the sulfonamidic group and the Zn^{2+}, while the benzyl group is deeply inserted in the lipophilic cavity of the enzyme, where it establishes Van der Waals interactions with some hydrophobic amino acid residues. This synthetic method allowed the building of a chiral piperazine scaffold, incorporating the aminoacid-skeleton. Starting from different amino acids, it should be possible to modify the nature of this substituent and at the same time to control its stereochemistry.

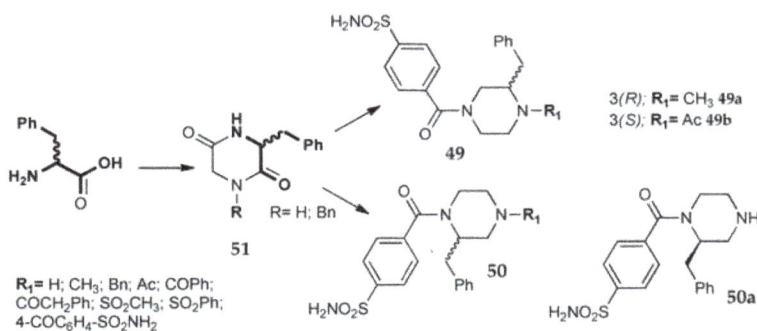

Figure 24. Piperazine derivatives studied by Chiaramonte et al. [23].

8. Conclusions

Amino acids are useful synthons. Thanks to their characteristics, there are several potential applications in chemical synthesis. The possibility of using both proteinogenic and non-proteinogenic amino acids allow the preparation of many related derivatives, where the different side chains modulate the properties. The research of new CAIs or the structural modification of existing inhibitors is a dynamic field with many perspectives but also a lot of still unsolved challenges, the most important being the lack of selectivity of sulfonamides. The use of amino acids, or their derivatives, could sometimes provide a smart solution to both synthetic and pharmacological problems.

Acknowledgments: This work was supported by grants from the University of Florence (Fondo Ricerca Ateneo RICATEN17).

Conflicts of Interest: The authors declare no conflict of interest, except C.T. Supuran who declares a conflict of interest, being co-author of many patents describing CAIs.

Abbreviations

CA	carbonic anhydrase
hCA	human carbonic anhydrase
bCA	bovine carbonic anhydrase
CAIs	carbonic anhydrases inhibitors
MMPs	matrix metalloproteinases
IOP	intraocular pressure
Ala	alanine
Asn	asparagine
Asp	aspartic acid
Arg	arginine
β-Ala	beta-alanine
β-PhSer	beta-phenylserine
Boc	tert-butyloxycarbonyl
Cys	cysteine
DOPA	(3′,4′-dihydroxy)phenylalanine
Fmoc	fluorenylmethyloxycarbonyl
GABA	γ-aminobutyric acid
Gly	glycine
Gln	glutamine
Glu	glutamic acid
Ile	isoleucine
K_{cat}	turnover number
K_d	dissociation constant
K_i	inhibition constant
Leu	leucine
Lys	lysine
Met	Methionine
PDB	protein data bank
PET	positron emission tomography
Phe	phenylalanine
PhGly	phenylglycine
Pro	proline
Ser	serine
Thr	threonine
Trp	tryptophan
Tyr	tyrosine
Val	valine
ZBG	zinc binding group

References

1. Kupriyanova, E.; Pronina, N.; Los, D. Carbonic anhydrase—A universal enzyme of the carbon-based life. *Photosynthetica* **2017**, *55*, 3–19. [CrossRef]
2. Supuran, C.T. Structure and function of carbonic anhydrases. *Biochem. J.* **2016**, *473*, 2023–2032. [CrossRef] [PubMed]
3. Supuran, C.T.; De Simone, G. Carbonic Anhydrases: An Overview. In *Carbonic Anhydrases as Biocatalysts: From Theory to Medical and Industrial Applications*; Elsevier: Amsterdam, Netherlands, 2015; pp. 3–13.
4. D'Ambrosio, K.; De Simone, G.; Supuran, C.T. Human Carbonic Anhydrases: Catalytic Properties, Structural Features, and Tissue Distribution. In *Carbonic Anhydrases as Biocatalysts: From Theory to Medical and Industrial Applications*; Elsevier: Amsterdam, Netherlands, 2015; pp. 17–30.
5. Supuran, C.T. Carbonic anhydrases: Novel therapeutic applications for inhibitors and activators. *Nat. Rev. Drug Discov.* **2008**, *7*, 168–181. [CrossRef] [PubMed]

6. Supuran, C. Carbonic Anhydrases and Metabolism. *Metabolites* **2018**, *8*, 25. [CrossRef] [PubMed]
7. Smith, K.S.; Jakubzick, C.; Whittam, T.S.; Ferry, J.G. Carbonic anhydrase is an ancient enzyme widespread in prokaryotes. *Proc. Natl. Acad. Sci. USA* **1999**, *96*, 15184–15189. [CrossRef] [PubMed]
8. Khalifah, R.G. The Carbon Dioxide Hydration Activity of Carbonic Anhydrase. *J. Biol. Chem.* **1971**, *246*, 2561–2573. [PubMed]
9. Lindskog, S. Structure and mechanism of carbonic anhydrase. *Pharmacol. Ther.* **1997**, *74*, 1–20. [CrossRef]
10. Silverman, D.N.; Lindskog, S. The Catalytic Mechanism of Carbonic Anhydrase: Implications of a Rate-Limiting Protolysis of Water. *Acc. Chem. Res.* **1988**, *21*, 30–36. [CrossRef]
11. Alterio, V.; Di Fiore, A.; D'Ambrosio, K.; Supuran, C.T.; De Simone, G. Multiple binding modes of inhibitors to carbonic anhydrases: How to design specific drugs targeting 15 different isoforms? *Chem. Rev.* **2012**, *112*, 4421–4468. [CrossRef] [PubMed]
12. Masini, E.; Carta, F.; Scozzafava, A.; Supuran, C.T. Antiglaucoma carbonic anhydrase inhibitors: A patent review. *Expert Opin. Ther. Pat.* **2013**, *23*, 705–716. [CrossRef] [PubMed]
13. Neri, D.; Supuran, C.T. Interfering with pH regulation in tumours as a therapeutic strategy. *Nat. Rev. Drug Discov.* **2011**, *10*, 767–777. [CrossRef] [PubMed]
14. Supuran, C.T. Carbonic anhydrase inhibition and the management of hypoxic tumors. *Metabolites* **2017**, *7*, 48. [CrossRef] [PubMed]
15. Singh, S.; Lomelino, C.L.; Mboge, M.Y.; Frost, S.C. Cancer Drug Development of Carbonic Anhydrase Inhibitors beyond the Active Site. *Molecules* **2018**, *23*, 1045. [CrossRef] [PubMed]
16. Mboge, M.; Mahon, B.; McKenna, R.; Frost, S. Carbonic Anhydrases: Role in pH Control and Cancer. *Metabolites* **2018**, *8*, 19. [CrossRef] [PubMed]
17. Supuran, C.T. Carbonic anhydrase activators. *Future Med. Chem.* **2018**, *10*, 561–573. [CrossRef] [PubMed]
18. Carta, F.; Supuran, C.T. Diuretics with carbonic anhydrase inhibitory action: A patent and literature review (2005–2013). *Expert Opin. Ther. Pat.* **2013**, *23*, 681–691. [CrossRef] [PubMed]
19. Monti, S.M.; Supuran, C.T.; De Simone, G. Anticancer carbonic anhydrase inhibitors: A patent review (2008–2013). *Expert Opin. Ther. Pat.* **2013**, *23*, 737–749. [CrossRef] [PubMed]
20. Scozzafava, A.; Supuran, C.T.; Carta, F. Antiobesity carbonic anhydrase inhibitors: A literature and patent review. *Expert Opin. Ther. Pat.* **2013**, *23*, 725–735. [CrossRef] [PubMed]
21. Supuran, C.T. Advances in structure-based drug discovery of carbonic anhydrase inhibitors. *Expert Opin. Drug Discov.* **2017**, *12*, 61–88. [CrossRef] [PubMed]
22. De Simone, G.; Alterio, V.; Supuran, C.T. Exploiting the hydrophobic and hydrophilic binding sites for designing carbonic anhydrase inhibitors. *Expert Opin. Drug Discov.* **2013**, *8*, 793–810. [CrossRef] [PubMed]
23. Chiaramonte, N.; Bua, S.; Ferraroni, M.; Nocentini, A.; Bonardi, A.; Bartolucci, G.; Durante, M.; Lucarini, L.; Chiapponi, D.; Dei, S.; et al. 2-Benzylpiperazine: A new scaffold for potent human carbonic anhydrase inhibitors. Synthesis, enzyme inhibition, enantioselectivity, computational and crystallographic studies and in vivo activity for a new class of intraocular pressure lowering agents. *Eur. J. Med. Chem.* **2018**, *151*, 363–375. [CrossRef] [PubMed]
24. Supuran, C.T. How many carbonic anhydrase inhibition mechanisms exist? *J. Enzym. Inhib. Med. Chem.* **2016**, *31*, 345–360. [CrossRef] [PubMed]
25. Supuran, C.T. Structure-based drug discovery of carbonic anhydrase inhibitors. *J. Enzym. Inhib. Med. Chem.* **2012**, *27*, 759–772. [CrossRef] [PubMed]
26. Karioti, A.; Carta, F.; Supuran, C.T. Phenols and polyphenols as Carbonic anhydrase inhibitors. *Molecules* **2016**, *21*, 1649. [CrossRef] [PubMed]
27. Maresca, A.; Temperini, C.; Pochet, L.; Masereel, B.; Scozzafava, A.; Supuran, C.T. Deciphering the mechanism of carbonic anhydrase inhibition with coumarins and thiocoumarins. *J. Med. Chem.* **2010**, *53*, 335–344. [CrossRef] [PubMed]
28. D'Ambrosio, K.; Carradori, S.; Monti, S.M.; Buonanno, M.; Secci, D.; Vullo, D.; Supuran, C.T.; De Simone, G. Out of the active site binding pocket for carbonic anhydrase inhibitors. *Chem. Commun.* **2015**, *51*, 302–305. [CrossRef] [PubMed]
29. Bischoff, R.; Schlüter, H. Amino acids: Chemistry, functionality and selected non-enzymatic post-translational modifications. *J. Proteomics* **2012**, *75*, 2275–2296. [CrossRef] [PubMed]
30. Jaradat, D.M.M. Thirteen decades of peptide synthesis: Key developments in solid phase peptide synthesis and amide bond formation utilized in peptide ligation. *Amino Acids* **2018**, *50*, 39–68. [CrossRef] [PubMed]

31. Schmidl, D.; Schmetterer, L.; Garhöfer, G.; Popa-Cherecheanu, A. Pharmacotherapy of Glaucoma. *J. Ocul. Pharmacol. Ther.* **2015**, *31*, 63–77. [CrossRef] [PubMed]
32. Sugrue, M.F. Pharmacological and ocular hypotensive properties of topical carbonic anhydrase inhibitors. *Prog. Retin. Eye Res.* **2000**, *19*, 87–112. [CrossRef]
33. Supuran, C.T.; Casini, A.; Scozzafava, A. Development of Sulfonamide Carbonic Anhydrase Inhibitors. In *Carbonic Anhydrase*; CRC Press: Boca Raton, FL, USA, 2004; pp. 67–147.
34. Maren, T.H. Carbonic anhydrase: Chemistry, physiology, and inhibition. *Physiol. Rev.* **1967**, *47*, 595–781. [CrossRef] [PubMed]
35. Casini, A.; Scozzafava, A.; Mincione, F.; Menabuoni, L.; Ilies, M.A.; Supuran, C.T. Carbonic anhydrase inhibitors: Water-soluble 4-sulfamoylphenylthioureas as topical intraocular pressure-lowering agents with long-lasting effects. *J. Med. Chem.* **2000**, *43*, 4884–4892. [CrossRef] [PubMed]
36. Scozzafava, A.; Briganti, F.; Mincione, G.; Menabuoni, L.; Mincione, F.; Supuran, C.T. Carbonic anhydrase inhibitors: Synthesis of water-soluble, aminoacyl/dipeptidyl sulfonamides possessing long-lasting intraocular pressure-lowering properties via the topical route. *J. Med. Chem.* **1999**, *42*, 3690–3700. [CrossRef] [PubMed]
37. Antonaroli, S.; Bianco, A.; Brufani, M.; Cellai, L.; Baido, G.L.; Potier, E.; Bonomi, L.; Perfetti, S.; Fiaschi, A.I.; Segre, G. Acetazolamide-like Carbonic Anhydrase Inhibitors with Topical Ocular Hypotensive Activity. *J. Med. Chem.* **1992**, *35*, 2697–2703. [CrossRef] [PubMed]
38. Jayaweera, G.D.S.A.; MacNeil, S.A.; Trager, S.F.; Blackburn, G.M. Synthesis of 2-substituted-1,3,4-thiadiazole-5-sulphonamides as novel water-soluble inhibitors of carbonic anhydrase. *Bioorg. Med. Chem. Lett.* **1991**, *1*, 407–410. [CrossRef]
39. Barboiu, M.; Supuran, C.T.; Menabuoni, L.; Scozzafava, A.; Mincione, F.; Briganti, F.; Mincione, G. Carbonic anhydrase inhibitors. Synthesis of topically effective intraocular pressure lowering agents derived from 5-(omega-aminoalkylcarboxamido)-1,3,4-thiadiazole-2-sulfonamide. *J. Enzym. Inhib.* **1999**, *15*, 23–46. [PubMed]
40. Supuran, C.T.; Briganti, F.; Menabuoni, L.; Mincione, G.; Mincione, F.; Scozzafava, A. Carbonic anhydrase inhibitors—Part 78. Synthesis of water-soluble sulfonamides incorporating β-alanyl moieties, possessing long lasting- intraocular pressure lowering properties via the topical route. *Eur. J. Med. Chem.* **2000**, *35*, 309–321. [CrossRef]
41. Maren, T.H.; Brechue, W.F.; Bar-Ilan, A. Relations among IOP reduction, ocular disposition and pharmacology of the carbonic anhydrase inhibitor ethoxzolamide. *Exp. Eye Res.* **1992**, *55*, 73–79. [CrossRef]
42. Brechue, W.F.; Maren, T.H. pH and drug ionization affects ocular pressure lowering of topical carbonic anhydrase inhibitors. *Investig. Ophthalmol. Vis. Sci.* **1993**, *34*, 2581–2587.
43. Ceruso, M.; Del Prete, S.; Alothman, Z.; Osman, S.M.; Scozzafava, A.; Capasso, C.; Supuran, C.T. Synthesis of sulfonamides with effective inhibitory action against Porphyromonas gingivalis γ-carbonic anhydrase. *Bioorg. Med. Chem. Lett.* **2014**, *24*, 4006–4010. [CrossRef] [PubMed]
44. Ceruso, M.; Bragagni, M.; Alothman, Z.; Osman, S.M.; Supuran, C.T. New series of sulfonamides containing amino acid moiety act as effective and selective inhibitors of tumor-associated carbonic anhydrase XII. *J. Enzym. Inhib. Med. Chem.* **2015**, *30*, 430–434. [CrossRef] [PubMed]
45. Casin, A.; Scozzafava, A.; Mincione, F.; Menabuoni, L.; Supuran, C.T. Carbonic anhydrase inhibitors: Synthesis of water soluble sulfonamides incorporating a 4-sulfamoylphenylmethylthiourea scaffold, with potent intraocular pressure lowering properties. *J. Enzym. Inhib. Med. Chem.* **2002**, *17*, 333–343. [CrossRef] [PubMed]
46. Mincione, F.; Starnotti, M.; Menabuoni, L.; Scozzafava, A.; Casini, A.; Supuran, C.T. Carbonic anhydrase inhibitors: 4-Sulfamoyl-benzenecarboxamides and 4-chloro-3-sulfamoyl-benzenecarboxamides with strong topical antiglaucoma properties. *Bioorg. Med. Chem. Lett.* **2001**, *11*, 1787–1791. [CrossRef]
47. Boriack, P.A.; Christianson, D.W.; Kingery-Wood, J.; Whitesides, G.M. Secondary Interactions Significantly Removed from the Sulfonamide Binding Pocket of Carbonic Anhydrase II Influence Inhibitor Binding Constants. *J. Med. Chem.* **1995**, *38*, 2286–2291. [CrossRef] [PubMed]
48. Jain, A.; Huang, S.G.; Whitesides, G.M. Lack of Effect of the Length of Oligoglycine- and Oligo(ethylene glycol)-Derived para-Substituents on the Affinity of Benzenesulfonamides for Carbonic Anhydrase II in Solution. *J. Am. Chem. Soc.* **1994**, *116*, 5057–5062. [CrossRef]

49. Jain, A.; Whitesides, G.M.; Alexander, R.S.; Christianson, D.W. Identification of Two Hydrophobic Patches in the Active-Site Cavity of Human Carbonic Anhydrase II by Solution-Phase and Solid-State Studies and Their Use in the Development of Tight-Binding Inhibitors. *J. Med. Chem.* **1994**, *37*, 2100–2105. [CrossRef] [PubMed]

50. Garaj, V.; Puccetti, L.; Fasolis, G.; Winum, J.Y.; Montero, J.L.; Scozzafava, A.; Vullo, D.; Innocenti, A.; Supuran, C.T. Carbonic anhydrase inhibitors: Novel sulfonamides incorporating 1,3,5-triazine moieties as inhibitors of the cytosolic and tumour-associated carbonic anhydrase isozymes I, II and IX. *Bioorg. Med. Chem. Lett.* **2005**, *15*, 3102–3108. [CrossRef] [PubMed]

51. Carta, F.; Garaj, V.; Maresca, A.; Wagner, J.; Avvaru, B.S.; Robbins, A.H.; Scozzafava, A.; McKenna, R.; Supuran, C.T. Sulfonamides incorporating 1,3,5-triazine moieties selectively and potently inhibit carbonic anhydrase transmembrane isoforms IX, XII and XIV over cytosolic isoforms I and II: Solution and X-ray crystallographic studies. *Bioorg. Med. Chem.* **2011**, *19*, 3105–3119. [CrossRef] [PubMed]

52. Innocenti, A.; Durdagi, S.; Doostdar, N.; Amanda Strom, T.; Barron, A.R.; Supuran, C.T. Nanoscale enzyme inhibitors: Fullerenes inhibit carbonic anhydrase by occluding the active site entrance. *Bioorg. Med. Chem.* **2010**, *18*, 2822–2828. [CrossRef] [PubMed]

53. Nosik, D.N.; Lyalina, I.K.; Kalnlna, L.B.; Lobach, O.A.; Chataeva, M.S.; Rasnetsov, L.D. The antiretroviral agent Fullevir. *Vopr. Virusol.* **2009**, *54*, 41–43. [PubMed]

54. Sijbesma, R.; Srdanov, G.; Wudl, F.; Castoro, J.A.; Wilkins, C.; Friedman, S.H.; DeCamp, D.L.; Kenyon, G.L. Synthesis of a Fullerene Derivative for the Inhibition of HIV Enzymes. *J. Am. Chem. Soc.* **1993**, *115*, 6510–6512. [CrossRef]

55. Kang, B.; Yu, D.; Dai, Y.; Chang, S.; Chen, D.; Ding, Y. Cancer-cell targeting and photoacoustic therapy using carbon nanotubes as "bomb" agents. *Small* **2009**, *5*, 1292–1301. [CrossRef] [PubMed]

56. Horie, M.; Fukuhara, A.; Saito, Y.; Yoshida, Y.; Sato, H.; Ohi, H.; Obata, M.; Mikata, Y.; Yano, S.; Niki, E. Antioxidant action of sugar-pendant C_{60} fullerenes. *Bioorg. Med. Chem. Lett.* **2009**, *19*, 5902–5904. [CrossRef] [PubMed]

57. Chaudhuri, P.; Paraskar, A.; Soni, S.; Mashelkar, R.A.; Sengupta, S. Fullerenol-cytotoxic conjugates for cancer chemotherapy. *ACS Nano* **2009**, *3*, 2505–2514. [CrossRef] [PubMed]

58. Alterio, V.; Di Flore, A.; D'Ambrosio, K.; Supuran, C.T.; De Simone, G. *X-ray Crystallography of Carbonic Anhydrase Inhibitors and Its Importance in Drug Design*; Wiley: Hoboken, NJ, USA, 2009; pp. 73–138.

59. Küçükbay, F.Z.; Küçükbay, H.; Tanc, M.; Supuran, C.T. Synthesis and carbonic anhydrase I, II, IV and XII inhibitory properties of N-protected amino acid-sulfonamide conjugates. *J. Enzym. Inhib. Med. Chem.* **2016**, *31*, 1476–1483. [CrossRef] [PubMed]

60. Küçükbay, F.Z.; Küçükbay, H.; Tanc, M.; Supuran, C.T. Synthesis and carbonic anhydrase inhibitory properties of amino acid–coumarin/quinolinone conjugates incorporating glycine, alanine and phenylalanine moieties. *J. Enzym. Inhib. Med. Chem.* **2016**, *31*, 1198–1202. [CrossRef] [PubMed]

61. Küçükbay, F.Z.; Buğday, N.; Küçükbay, H.; Tanc, M.; Supuran, C.T. Synthesis, characterization and carbonic anhydrase inhibitory activity of novel benzothiazole derivatives. *J. Enzym. Inhib. Med. Chem.* **2016**, *31*, 1221–1225. [CrossRef] [PubMed]

62. Lomelino, C.L.; Supuran, C.T.; McKenna, R. Non-classical inhibition of carbonic anhydrase. *Int. J. Mol. Sci.* **2016**, *17*, 1150. [CrossRef] [PubMed]

63. Žalubovskis, R. In a search for selective inhibitors of carbonic anhydrases: Coumarin and its bioisosteres—Synthesis and derivatization. *Chem. Heterocycl. Compd.* **2015**, *51*, 607–612. [CrossRef]

64. Kolb, H.C.; Walsh, J.C.; Kasi, D.; Mocharla, V.; Wang, B.; Gangadharmath, U.B.; Duclos, B.A.; Chen, K.; Zhang, W.; Chen, G.; et al. Development of Triazole Derivatives as Molecular Imaging Probes for Carbonic Anhydrase-IX Using Click Chemistry. PCT International Application EP20080745260, 7 April 2008. 188p.

65. Peeters, S.G.J.A.; Dubois, L.; Lieuwes, N.G.; Laan, D.; Mooijer, M.; Schuit, R.C.; Vullo, D.; Supuran, C.T.; Eriksson, J.; Windhorst, A.D.; et al. [^{18}F]VM$_{4-037}$ MicroPET Imaging and Biodistribution of Two In Vivo CAIX-Expressing Tumor Models. *Mol. Imaging Biol.* **2015**, *17*, 615–619. [CrossRef] [PubMed]

66. Doss, M.; Kolb, H.C.; Walsh, J.C.; Mocharla, V.P.; Zhu, Z.; Haka, M.; Alpaugh, R.K.; Chen, D.Y.T.; Yu, J.Q. Biodistribution and radiation dosimetry of the carbonic anhydrase IX imaging agent [^{18}F]VM$_{4-037}$ determined from PET/CT scans in healthy volunteers. *Mol. Imaging Biol.* **2014**, *16*, 739–746. [CrossRef] [PubMed]

67. Rutkauskas, K.; Zubrienė, A.; Tumosienė, I.; Kantminienė, K.; Kažemėkaitė, M.; Smirnov, A.; Kazokaitė, J.; Morkūnaitė, V.; Čapkauskaitė, E.; Manakova, E.; et al. 4-Amino-substituted Benzenesulfonamides as Inhibitors of Human Carbonic Anhydrases. *Molecules* **2014**, *19*, 17356–17380. [CrossRef] [PubMed]

68. Vaškevičienė, I.; Paketurytė, V.; Zubrienė, A.; Kantminienė, K.; Mickevičius, V.; Matulis, D. *N*-Sulfamoylphenyl- and *N*-sulfamoylphenyl-*N*-thiazolyl-β-alanines and their derivatives as inhibitors of human carbonic anhydrases. *Bioorg. Chem.* **2017**, *75*, 16–29. [CrossRef] [PubMed]

69. Ceruso, M.; Antel, S.; Vullo, D.; Scozzafava, A.; Supuran, C.T. Inhibition studies of new ureido-substituted sulfonamides incorporating a GABA moiety against human carbonic anhydrase isoforms I-XIV. *Bioorg. Med. Chem.* **2014**, *22*, 6768–6775. [CrossRef] [PubMed]

70. Moeker, J.; Mahon, B.P.; Bornaghi, L.F.; Vullo, D.; Supuran, C.T.; McKenna, R.; Poulsen, S.A. Structural insights into carbonic anhydrase IX isoform specificity of carbohydrate-based sulfamates. *J. Med. Chem.* **2014**, *57*, 8635–8645. [CrossRef] [PubMed]

71. Shay, G.; Lynch, C.C.; Fingleton, B. Moving targets: Emerging roles for MMPs in cancer progression and metastasis. *Matrix Biol.* **2015**, *44–46*, 200–206. [CrossRef] [PubMed]

72. Zhong, Y.; Lu, Y.T.; Sun, Y.; Shi, Z.H.; Li, N.G.; Tang, Y.P.; Duan, J.A. Recent opportunities in matrix metalloproteinase inhibitor drug design for cancer. *Expert Opin. Drug Discov.* **2018**, *13*, 75–87. [CrossRef] [PubMed]

73. Bonnans, C.; Chou, J.; Werb, Z. Remodelling the extracellular matrix in development and disease. *Nat. Rev. Mol. Cell Biol.* **2014**, *15*, 786–801. [CrossRef] [PubMed]

74. Hidalgo, M.; Eckhardt, S.G. Development of Matrix Metalloproteinase Inhibitors in Cancer Therapy. *JNCI J. Natl. Cancer Inst.* **2001**, *93*, 178–193. [CrossRef] [PubMed]

75. Scozzafava, A.; Supuran, C.T. Carbonic anhydrase and matrix metalloproteinase inhibitors: Sulfonylated amino acid hydroxamates with MMP inhibitory properties act as efficient inhibitors of CA isozymes I, II, and IV, and N-hydroxysulfonamides inhibit both these zinc enzymes. *J. Med. Chem.* **2000**, *43*, 3677–3687. [CrossRef] [PubMed]

76. Nuti, E.; Orlandini, E.; Nencetti, S.; Rossello, A.; Innocenti, A.; Scozzafava, A.; Supuran, C.T. Carbonic anhydrase and matrix metalloproteinase inhibitors. Inhibition of human tumor-associated isozymes IX and cytosolic isozyme I and II with sulfonylated hydroxamates. *Bioorg. Med. Chem.* **2007**, *15*, 2298–2311. [CrossRef] [PubMed]

77. Korkmaz, N.; Obaidi, O.A.; Senturk, M.; Astley, D.; Ekinci, D.; Supuran, C.T. Synthesis and biological activity of novel thiourea derivatives as carbonic anhydrase inhibitors. *J. Enzym. Inhib. Med. Chem.* **2015**, *30*, 75–80. [CrossRef] [PubMed]

78. Mollica, A.; Costante, R.; Akdemir, A.; Carradori, S.; Stefanucci, A.; Macedonio, G.; Ceruso, M.; Supuran, C.T. Exploring new Probenecid-based carbonic anhydrase inhibitors: Synthesis, biological evaluation and docking studies. *Bioorg. Med. Chem.* **2015**, *23*, 5311–5318. [CrossRef] [PubMed]

79. Fidan, I.; Salmas, R.E.; Arslan, M.; Şentürk, M.; Durdagi, S.; Ekinci, D.; Şentürk, E.; Coşgun, S.; Supuran, C.T. Carbonic anhydrase inhibitors: Design, synthesis, kinetic, docking and molecular dynamics analysis of novel glycine and phenylalanine sulfonamide derivatives. *Bioorg. Med. Chem.* **2015**, *23*, 7353–7358. [CrossRef] [PubMed]

metabolites

MDPI

Article

Benzamide-4-Sulfonamides Are Effective Human Carbonic Anhydrase I, II, VII, and IX Inhibitors

Morteza Abdoli [1,2], Murat Bozdag [1,*] , Andrea Angeli [1] and Claudiu T. Supuran [1,*]

1   Dipartimento Neurofarba, Sezione di Scienze Farmaceutiche e Nutraceutiche, Università degli Studi di Firenze, Via U. Schiff 6, Sesto Fiorentino, 50019 Florence, Italy; mortezaabdoli1987@gmail.com (M.A.); andrea.angeli@unifi.it (A.A.)
2   Department of Chemistry, Faculty of Science, Lorestan University, Khorramabad 6813833946, Iran
*   Correspondences: bozdag.murat@unifi.it (M.B.); claudiu.supuran@unifi.it (C.T.S.);
    Tel.: +39-055-457-3666 (M.B.); +39-055-457-3729 (C.T.S.)

Received: 11 May 2018; Accepted: 30 May 2018; Published: 1 June 2018

Abstract: A series of benzamides incorporating 4-sulfamoyl moieties were obtained by reacting 4-sulfamoyl benzoic acid with primary and secondary amines and amino acids. These sulfonamides were investigated as inhibitors of the metalloenzyme carbonic anhydrase (CA, EC 4.2.1.1). The human (h) isoforms hCA II, VII, and IX were inhibited in the low nanomolar or subnanomolar ranges, whereas hCA I was slightly less sensitive to inhibition (K_Is of 5.3–334 nM). The β- and γ-class CAs from pathogenic bacteria and fungi, such as *Vibrio cholerae* and *Malassezia globosa*, were inhibited in the micromolar range by the sulfonamides reported in the paper. The benzamide-4-sulfonamides are a promising class of highly effective CA inhibitors.

Keywords: carbonic anhydrase; human isoform; sulfonamide; benzamide; pathogens

1. Introduction

Benzamides incorporating 3- or 4-sulfamoyl moieties, such as derivatives **A** and **B** (Figure 1) were investigated [1,2] as inhibitors of the zinc metallo-enzyme carbonic anhydrase (CA, EC 4.2.1.1) [3–12] in this study, in the search of agents with intraocular pressure lowering effects [1,2]. The incorporation of a wide range of amino acid (AA) or dipeptide AA moieties in molecules **A** and **B** led to enhanced water solubility for topical administration within the eye. These compounds showed remarkable in vitro inhibitory effects, assayed by an esterase method with 4-nitrophenyl acetate as substrate, against isoforms hCA II and IV, involved in aqueous humor production within the eye [1–12].

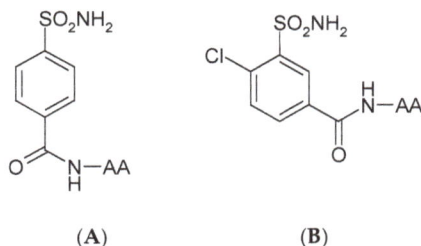

Figure 1. (**A,B**) Sulfonamides incorporating benzamide moieties, amino acid (AA) and dipeptide AA moieties [1,2].

The CA inhibitors (CAIs) belonging to the sulfonamide and sulfamate types have been used clinically for several decades as diuretics [13,14], antiglaucoma agents [15], and anti-obesity drugs [16,17]. More recently, a large number of studies showed that CA inhibition has profound antitumor effects by inhibiting hypoxia-inducible isoforms hCA IX and XII, overexpressed in many hypoxic tumors [18–22]. Furthermore, several proof-of-concept studies demonstrated the involvement of some CA isoforms in neuropathic pain [23,24] and arthritis [25,26], with the CAIs of sulfonamide and coumarin [27–30] types demonstrating significant in vivo effects in animal models of these diseases. Thus, the field of drug design, synthesis, and in vivo investigations of various types of CAIs is highly dynamic, with the action of a large number of interesting new chemotypes on these widespread enzymes being constantly studied [27–39]. As they catalyze the interconversion between carbon dioxide (CO_2) and bicarbonate with the formation of a proton, CAs are widespread in organisms all over the phylogenetic tree as seven distinct genetic families: the α-, β-, γ-, δ-, η-, ɛ-, and θ-CAs [3–12,40–47]. CAs participate in crucial physiologic processes connected to pH homeostasis, metabolism, transport of gases and ions, and secretion of electrolytes in virtually all living beings [3–12,40–47].

Apart from the inhibition of human (h) or other vertebrate CA isoforms, the interest in inhibiting such enzymes present in various pathogenic organisms (bacteria, fungi, protozoa, or worms) has presented the possibility of designing anti-infective agents with a novel mechanism of action [40–51]. Thus, in this paper, we explored novel CAIs belonging to the sulfonamide class, incorporating benzamide moieties similar to compounds reported earlier, but that were investigated for the inhibition of isoforms involved in important diseases, such as glaucoma (hCA II), neuropathic pain (hCA VII), or tumors (hCA IX), and ubiquitous off target isoform hCA I. Furthermore, we investigated whether this chemotype shows inhibitory effects against β- and γ-class CAs from pathogenic bacteria (*Vibrio cholerae*) or fungi (*Malassezia globosa*).

2. Results

2.1. Chemistry

The classical coupling of carboxylic acid **1** with amines, in the presence of carbodiimides (EDCI) and hydroxybenzotriazole has been used for synthesis, as reported previously [1,2] (Scheme 1).

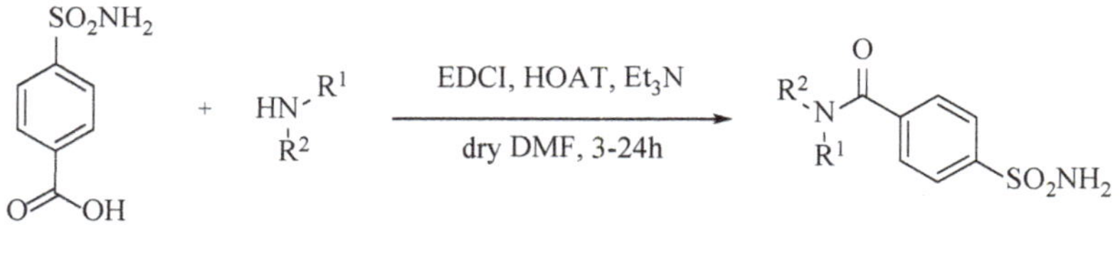

3a: R^1=H, R^2= Et
3b: R^1=H, R^2= *n*-Pr
3c: R^1=H, R^2= Propargyl
3d: R^1= O[(CH$_2$CH$_2$)]$_2$, R^2 = ⁻
3e: R^1= [C$_2$H$_4$C$_2$H$_4$CH$_2$], R^2 = ⁻
3f: R^1=H, R^2= methyl *DL*-valinate

3g, R^1=H, R^2= dimethyl *D*-glutamate
3h, R^1=H, R^2= methyl *L*-leucinate
3i, R^1=H, R^2= dimethyl *D*-aspartate
3j, R^1=H, R^2= methyl D*L*-alaninate
3k, R^1=H, R^2= ethyl 4-aminobutanoate
3l, R^1=H, R^2= methyl *L*-phenylalaninate

Scheme 1. Synthesis of compounds **3a–l**.

Compound **1** was condensed with compounds **3a–e** that possess primary or secondary amines as well amino acid derivatives **3f–l** in the presence of EDCI and 1-hydroxy-7-azabenzotriazole (HOAT) to obtain their corresponding amides (Scheme 1). By choosing variously substituted amines and amino acids, incorporating both simple aliphatic and heterocyclic scaffolds (for the amine) and aliphatic and aromatic amino acids, the physico-chemical properties and enzyme inhibitory properties of the new compounds could be modulated. For example, the amino acid derivatives **3f**, **3g**, **3h**, **3j**, and **3l** may form sodium salts leading to water soluble CAIs.

2.2. Carbonic Anhydrase Inhibition

Sulfonamides **3a–3l** were tested as inhibitors of four hCAs involved in various pathologies, hCA I, II, VII, and IX, as well as three β- and γ-CAs from pathogenic organisms: the β-CAs from the bacterium *Vibrio cholerae* (VchCAβ) and the fungus *Malassezia globosa* (MgCA), and the γ-CA from the same pathogenic bacterium, VchCAγ–enzymes recently cloned and characterized by our group as potential anti-infective targets [52–59] (Table 1).

Table 1. Inhibition data of human carbonic anhydrase (CA) isoforms hCA I, II, VII, IX, and pathogenic bacteria and fungi β- and γ-CAs with compounds **3a–3l** in comparison with the standard sulfonamide inhibitor **AAZ** by a stopped flow carbon dioxide (CO_2) hydrase assay [60].

Cpd	K_I (nM) [a]						
	hCA I	hCA II	hCA VII	hCA IX	VchCAβ	MgCA	VchCAγ
3a	334	5.3	26.7	15.9	7082	7669	929
3b	8.2	3.5	0.4	26.0	7680	3921	636
3c	67.6	1.9	0.6	22.9	741	5781	383
3d	8.7	6.2	0.8	10.7	8587	5880	693
3e	29.7	7.0	6.2	18.1	749	3985	453
3f	57.8	4.5	3.7	16.0	8172	5500	4458
3g	8.2	5.2	0.6	19.7	862	632	503
3h	5.6	3.7	0.4	8.0	719	763	891
3i	75.7	6.1	0.7	12.1	910	6946	744
3j	85.3	6.1	3.7	21.5	412	87.3	271
3k	5.3	4.0	0.4	9.3	953	6695	756
3l	5.6	3.3	0.5	19.2	663	517	409
AAZ	250.0	12.1	5.7	25.8	451	74000	473

[a] Mean from three different assay using a stopped flow technique. Errors were in the range of ±5% to 10% of the reported values.

3. Discussion

The following structure-activity relationship (SAR) were determined from the data of Table 1, in which the standard sulfonamide inhibitor acetazolamide (**AAZ**) was also included for comparison.

The slow cytosolic isoform hCAI, involved in some ocular diseases (not glaucoma) [3–7], was inhibited by sulfonamides **3a–l** reported here with K_Is in the range of 5.3 to 334 nM. The ethyl- (**3a**) derivative was the weakest inhibitor, whereas **3c**, **3f**, **3i**, and **3j** showed medium potency inhibitory action, with a K_Is in the range of 57.8 to 85.3. These compounds incorporate propargyl, valyl, aspartyl, and alanyl moieties. The remaining derivatives, **3b**, **3d**, **3e**, **3g**, **3h**, **3k**, and **3l** showed very effective hCA I inhibitory properties, with a K_Is in the range of 5.3 to 29.7 nM, being CAIs an order of magnitude better compared to acetazolamide (Table 1). Small changes in the scaffold (compare **3a** and **3b**) led to dramatic changes in the hCA I inhibitory effects, with the propyl derivative **3b** being 40.7 times more effective an inhibitor compared with the ethyl derivative **3a**.

All sulfonamides **3a–l** reported here were excellent hCA II inhibitors, with a K_Is in the range of 1.9 to 7.0 nM, thus being more effective than **AAZ** (Table 1). With this highly effective inhibition and small range in the variation of the K_Is, the SAR is flat and the only conclusion is that all the explored substitution patterns led to highly effective hCA II inhibitors. This is also the dominant cytosolic isoform, involved in glaucoma, diuresis, respiration, and electrolyte secretion in a multitude of tissues [3–12], meaning these results are highly significant.

The third cytosolic isoform investigated here, hCA VII, predominantly found in the brain and involved in epileptogenesis and neuropathic pain [16–24], was also effectively inhibited by sulfonamides **3a–l**, which showed a K_Is in the range of 0.4 to 26.7 nM. Most of these compounds were sub-nanomolar hCA VII inhibitors (e.g., **3b–3d, 3g–3i, 3k, 3l**), being more effective by an order of magnitude compared with the standard **AAZ**, whereas few of them showed the same potency as **AAZ** (**3e, 3f, 3j**) and only the ethyl derivative **3a** was a less effective inhibitor compared to **AAZ**, with a K_I of 26.7 nM. Overall, the SAR is extremely simple, and except for the ethyl derivative mentioned above, all the substitution patterns from derivatives **3b–3l** indicated all compounds are highly effective hCA VII inhibitors.

The tumor-associated, hypoxia-inducible isoform hCA IX was effectively inhibited by sulfonamides **3a–l**, with a K_Is in the range of 8.0 to 26. 0 nMh. **AAZ** has an inhibition constant of 25.8 nM against this isoform. The most effective inhibitors, **3h** and **3k**, with a K_Is of 8.0–9.3 nM, incorporated amino acyl moieties, but all substitution patterns present in compound **3**, of the amine or amino acid type, led to highly effective hCA IX inhibition.

Conversely, the β- and γ-CAs from pathogenic organisms investigated here were poorly inhibited by these compounds, which showed activity in the micromolar range, with few exceptions (Table 1). Thus, for VchCAβ, the K_Is was in the range of 0.41 to 8.58 μM; for MgCA, in the range of 87.3 nM to 7.67 μM; and for VchCAγ, in the range of 0.27 to 4.45 μM. Notably, **3j** compounds, which incorporate the alanyl moiety, showed a good inhibitory effect against the *Malassezia* enzyme, one of the causative agents of dandruff. Acetazolamide is a highly ineffective MgCA inhibitor, and most other sulfonamides investigated here, although less effective than **3j**, showed a better activity compared with the standard sulfonamide CAI. Overall, β- and γ-CAs are less sensitive to inhibition with sulfonamides compared with α-CAs [3–14].

4. Materials and Methods

4.1. Chemistry

Amines, 4-sulfamoyl-benzoic acid, buffers, solvents, and acetazolamide (**AAZ**) were commercially available, obtained as highest purity reagents from Sigma-Aldrich/Merck, Milan, Italy. Nuclear magnetic resonance (^{1}H NMR, ^{13}C NMR) spectra were recorded using a Bruker Avance III 400 MHz spectrometer (Bruker, Billerica, MA, USA) in dimethyl sulfoxide (DMSO-d_{6l}). Chemical shifts are reported in parts per million (ppm) and the coupling constants (J) are expressed in Hertz (Hz). Splitting patterns were designated as follows: s, singlet; d, doublet; t, triplet; m, multiplet; brs, broad singlet; and dd, double of doubles. The assignment of exchangeable protons (OH and NH) was confirmed by the addition of D_2O. Analytical thin-layer chromatography (TLC) was performed on Merck silica gel F-254 plates. Flash chromatography purifications were performed on Merck Silica gel 60 (230–400 mesh ASTM) as the stationary phase and MeOH/DCM were used as eluents.

4.1.1. General Procedure to Synthesize Compounds **3a–l**

A solution of 4-carboxybenzene sulfonamide **1** (1.0 eq) in dry dimethylformamide (DMF, 3–5 mL) was treated with primary or secondary amines or amino acids **2a–l** (1.2 eq), then followed by addition of *N*-(3-dimethylaminopropyl)-*N'*-ethylcarbodiimide hydrochloride (EDCI, 1.5 eq.), 1-hydroxy-7-azabenzotriazole (HOAT, 1.5 eq), and triethylamine (Et$_3$N, 3 eq). The reaction continued until the consumption of starting materials (TLC monitoring, 3–24 h) and quenched with water. The title compounds were either obtained from filtration of the precipitates formed followed by washing with water (**3a–3e, 3h, 3k–l**) or extracted from ethyl acetate (EtOAc). In the latter, the combined organic layers were washed with H_2O (3 × 20 mL), dried over sodium sulfate, filtered, and concentrated in a vacuum to provide a residue that was triturated from dichloromethane (**3f–g, 3i–j**).

4.1.2. Characterization of Synthesized Compounds (**3a–l**)

N-Ethyl-4-Sulfamoylbenzamide (**3a**): 140 mg white solid, yield 83%; δ_H (400 MHz, DMSO-d_6) 1.17 (3H, t, *J* 7.2), 3.33 (2H, m), 7.51 (2H, s, exchange with D_2O, SO_2NH_2), 7.93 (2H, d, *J* 8.4), 8.02 (2H, d, *J* 8.4), 8.68 (1H, t, *J* 7.2 exchange with D_2O, NH); δ_C (100 MHz, DMSO-d_6) 15.5, 35.1, 126.5, 128.6, 138.5, 147.0, and 165.8; *m/z* (ESI positive) 229.0 [M + H]$^+$.

N-Propyl-4-Sulfamoylbenzamide (**3b**): 120 mg white solid, yield 80%; δ_H (400 MHz, DMSO-d_6) 0.93 (3H, t, *J* 7.2), 1.58 (2H, m), 3.27 (2H, q, *J* 7.2), 7.50 (2H, s, exchange with D_2O, SO_2NH_2), 7.93 (2H, d, *J* 8.4), 8.02 (2H, d, *J* 8.4), 8.66 (1H, t, *J* 7.2, exchange with D_2O, NH); δ_C (100 MHz, DMSO-d_6) 12.3, 23.2, 42.0, 126.5, 128.7, 138.5, 147.0, 166.0; *m/z* (ESI positive) 243.1 [M + H]$^+$.

N-(Prop-2-Yn-1-Yl)-4-Sulfamoylbenzamide (**3c**): 120 mg yellow solid, yield 74%; δ_H (400 Mhz, DMSO-d_6) 3.18 (1H, T, *J* 2.5), 4.12 (2H, dd, *J* 5.5, 2.5), 7.52 (2H, s, exchange with D_2O, SO_2NH_2), 7.94 (2H, d, *J* 8.8), 8.04 (2H, d, *J* 8.8), 9.16 (1H, t, *J* 5.5, exchange with D_2O, N*H*); δ_c (100 Mhz, DMSO-d_6) 29.5, 73.9, 81.9, 126.5, 128.8, 137.6, 147.3, 165.8; *m/z* (ESI Positive) 239.0 [M + H]$^+$. Experimental data in agreement with reported data [61].

4-(Morpholine-4-Carbonyl)Benzenesulfonamide (**3d**): 10 mg pale yellow solid; 9% yield; δ_H (400 MHz, DMSO-d_6) 3.65 (8H, m), 7.44 (2H, s, exchange with D_2O, SO_2NH_2), 7.63 (2H, d, *J* 8.0), 7.92 (2H, d, *J* 8.0); δ_C (100 MHz, DMSO-d_6) 66.9, 66.9, 126.7, 128.5, 139.7, 145.8, 168.8; *m/z* (ESI positive) 271.1 [M + H]$^+$.

4-(Piperidine-1-Carbonyl)Benzenesulfonamide (**3e**): 12 mg yellow solid, yield 18%; δ_H (400 MHz, DMSO-d_6) 1.50 (2H, m), 1.65 (4H, m), 3.25 (2H, m), 3.63 (2H, m), 7.48 (2H, s, exchange with D_2O, SO_2NH_2), 7.59 (2H, d, *J* 8.0), 7.91 (2H, d, *J* 8.0); δ_C (100 MHz, DMSO-d_6) 26.1, 26.7, 48.8, 126.7, 128.0, 140.7, 145.4, 168.5; *m/z* (ESI positive) 269.1 [M + H]$^+$.

Methyl (4-Sulfamoylbenzoyl)-DL-Valinate (**3f**): 12 mg pale yellow solid, yield 10%; δ_H (400 MHz, DMSO-d_6) 0.98 (3H, d, *J* 6.8), 1.02 (3H, d, *J* 6.8), 2.23 (1H, m), 3.70 (3H, s), 4.36 (1H, t, *J* 6.8), 7.53 (2H, s, exchange with D_2O, SO_2NH_2), 7.94 (2H, d, *J* 8.4), 8.05 (2H, d, *J* 8.4), 8.83 (1H, d, *J* 6.8, exchange with D_2O, NH); δ_C (100 MHz, DMSO-d_6) 19.9, 20.0, 30.5, 52.6, 59.6, 126.4, 129.2, 137.7, 147.4, 167.0, 172.9; *m/z* (ESI positive) 315.0 [M + H]$^+$.

Dimethyl (4-Sulfamoylbenzoyl)-D-Glutamate (**3g**): 14 mg pale yellow solid, yield 16%; δ_H (400 Mhz, DMSO-d_6) 2.07 (2H, m), 2.16 (2H, m), 3.63 (3H, s), 3.70 (3H, s), 4.53 (1H, m), 7.52 (2H, s, exchange with D_2O, SO_2NH_2), 7.94 (2H, d, *J* 8.8), 8.05 (2H, d, *J* 8.8), 8.83 (1H, d, *J* 7.3, exchange with D_2O, NH); δ_C (100 Mhz, DMSO-d_6) 26.6, 30.8, 52.3, 52.9, 53.0, 126.5, 129.0, 137.4, 147.5, 166.6, 172.9, 173.5; *m/z* (ESI Positive) 359.1 [M + H]$^+$.

Methyl (4-Sulfamoylbenzoyl)-L-Leucinate (**3h**): 37 mg white solid, yield 25%; δ_H (400 Mhz, DMSO-d_6) 0.92 (3H, d, *J* 6.4), 0.97 (3H, d, *J* 6.4), 1.63 (1H, m), 1.70–1.86 (2H, m), 3.69 (3H, s), 4.56 (1H, m), 7.52 (2H, s, exchange with D_2O, SO_2NH_2), 7.95 (2H, d, *J* 8.3), 8.06 (2H, d, *J* 8.3), 8.94 (1H, d, *J* 6.4, exchange with D_2O, NH); δ_C (100 Mhz, DMSO-d_6) 22.1, 23.7, 25.3, 40.2, 51.9, 52.8, 126.5, 129.0, 137.5, 147.4, 166.5, 173.8; *m/z* (ESI Positive) 329.01 [M + H]$^+$.

Dimethyl (4-Sulfamoylbenzoyl)-D-Aspartate (**3i**): 35 mg yellow solid, yield 50%; δ_H (400 Mhz, DMSO-d_6) 2.86–3.04 (2H, m), 3.67 (3H, s), 3.70 (3H, s), 4.89 (1H, m), 7.52 (2H, s, exchange with D_2O, SO_2NH_2), 7.96 (2H, d, *J* 8.7), 8.03 (2H, d, *J* 8.7), 9.13 (1H, d, *J* 7.6, exchange with D_2O, NH); δ_C (100 Mhz, DMSO-d_6) 36.2, 50.2, 52.6, 53.2, 126.6, 129.0, 137.2, 147.6, 166.1, 171.3, 171.9; *m/z* (ESI Positive) 345.0 [M + H]$^+$.

Methyl (4-Sulfamoylbenzoyl)-DL-Alaninate (**3j**): 14 mg white solid, yield 13%; δ_H (400 Mhz, DMSO-d_6) 1.46 (3H, d, *J* 7.3), 3,69 (3H, s), 4.54 (1H, m), 7.51 (2H, s, exchange with D_2O, SO_2NH_2), 7.96 (2H, d, *J* 8.4), 8.06 (2H, d, *J* 8.4), 9.01 (1H, d, *J* 7.3, exchange with D_2O, NH); δ_C (100 Mhz, DMSO-d_6) 17.6, 49.3, 52.8, 126.5, 129.0, 137.4, 147.4, 166.2, 173.8; *m/z* (ESI Positive) 287.0 [M + H]$^+$.

Ethyl 4-(4-Sulfamoylbenzamido)Butanoate (**3k**): 80 mg white solid, yield 58%; δ_H (400 Mhz, DMSO-d_6) 1.20 (3H, t, J 7.2), 1.83 (2H, pent, J 6.8), 2.39 (2H, t, J 6.8), 3.31 (2H, m), 4.09 (2H, q, J 7.2), 7.47 (2H, s, exchange with D_2O, SO_2NH_2), 7.91 (2H, d, J 8.0), 8.01 (2H, d, J 8.0), 8.66 (1H, t, J 6.8, exchange with D_2O, NH); δ_C (100 Mhz, DMSO-d_6) 15.0, 25.3, 31.9, 39.6, 60.6, 126.5, 128.7, 138.3, 147.1, 166.1, 173.5; *m/z* (ESI Positive) 315.0 [M + H]$^+$.

Methyl (4-Sulfamoylbenzoyl)-L-Phenylalaninate (**3l**): 130 mg white solid, yield 72%; δ_H (400 Mhz, DMSO-d_6) 3.10–3.25 (2H, m), 3.69 (3H, s), 4.70–4.76 (1H, m), 7.24 (1H, m), 7.32 (4H, m), 7.52 (2H, s, exchange with D_2O, SO_2NH_2), 7.93 (2H, d, J 8.4), 7.98 (2H, d, J 8.4), 9.08 (1H, d, J 7.9, exchange with D_2O, NH); δ_C (100 Mhz, DMSO-d_6) 37.1, 52.9, 55.2, 126.5, 127.4, 128.9, 129.2, 130.0, 137.4, 138.4, 147.4, 166.3, 172.8; *m/z* (ESI Positive) 363.0 [M + H]$^+$.

4.2. CA Enzyme Inhibition Assay

An Sx.18Mv-R Applied Photophysics (Oxford, U.K.) stopped-flow instrument was used to assay he catalytic activity of various CA isozymes for CO_2 hydration reaction [60]. Phenol red, at a concentration of 0.2 mM, was used as an indicator, working at the absorbance maximum of 557 nm, with 10 mM Hepes (pH 7.5, for α-CAs) or TRIS (pH 8.3, for β- and γ-CAs) as buffers, 0.1 M sodium sulfate (Na_2SO_4) (for maintaining constant ionic strength), following the CA-catalyzed CO_2 hydration reaction for a period of 10 s at 25 °C. The CO_2 concentrations ranged from 1.7 to 17 mM for the determination of the kinetic parameters and inhibition constants. For each inhibitor, at least six traces of the initial 5–10% of the reaction were used for determining the initial velocity. The uncatalyzed rates were determined in the same manner and subtracted from the total observed rates. Stock solutions of inhibitors (10 mM) were prepared in distilled-deionized water. Dilutions up to 1 nM were performed thereafter with the assay buffer. Enzyme and inhibitor solutions were pre-incubated together for 15 min (standard assay at room temperature) prior to assay, to allow for the formation of the enzyme–inhibitor complex. The inhibition constants were obtained by non-linear least-squares methods using PRISM 3 and the Cheng-Prusoff equation, as reported earlier [62–75]. All CAs were recombinant proteins produced as reported earlier by our groups [52–76].

5. Conclusions

We report a series of benzamides incorporating 4-sulfamoyl moieties, which were obtained by reacting 4-sulfamoyl benzoic acid with primary and secondary amines and amino acids. These sulfonamides were investigated as inhibitors of several enzymes, including the human (h) isoforms hCA II, VII, and IX, involved in severe pathologies, such as glaucoma, epilepsy, neuropathic pain and cancer; and β- and γ-class CAs from pathogenic bacteria and fungi. hCA II, VII, and IX were inhibited in the low nanomolar or subnanomolar ranges by all investigated sulfonamides, whereas hCA I was slightly less sensitive to inhibition (K_Is of 5.3–334 nM). The *Vibrio cholerae* and *Malassezia globosa* CAs were generally inhibited in the micromolar range by the sulfonamides reported in the paper. The benzamide-4-sulfonamides constitute a promising class of highly effective CA inhibitors. Further investigations will focus on extending the series of sulfanilamide possessing aliphatic tails with carbamide linkers, such as cyclic and aliphatic and aromatic, to investigate and obtain isoform selective inhibitors for their profiling and possible in vivo applications.

Author Contributions: M.A. synthesized and characterized the CAIs, A.A. performed the stopped-flow analysis. M.B. supervised to study and edited the manuscript. C.T.S. wrote and edited the manuscript. All authors read and approved the final paper

Funding: This research received no external funding

Conflicts of Interest: The authors declare no competing financial interest.

References

1. Mincione, F.; Starnotti, M.; Menabuoni, L.; Scozzafava, A.; Casini, A.; Supuran, C.T. Carbonic anhydrase inhibitors: 4-sulfamoyl-benzenecarboxamides and 4-chloro-3-sulfamoyl-benzenecarboxamides with strong topical antiglaucoma properties. *Bioorg. Med. Chem. Lett.* **2001**, *11*, 1787–1791. [CrossRef]

2. Casini, A.; Scozzafava, A.; Mincione, F.; Menabuoni, L.; Starnotti, M.; Supuran, C.T. Carbonic anhydrase inhibitors: Topically acting antiglaucoma sulfonamides incorporating esters and amides of 3- and 4-carboxybenzolamide. *Bioorg. Med. Chem. Lett.* **2003**, *13*, 2867–2873. [CrossRef]

3. Supuran, C.T. Carbonic anhydrases: From biomedical applications of the inhibitors and activators to biotechnological use for CO_2 capture. *J. Enzyme Inhib. Med. Chem.* **2013**, *28*, 229–230. [CrossRef] [PubMed]

4. Supuran, C.T. How many carbonic anhydrase inhibition mechanisms exist? *J. Enzyme Inhib. Med. Chem.* **2016**, *31*, 345–360. [CrossRef] [PubMed]

5. Alterio, V.; Di Fiore, A.; D'Ambrosio, K.; Supuran, C.T.; De Simone, G. Multiple binding modes of inhibitors to carbonic anhydrases: How to design specific drugs targeting 15 different isoforms? *Chem. Rev.* **2012**, *112*, 4421–4468. [CrossRef] [PubMed]

6. Abbate, F.; Winum, J.Y.; Potter, B.V.; Casini, A.; Montero, J.L.; Scozzafava, A.; Supuran, C.T. Carbonic anhydrase inhibitors: X-ray crystallographic structure of the adduct of human isozyme II with EMATE, a dual inhibitor of carbonic anhydrases and steroid sulfatase. *Bioorg. Med. Chem. Lett.* **2004**, *14*, 231–234. [CrossRef] [PubMed]

7. Capasso, C.; Supuran, C.T. An overview of the alpha-, beta-and gamma-carbonic anhydrases from Bacteria: Can bacterial carbonic anhydrases shed new light on evolution of bacteria? *J. Enzyme Inhib. Med. Chem.* **2015**, *30*, 325–332. [CrossRef] [PubMed]

8. Supuran, C.T. Advances in structure-based drug discovery of carbonic anhydrase inhibitors. *Expert Opin. Drug Discov.* **2017**, *12*, 61–88. [CrossRef] [PubMed]

9. Supuran, C.T. Structure and function of carbonic anhydrases. *Biochem. J.* **2016**, *473*, 2023–2032. [CrossRef] [PubMed]

10. Supuran, C.T. Carbonic anhydrases: Novel therapeutic applications for inhibitors and activators. *Nat. Rev. Drug Discov.* **2008**, *7*, 168–181. [CrossRef] [PubMed]

11. Neri, D.; Supuran, C.T. Interfering with pH regulation in tumours as a therapeutic strategy. *Nat. Rev. Drug Discov.* **2011**, *10*, 767–777. [CrossRef] [PubMed]

12. Supuran, C.T.; Vullo, D.; Manole, G.; Casini, A.; Scozzafava, A. Designing of novel carbonic anhydrase inhibitors and activators. *Curr. Med. Chem. Cardiovasc. Hematol. Agents* **2004**, *2*, 49–68. [CrossRef] [PubMed]

13. Carta, F.; Supuran, C.T. Diuretics with carbonic anhydrase inhibitory action: A patent and literature review (2005–2013). *Expert Opin. Ther. Pat.* **2013**, *23*, 681–691. [CrossRef] [PubMed]

14. Temperini, C.; Cecchi, A.; Scozzafava, A.; Supuran, C.T. Carbonic anhydrase inhibitors. Sulfonamide diuretics revisited—Old leads for new applications? *Org. Biomol. Chem.* **2008**, *6*, 2499–2506. [CrossRef] [PubMed]

15. Masini, E.; Carta, F.; Scozzafava, A.; Supuran, C.T. Antiglaucoma carbonic anhydrase inhibitors: A patent review. *Expert Opin. Ther. Pat.* **2013**, *23*, 705–716. [CrossRef] [PubMed]

16. Scozzafava, A.; Supuran, C.T.; Carta, F. Antiobesity carbonic anhydrase inhibitors: A literature and patent review. *Expert Opin. Ther. Pat.* **2013**, *23*, 725–735. [CrossRef] [PubMed]

17. Supuran, C.T. Carbonic anhydrases and metabolism. *Metabolites* **2018**, *8*, E25. [CrossRef] [PubMed]

18. Monti, S.M.; Supuran, C.T.; De Simone, G. Anticancer carbonic anhydrase inhibitors: A patent review (2008–2013). *Expert Opin. Ther. Pat.* **2013**, *23*, 737–749. [CrossRef] [PubMed]

19. Supuran, C.T. Carbonic Anhydrase Inhibition and the Management of Hypoxic Tumors. *Metabolites* **2017**, *7*, E48. [CrossRef] [PubMed]

20. Ward, C.; Langdon, S.P.; Mullen, P.; Harris, A.L.; Harrison, D.J.; Supuran, C.T.; Kunkler, I.H. New strategies for targeting the hypoxic tumour microenvironment in breast cancer. *Cancer Treat. Rev.* **2013**, *39*, 171–179. [CrossRef] [PubMed]

21. Garaj, V.; Puccetti, L.; Fasolis, G.; Winum, J.Y.; Montero, J.L.; Scozzafava, A.; Vullo, D.; Innocenti, A.; Supuran, C.T. Carbonic anhydrase inhibitors: Novel sulfonamides incorporating 1,3,5-triazine moieties as inhibitors of the cytosolic and tumour-associated carbonic anhydrase isozymes I, II and IX. *Bioorg. Med. Chem. Lett.* **2005**, *15*, 3102–3108. [CrossRef] [PubMed]

22. Casey, J.R.; Morgan, P.E.; Vullo, D.; Scozzafava, A.; Mastrolorenzo, A.; Supuran, C.T. Carbonic anhydrase inhibitors. Design of selective, membrane-impermeant inhibitors targeting the human tumor-associated isozyme IX. *J. Med. Chem.* **2004**, *47*, 2337–2347. [CrossRef] [PubMed]

23. Supuran, C.T. Carbonic anhydrase inhibition and the management of neuropathic pain. *Expert Rev. Neurother.* **2016**, *16*, 961–968. [CrossRef] [PubMed]

24. Di Cesare Mannelli, L.; Micheli, L.; Carta, F.; Cozzi, A.; Ghelardini, C.; Supuran, C.T. Carbonic anhydrase inhibition for the management of cerebral ischemia: In vivo evaluation of sulfonamide and coumarin inhibitors. *J. Enzyme Inhib. Med. Chem.* **2016**, *31*, 894–899. [CrossRef] [PubMed]

25. Margheri, F.; Ceruso, M.; Carta, F.; Laurenzana, A.; Maggi, L.; Lazzeri, S.; Simonini, G.; Annunziato, F.; Del Rosso, M.; Supuran, C.T.; et al. Overexpression of the transmembrane carbonic anhydrase isoforms IX and XII in the inflamed synovium. *J. Enzyme Inhib. Med. Chem.* **2016**, *31*, 60–63. [CrossRef] [PubMed]

26. Bua, S.; Di Cesare Mannelli, L.; Vullo, D.; Ghelardini, C.; Bartolucci, G.; Scozzafava, A.; Supuran, C.T.; Carta, F. Design and Synthesis of Novel Nonsteroidal Anti-Inflammatory Drugs and Carbonic Anhydrase Inhibitors Hybrids (NSAIDs-CAIs) for the Treatment of Rheumatoid Arthritis. *J. Med. Chem.* **2017**, *60*, 1159–1170. [CrossRef] [PubMed]

27. Maresca, A.; Temperini, C.; Vu, H.; Pham, N.B.; Poulsen, S.A.; Scozzafava, A.; Quinn, R.J.; Supuran, C.T. Non-zinc mediated inhibition of carbonic anhydrases: Coumarins are a new class of suicide inhibitors. *J. Am. Chem. Soc.* **2009**, *131*, 3057–3062. [CrossRef] [PubMed]

28. Maresca, A.; Temperini, C.; Pochet, L.; Masereel, B.; Scozzafava, A.; Supuran, C.T. Deciphering the mechanism of carbonic anhydrase inhibition with coumarins and thiocoumarins. *J. Med. Chem.* **2010**, *53*, 335–344. [CrossRef] [PubMed]

29. Carta, F.; Maresca, A.; Scozzafava, A.; Supuran, C.T. Novel coumarins and 2-thioxo-coumarins as inhibitors of the tumor-associated carbonic anhydrases IX and XII. *Bioorg. Med. Chem.* **2012**, *20*, 2266–2273. [CrossRef] [PubMed]

30. Supuran, C.T. Carbonic anhydrase activators. *Future Med. Chem.* **2018**, *10*, 561–573. [CrossRef] [PubMed]

31. Alterio, V.; Cadoni, R.; Esposito, D.; Vullo, D.; Fiore, A.D.; Monti, S.M.; Caporale, A.; Ruvo, M.; Sechi, M.; Dumy, P.; et al. Benzoxaborole as a new chemotype for carbonic anhydrase inhibition. *Chem. Commun.* **2016**, *52*, 11983–11986. [CrossRef] [PubMed]

32. Nocentini, A.; Cadoni, R.; Del Prete, S.; Capasso, C.; Dumy, P.; Gratteri, P.; Supuran, C.T.; Winum, J.Y. Benzoxaboroles as Efficient Inhibitors of the β-Carbonic Anhydrases from Pathogenic Fungi: Activity and Modeling Study. *ACS Med. Chem. Lett.* **2017**, *8*, 1194–1198. [CrossRef] [PubMed]

33. Tars, K.; Vullo, D.; Kazaks, K.; Leitans, J.; Lends, A.; Grandane, A.; Zalubovskis, R.; Scozzafava, A.; Supuran, C.T. Sulfocoumarins (1,2-benzoxathiine-2,2-dioxides): A class of potent and isoform-selective inhibitors of tumor-associated carbonic anhydrases. *J. Med. Chem.* **2013**, *56*, 293–300. [CrossRef] [PubMed]

34. Métayer, B.; Mingot, A.; Vullo, D.; Supuran, C.T.; Thibaudeau, S. New superacid synthesized (fluorinated) tertiary benzenesulfonamides acting as selective hCA IX inhibitors: Toward a new mode of carbonic anhydrase inhibition by sulfonamides. *Chem. Commun.* **2013**, *49*, 6015–6017. [CrossRef] [PubMed]

35. Métayer, B.; Mingot, A.; Vullo, D.; Supuran, C.T.; Thibaudeau, S. Superacid synthesized tertiary benzenesulfonamides and benzofuzed sultams act as selective hCA IX inhibitors: Toward understanding a new mode of inhibition by tertiary sulfonamides. *Org. Biomol. Chem.* **2013**, *11*, 7540–7549. [CrossRef] [PubMed]

36. Supuran, C.T. Carbon-versus sulphur-based zinc binding groups for carbonic anhydrase inhibitors? *J. Enzyme Inhib. Med. Chem.* **2018**, *33*, 485–495. [CrossRef] [PubMed]

37. Di Fiore, A.; Maresca, A.; Supuran, C.T.; De Simone, G. Hydroxamate represents a versatile zinc binding group for the development of new carbonic anhydrase inhibitors. *Chem. Commun.* **2012**, *48*, 8838–8840. [CrossRef] [PubMed]

38. Marques, S.M.; Nuti, E.; Rossello, A.; Supuran, C.T.; Tuccinardi, T.; Martinelli, A.; Santos, M.A. Dual inhibitors of matrix metalloproteinases and carbonic anhydrases: Iminodiacetyl-based hydroxamate-benzenesulfonamide conjugates. *J. Med. Chem.* **2008**, *51*, 7968–7979. [CrossRef] [PubMed]

39. Bozdag, M.; Carta, F.; Angeli, A.; Osman, S.M.; Alasmary, F.A.S.; AlOthman, Z.; Supuran, C.T. Synthesis of N'-phenyl-N-hydroxyureas and investigation of their inhibitory activities on human carbonic anhydrases. *Bioorg. Chem.* **2018**, *78*, 1–6. [CrossRef] [PubMed]

40. Capasso, C.; Supuran, C.T. Bacterial, fungal and protozoan carbonic anhydrases as drug targets. *Expert Opin. Ther. Targets* **2015**, *19*, 1689–1704. [CrossRef] [PubMed]

41. Vermelho, A.B.; Da Silva Cardoso, V.; Ricci Junior, E.; Dos Santos, E.P.; Supuran, C.T. Nanoemulsions of sulfonamide carbonic anhydrase inhibitors strongly inhibit the growth of *Trypanosoma cruzi*. *J. Enzyme Inhib. Med. Chem.* **2018**, *33*, 139–146. [CrossRef] [PubMed]

42. de Menezes Dda, R.; Calvet, C.M.; Rodrigues, G.C.; de Souza Pereira, M.C.; Almeida, I.R.; de Aguiar, A.P.; Supuran, C.T.; Vermelho, A.B. Hydroxamic acid derivatives: A promising scaffold for rational compound optimization in Chagas disease. *J. Enzyme Inhib. Med. Chem.* **2016**, *31*, 964–973. [CrossRef] [PubMed]

43. Nocentini, A.; Cadoni, R.; Dumy, P.; Supuran, C.T.; Winum, J.Y. Carbonic anhydrases from Trypanosoma cruzi and Leishmania donovani chagasi are inhibited by benzoxaboroles. *J. Enzyme Inhib. Med. Chem.* **2018**, *33*, 286–289. [CrossRef] [PubMed]

44. Del Prete, S.; De Luca, V.; De Simone, G.; Supuran, C.T.; Capasso, C. Cloning, expression and purification of the complete domain of the η-carbonic anhydrase from *Plasmodium falciparum*. *J. Enzyme Inhib. Med. Chem.* **2016**, *31*, 54–59. [CrossRef] [PubMed]

45. Supuran, C.T.; Capasso, C. The η-class carbonic anhydrases as drug targets for antimalarial agents. *Expert Opin. Ther. Targets* **2015**, *19*, 551–563. [CrossRef] [PubMed]

46. Vullo, D.; Del Prete, S.; Fisher, G.M.; Andrews, K.T.; Poulsen, S.A.; Capasso, C.; Supuran, C.T. Sulfonamide inhibition studies of the η-class carbonic anhydrase from the malaria pathogen *Plasmodium Falciparum*. *Bioorg. Med. Chem.* **2015**, *23*, 526–531. [CrossRef] [PubMed]

47. De Simone, G.; Di Fiore, A.; Capasso, C.; Supuran, C.T. The zinc coordination pattern in the η-carbonic anhydrase from Plasmodium falciparum is different from all other carbonic anhydrase genetic families. *Bioorg. Med. Chem. Lett.* **2015**, *25*, 1385–1389. [CrossRef] [PubMed]

48. Modak, J.K.; Liu, Y.C.; Supuran, C.T.; Roujeinikova, A. Structure-Activity Relationship for Sulfonamide Inhibition of Helicobacter pylori α-Carbonic Anhydrase. *J. Med. Chem.* **2016**, *59*, 11098–11109. [CrossRef] [PubMed]

49. Buzás, G.M.; Supuran, C.T. The history and rationale of using carbonic anhydrase inhibitors in the treatment of peptic ulcers. In memoriam Ioan Puşcaş (1932–2015). *J. Enzyme Inhib. Med. Chem.* **2016**, *31*, 527–533. [CrossRef] [PubMed]

50. Supuran, C.T. Bacterial carbonic anhydrases as drug targets: Toward novel antibiotics? *Front. Pharmacol.* **2011**, *2*, 34. [CrossRef] [PubMed]

51. Nishimori, I.; Onishi, S.; Takeuchi, H.; Supuran, C.T. The alpha and beta classes carbonic anhydrases from *Helicobacter pylori* as novel drug targets. *Curr. Pharm. Des.* **2008**, *14*, 622–630. [CrossRef] [PubMed]

52. De Vita, D.; Angeli, A.; Pandolfi, F.; Bortolami, M.; Costi, R.; Di Santo, R.; Suffredini, E.; Ceruso, M.; Del Prete, S.; Capasso, C.; et al. Inhibition of the α-carbonic anhydrase from Vibrio cholerae with amides and sulfonamides incorporating imidazole moieties. *J. Enzyme Inhib. Med. Chem.* **2017**, *32*, 798–804. [CrossRef] [PubMed]

53. Del Prete, S.; Vullo, D.; De Luca, V.; Carginale, V.; Ferraroni, M.; Osman, S.M.; AlOthman, Z.; Supuran, C.T.; Capasso, C. Sulfonamide inhibition studies of the β-carbonic anhydrase from the pathogenic bacterium Vibrio cholerae. *Bioorg. Med. Chem.* **2016**, *24*, 1115–1120. [CrossRef] [PubMed]

54. Del Prete, S.; Isik, S.; Vullo, D.; De Luca, V.; Carginale, V.; Scozzafava, A.; Supuran, C.T.; Capasso, C. DNA cloning, characterization, and inhibition studies of an α-carbonic anhydrase from the pathogenic bacterium Vibrio cholerae. *J. Med. Chem.* **2012**, *55*, 10742–10748. [CrossRef] [PubMed]

55. Del Prete, S.; Vullo, D.; De Luca, V.; Carginale, V.; di Fonzo, P.; Osman, S.M.; AlOthman, Z.; Supuran, C.T.; Capasso, C. Anion inhibition profiles of α-, β- and γ-carbonic anhydrases from the pathogenic bacterium Vibrio cholerae. *Bioorg. Med. Chem.* **2016**, *24*, 3413–3417. [CrossRef] [PubMed]

56. Angeli, A.; Del Prete, S.; Osman, S.M.; Alasmary, F.A.S.; AlOthman, Z.; Donald, W.A.; Capasso, C.; Supuran, C.T. Activation studies of the α- and β-carbonic anhydrases from the pathogenic bacterium Vibrio cholerae with amines and amino acids. *J. Enzyme Inhib. Med. Chem.* **2018**, *33*, 227–233. [CrossRef] [PubMed]

57. Del Prete, S.; De Luca, V.; Vullo, D.; Osman, S.M.; AlOthman, Z.; Carginale, V.; Supuran, C.T.; Capasso, C. A new procedure for the cloning, expression and purification of the β-carbonic anhydrase from the pathogenic yeast Malassezia globosa, an anti-dandruff drug target. *J. Enzyme Inhib. Med. Chem.* **2016**, *31*, 1156–1161. [CrossRef] [PubMed]

58. Nocentini, A.; Vullo, D.; Del Prete, S.; Osman, S.M.; Alasmary, F.A.S.; AlOthman, Z.; Capasso, C.; Carta, F.; Gratteri, P.; Supuran, C.T. Inhibition of the β-carbonic anhydrase from the dandruff-producing fungus Malassezia globosa with monothiocarbamates. *J. Enzyme Inhib. Med. Chem.* **2017**, *32*, 1064–1070. [CrossRef] [PubMed]

59. Angiolella, L.; Carradori, S.; Maccallini, C.; Giusiano, G.; Supuran, C.T. Targeting Malassezia species for Novel Synthetic and Natural Antidandruff Agents. *Curr. Med. Chem.* **2017**, *24*, 2392–2412. [CrossRef] [PubMed]

60. Khalifah, R.G. The carbon dioxide hydration activity of carbonic anhydrase. I. Stop-flow kinetic studies on the native human isoenzymes B and C. *J. Biol. Chem.* **1971**, *246*, 2561–2573. [PubMed]

61. Wilkinson, B.L.; Bornaghi, L.F.; Houston, T.A.; Innocenti, A.; Supuran, C.T.; Poulsen, S.A. A novel class of carbonic anhydrase inhibitors: Glycoconjugate benzene sulfonamides prepared by "click-tailing". *J. Med. Chem.* **2006**, *49*, 6539–6548. [CrossRef] [PubMed]

62. Diaz, J.R.; Fernández Baldo, M.; Echeverría, G.; Baldoni, H.; Vullo, D.; Soria, D.B.; Supuran, C.T.; Camí, G.E. A substituted sulfonamide and its Co (II), Cu (II), and Zn (II) complexes as potential antifungal agents. *J. Enzyme Inhib. Med. Chem.* **2016**, *31*, 51–62. [CrossRef] [PubMed]

63. Menchise, V.; De Simone, G.; Alterio, V.; Di Fiore, A.; Pedone, C.; Scozzafava, A.; Supuran, C.T. Carbonic anhydrase inhibitors: Stacking with Phe131 determines active site binding region of inhibitors as exemplified by the X-ray crystal structure of a membrane-impermeant antitumor sulfonamide complexed with isozyme II. *J. Med. Chem.* **2005**, *48*, 5721–5727. [CrossRef] [PubMed]

64. Supuran, C.T.; Mincione, F.; Scozzafava, A.; Briganti, F.; Mincione, G.; Ilies, M.A. Carbonic anhydrase inhibitors—Part 52. Metal complexes of heterocyclic sulfonamides: A new class of strong topical intraocular pressure-lowering agents in rabbits. *Eur. J. Med. Chem.* **1998**, *33*, 247–254. [CrossRef]

65. Şentürk, M.; Gülçin, İ.; Beydemir, Ş.; Küfrevioğlu, O.İ.; Supuran, C.T. In vitro inhibition of human carbonic anhydrase I and II isozymes with natural phenolic compounds. *Chem. Biol. Drug Des.* **2011**, *77*, 494–499. [CrossRef] [PubMed]

66. Fabrizi, F.; Mincione, F.; Somma, T.; Scozzafava, G.; Galassi, F.; Masini, E.; Impagnatiello, F.; Supuran, C.T. A new approach to antiglaucoma drugs: Carbonic anhydrase inhibitors with or without NO donating moieties. Mechanism of action and preliminary pharmacology. *J. Enzyme Inhib. Med. Chem.* **2012**, *27*, 138–147. [CrossRef] [PubMed]

67. Krall, N.; Pretto, F.; Decurtins, W.; Bernardes, G.J.; Supuran, C.T.; Neri, D. A Small-Molecule Drug Conjugate for the Treatment of Carbonic Anhydrase IX Expressing Tumors. *Angew. Chem. Int. Ed. Engl.* **2014**, *53*, 4231–4235. [CrossRef] [PubMed]

68. Rehman, S.U.; Chohan, Z.H.; Gulnaz, F.; Supuran, C.T. In-vitro antibacterial, antifungal and cytotoxic activities of some coumarins and their metal complexes. *J. Enzyme Inhib. Med. Chem.* **2005**, *20*, 333–340. [CrossRef] [PubMed]

69. Clare, B.W.; Supuran, C.T. Carbonic anhydrase activators. 3: Structure-activity correlations for a series of isozyme II activators. *J. Pharm. Sci.* **1994**, *83*, 768–773. [CrossRef] [PubMed]

70. Dubois, L.; Peeters, S.; Lieuwes, N.G.; Geusens, N.; Thiry, A.; Wigfield, S.; Carta, F.; McIntyre, A.; Scozzafava, A.; Dogné, J.M.; et al. Specific inhibition of carbonic anhydrase IX activity enhances the in vivo therapeutic effect of tumor irradiation. *Radiother. Oncol.* **2011**, *99*, 424–431. [CrossRef] [PubMed]

71. Chohan, Z.H.; Munawar, A.; Supuran, C.T. Transition metal ion complexes of Schiff-bases. Synthesis, characterization and antibacterial properties. *Met. Based Drugs* **2001**, *8*, 137–143. [CrossRef] [PubMed]

72. Zimmerman, S.A.; Ferry, J.G.; Supuran, C.T. Inhibition of the archaeal β-class (Cab) and γ-class (Cam) carbonic anhydrases. *Curr. Top. Med. Chem.* **2007**, *7*, 901–908. [CrossRef] [PubMed]

73. Supuran, C.T.; Nicolae, A.; Popescu, A. Carbonic anhydrase inhibitors. Part 35. Synthesis of Schiff bases derived from sulfanilamide and aromatic aldehydes: The first inhibitors with equally high affinity towards cytosolic and membrane-bound isozymes. *Eur. J. Med. Chem.* **1996**, *31*, 431–438. [CrossRef]

74. Pacchiano, F.; Aggarwal, M.; Avvaru, B.S.; Robbins, A.H.; Scozzafava, A.; McKenna, R.; Supuran, C.T. Selective hydrophobic pocket binding observed within the carbonic anhydrase II active site accommodate different 4-substituted-ureido-benzenesulfonamides and correlate to inhibitor potency. *Chem. Commun.* **2010**, *46*, 8371–8373. [CrossRef] [PubMed]

75. Ozensoy Guler, O.; Capasso, C.; Supuran, C.T. A magnificent enzyme superfamily: Carbonic anhydrases, their purification and characterization. *J. Enzyme Inhib. Med. Chem.* **2016**, *31*, 689–694. [CrossRef] [PubMed]

76. De Simone, G.; Langella, E.; Esposito, D.; Supuran, C.T.; Monti, S.M.; Winum, J.Y.; Alterio, V. Insights into the binding mode of sulphamates and sulphamides to hCA II: Crystallographic studies and binding free energy calculations. *J. Enzyme Inhib. Med. Chem.* **2017**, *32*, 1002–1011. [CrossRef] [PubMed]

metabolites

MDPI

Review

An Update on the Metabolic Roles of Carbonic Anhydrases in the Model Alga *Chlamydomonas reinhardtii*

Ashok Aspatwar [1,*] , Susanna Haapanen [1] and Seppo Parkkila [1,2]

[1] Faculty of Medicine and Life Sciences, University of Tampere, FI-33014 Tampere, Finland; Haapanen.Susanna.E@student.uta.fi (S.H.); seppo.parkkila@staff.uta.fi (S.P.)
[2] Fimlab, Ltd., and Tampere University Hospital, FI-33520 Tampere, Finland
* Correspondence: ashok.aspatwar@staff.uta.fi; Tel.: +358-46-596-2117

Received: 11 January 2018; Accepted: 10 March 2018; Published: 13 March 2018

Abstract: Carbonic anhydrases (CAs) are metalloenzymes that are omnipresent in nature. CAs catalyze the basic reaction of the reversible hydration of CO_2 to HCO_3^- and H^+ in all living organisms. Photosynthetic organisms contain six evolutionarily different classes of CAs, which are namely: α-CAs, β-CAs, γ-CAs, δ-CAs, ζ-CAs, and θ-CAs. Many of the photosynthetic organisms contain multiple isoforms of each CA family. The model alga *Chlamydomonas reinhardtii* contains 15 CAs belonging to three different CA gene families. Of these 15 CAs, three belong to the α-CA gene family; nine belong to the β-CA gene family; and three belong to the γ-CA gene family. The multiple copies of the CAs in each gene family may be due to gene duplications within the particular CA gene family. The CAs of *Chlamydomonas reinhardtii* are localized in different subcellular compartments of this unicellular alga. The presence of a large number of CAs and their diverse subcellular localization within a single cell suggests the importance of these enzymes in the metabolic and biochemical roles they perform in this unicellular alga. In the present review, we update the information on the molecular biology of all 15 CAs and their metabolic and biochemical roles in *Chlamydomonas reinhardtii*. We also present a hypothetical model showing the known functions of CAs and predicting the functions of CAs for which precise metabolic roles are yet to be discovered.

Keywords: carbonic anhydrases; CA gene family; *Chlamydomonas reinhardtii*; model alga; metabolic role; photosynthesis

---

1. Introduction

Carbonic anhydrases (EC 4.2.1.1) (CAs) are metalloenzymes that catalyze the reversible reaction of hydration of carbon dioxide to bicarbonate ($CO_2 + H_2O \Leftrightarrow HCO_3^- + H^+$). The CAs belong to seven evolutionarily unrelated CA-gene families (α-, β-, γ-, δ-, ζ-, η-, and θ-CAs) [1–5].

The CAs are widespread in nature, and are found abundantly in plants, animals, and microorganisms, suggesting that the CAs have many diverse metabolic roles in living organisms [6–8]. Vertebrates and mammals have only α-CAs and contain multiple isoforms of the enzyme. In contrast, multicellular plants and unicellular photosynthetic organisms seem to have members of six CA gene families, often multiple isoforms of CAs from each gene family [4,9]. The *Chlamydomonas reinhardtii* genome analysis has revealed the presence of at least 15 CA genes encoding three different families of CAs. The number of CAs in *C. reinhardtii* is thus much higher than previously thought for a unicellular cell alga. Interestingly, a recent study showed that the limiting CO_2-inducible B protein (LCIB) family belongs to the β-CAs [10]. The amino acid sequences of these CA families are different, but most of these CA families have a Zn^{2+} atom at the active site [11]. In this alga, CAs have been found in the mitochondria, chloroplast thylakoid, cytoplasm, and periplasmic space [12,13]. A recent study showed that CAH6 is localized in the flagella instead of the pyrenoid stroma, as previously reported [14].

The downregulation of CA activity using molecular techniques and chemical inhibitors has shown reduced lipid biosynthesis in chloroplasts compared with chloroplasts from wild-type plants [15]. CAs are indirectly involved in lipid synthesis (and perhaps other HCO_3^--requiring pathways in plastids), serving to "concentrate" CO_2 in plastids as HCO_3^-, and reduce the rate of CO_2 diffusion out of plastids [15]. The CA might indirectly influence fatty acid synthesis in plastids by modulating plastidial pH, as the enzyme fatty acid synthase activity requires an optimal pH for fatty acid synthesis [15].

The role of CAs in pH regulation is well known in animal cells. However, the roles of CAs in pH regulation in this model alga are not known, and need to be investigated. The presence of 15 CAs in *C. reinhardtii* suggests that they are involved in several other metabolic functions in addition to a CO_2-concentrating mechanism (CCM), which is attributed to the evolutionarily conserved enzymes in plants. In *C. reinhardtii*, CAs are involved in many metabolic functions that involve carboxylation or decarboxylation reactions, including both photosynthesis and respiration. In addition, it has been clearly shown that CAs also participate in the transport of inorganic carbon to actively photosynthesizing cells and away from nonphotosynthesizing, respiratory cells [12,16].

In the current article, we will review the information on CAs of *C. reinhardtii*, a unicellular model alga. We will describe the information that is available on the molecular biology, and present the data for the metabolic and biochemical roles of the three CA gene families. For each CA enzyme from the three CA families, we will highlight the current research and questions that have been addressed by researchers in the field. We will also present a hypothetical model showing the known functions of CAs and predicting the functions of CAs for which precise metabolic roles are yet to be discovered. Finally, we present future directions in the field of *C. reinhardtii* CA research in order to study the precise metabolic and physiological roles of CAs from this alga.

2. General Aspects of Carbonic Anhydrases

Carbon dioxide (CO_2) is a very important molecule that is found in all living organisms. CO_2 is soluble in lipid membranes, and freely diffusible in and out of the cell [17–19]. Carbon dioxide and bicarbonate constitute the main buffer system for pH regulation in all living cells. CAs play a very important role in the transport of CO_2 and protons across cell membranes [20,21]. CA families differ in their preference for the metal ions used within the active site for performing the catalysis. The general enzyme catalytic mechanism of all of the CAs involves a reaction between a metal cofactor bound to OH^- and CO_2, giving rise to a HCO_3^- ion that is subsequently replaced from the metal with an H_2O molecule. This reaction is shown in Equation (1) below, where Enz indicates CA enzymes, and M indicates the metal cofactor. The regeneration of OH^- involves a transfer of H^+ from the metal bound to the H_2O molecule to the solvent, as shown in Equation (2) [22,23].

$$EnzM - OH + CO_2 \Leftrightarrow EnzM - HCO_3^- \overset{H_2O}{\Leftrightarrow} EnzM - H_2O + HCO_3^- \tag{1}$$

$$EnzM - H_2O \Leftrightarrow EnzM\text{–}OH^- + H^+ \tag{2}$$

CAs are metal-containing enzymes that are found in every living organism and have been studied extensively in the past. These enzymes are widely distributed among metabolically diverse species from all three domains of life. CAs perform various physiological functions, such as respiration, photosynthesis, pH regulation, ion transport, bone resorption, and the secretion of gastric juice, cerebrospinal fluid, and pancreatic juice [8,24,25]. CAs are also involved in electrolyte secretion, CO_2 and pH homeostasis, CO_2 fixation, and biosynthetic reactions, such as gluconeogenesis and ureagenesis [26–29].

Although the CA families are diverse and widespread, the current understanding of their biological role is mainly based on studies with several α and β-class CAs. Among the other CA classes, the δ and ζ classes have been reported only in diatoms and coccoliths [30,31]. Cam is the γ-class archetype isolated from *Methanosarcina thermophila,* an anaerobic methane-producing species from the Archaea domain [9]. The η and θ-class CAs have been identified recently from *Plasmodium falciparum* and marine diatoms, respectively [1,4]. The α-CAs are typically found as monomers and dimers; the β-CAs are typically found in many oligomerization states; and the γ-CAs are typically found as trimeric forms. Table 1 shows details of all of the CA gene families of enzymes, and Figure 1 depicts example structures of the major CA families: α, β, and γ.

Table 1. Details of the carbonic anhydrase (CA) gene family enzymes in living organisms.

CAs	Enzyme	Metal Ion	Organisms	Ref.
α	Monomeric, dimeric	Zn^{2+}	Animals, prokaryotes, fungi, and plants	[8,32]
β	Multimeric	Zn^{2+}	Plants, bacteria, and fungi	[8,32]
γ	Trimeric	Zn^{2+} or Fe, Co	Plants, archaea, fungi, and bacteria	[33,34]
ζ	Monomeric	Cd or Zn	Marine diatoms	[30,33,34]
δ	Monomeric	Co	Marine diatoms	[30,35,36]
η	Monomeric	Zn^{2+}	*Plasmodium* spp.	[1,37]
θ	Monomeric	Zn^{2+}	Marine diatoms	[4,9,10]

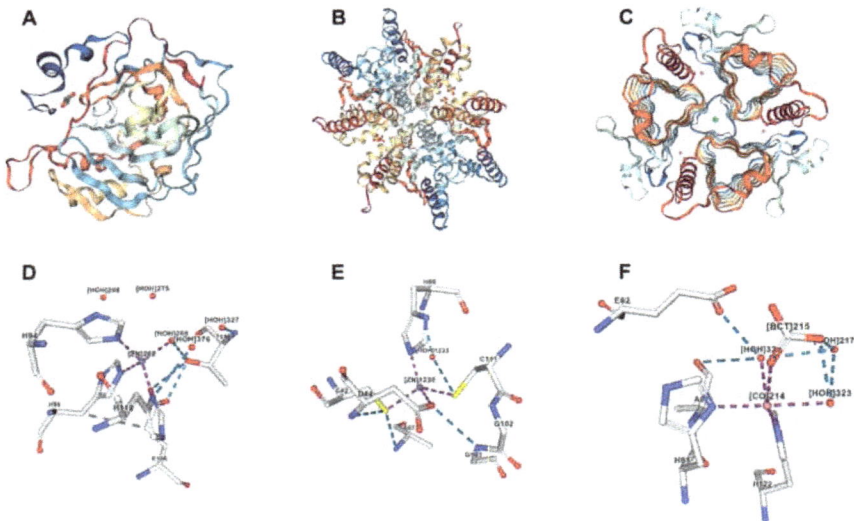

Figure 1. Representative structures of the α-CA, β-CA, and γ-CA families of enzymes and their ligand-binding sites. (**A**) Structure of human CAII enzymes retrieved from PDB 3U45. The human CAII monomer mostly consists of beta strands and contains a single active site with three zinc-coordinating histidine residues [38,39]; (**B**) Structure of *Haemophilus influenzae* β-CA retrieved from PDB 2A8C [40]; (**C**) Structure of γ-CA from *Methanosarcina thermophila* 1QRE [41,42]; (**D–F**) Metal at the active site coordinated with histidine residues (purple), hydrogen bonds (blue), halogen bonds (turquoise), hydrophobic contacts (gray), and pi interactions (orange, green). Images D and E show Zn^{2+} at the active site, and image F shows the Co^{2+} substitution in the structure.

The α-carbonic anhydrases: Among the CA gene families, the α-CA was first to be discovered in 1932 in erythrocytes [43,44]. Among the CA gene families, α-CAs are the most widely distributed CA family. α-CAs are also the most studied CA-group, probably because all of the human CAs belong to this enzyme family. In humans, there are 15 α-CA family isoforms; among these, 12 α-CA isoforms are catalytically active, and three CAs do not have catalytic activity, and are designated carbonic anhydrase-related proteins (CARPs) [45]. The α-CAs are monomers that are found in every type of tissue; furthermore, these enzymes have unique localizations within the cell. These enzymes have Zn^{2+} as a metal ion at their active site, and are coordinated by three histidine residues. The α-CAs play a wide variety of roles in living organisms.

The β-carbonic anhydrases: The first β-CA was discovered in 1938, and subsequently, many β-CAs have been reported from bacteria, archaea, yeasts, algae, and plants [32,46.47]. The β-CAs contain Zn^{2+} as the metal ion at the active site of the enzyme. In β-CAs, the Zn^{2+} atom is coordinated by two cysteines and one histidine, unlike the three histidines found in α and γ-CAs [48]. Some of the enzymes in the β-CA class have four zinc ligands, that is, one His residue, two Cys residues, and one Asp residue coordinated to Zn^{2+} [49]. β-CAs are found in many oligomerization states due to the presence of an α/β-fold that promotes an association with the formation of dimers. Several crystal structures of β-CAs have already been reported [50–52]. The β-CAs play important roles in CO_2 fixation in plants and other photosynthetic organisms, as well as in pH regulation, survival, virulence, and invasion in bacteria [53–55].

The γ-carbonic anhydrases: The γ-CAs were discovered in 1994, and are found in *Archaea, Bacteria,* and plants [56,57]. These enzymes are trimers, unlike the other CA gene families. The γ-CAs contain Zn^{2+} at the active site, and in some cases, the enzymes contain Fe^{2+} instead of the Zn^{2+} that is found in anaerobic *Archaea* [58]. The active site metal is coordinated by three histidine ligands and one water molecule [42]. The kinetic studies of γ-CA from *M. thermophila* showed that the hydration of CO_2 is a two-step process similar to that in α-CAs [59,60].

The δ-carbonic anhydrases: The first δ-CA was characterized from *Thalassiosira weissflogii,* a unicellular microalga representing a species of centric diatoms [61]. The δ-CAs are also present in many eukaryotic marine phytoplanktons [8,61]. The crystal structure of δ-CAs has not been resolved, but studies using X-ray absorption spectroscopy methods have shown that the metal ion Zn is coordinated by histidine ligands similar to those in the α and γ classes of CAs [48].

The ζ-carbonic anhydrases: Xu et al. described the ζ-CAs in marine diatoms in 2008 [8]. The ζ-CAs contain cadmium at the active site of the enzyme as an alternative metal cofactor [33]. The structure of CdCA1 has been resolved, and it is reported that the metal-binding region is repeated three times in this enzyme [62]. The ζ-CAs have been shown to bind to other metals apart from Cd and remain active. The reason for having Cd at their active site is possibly adaptation in the ocean, where Zn levels are often low.

The η-carbonic anhydrases: η-CA is a novel class of CA that Del Prete et al. recently discovered recently [1]. The η-CA is found in the malaria-causing protozoan parasite *Plasmodium falciparum.* The active site of η-CA contains Zn^{2+}, and is coordinated by two Hi residues and one Gln residue, in addition to the water molecule/hydroxide ion acting as a nucleophile in the catalytic cycle [37]. This arrangement is unique compared with that in the other CA gene families [37].

The θ-carbonic anhydrases: The θ-CA family is a recent addition to the existing CA gene family, as reported by Kikutani et al. [4] in 2016. These CAs have also been described in a cyanobacterium by Jin et al. [10]. The θ-CAs are implicated in the formation of CO_2 from HCO_3^- in the chloroplast.

2.1. Carbonic Anhydrases in Photosynthetic Organisms

Photosynthetic organisms contain CAs that belong to six different CA gene families, namely: α-CAs, β-CAs, γ-CAs, δ-CAs, ζ-CAs, and θ-CAs. Each of the at least three gene families of α-CAs, β-CAs, and γ-CAs are represented by multiple isoforms in all of the species. The γ-CAs are also found in photosynthetic bacteria [63,64]. θ-CA has been recently discovered in the thylakoid lumen of the marine diatom *Phaeodactylum tricornutum* [4]. The four CA gene families (α-, β-, γ-, and θ-CAs) that are found in photosynthetic organisms contain zinc as a metal ion at the active site of the enzymes [4,8]. Due to the alternative splicing of CA transcripts, the number of functional CA isoforms in many of the species is greater than the number of genes that encode a particular CA enzyme. In photosynthetic organisms, CAs are expressed in different cellular compartments, and are most prevalent in chloroplasts, cytosol, and mitochondria. The diversity in location suggests the importance of the CAs in the many physiological and biochemical roles they may play in photosynthetic organisms.

2.2. Carbonic Anhydrases in Chlamydomonas Reinhardtii

The model alga *Chlamydomonas reinhardtii* is a unicellular photosynthetic eukaryote that contains multiple genes encoding CAs for three different gene families. The α-CAs were discovered in the 1980s and 1990s in *C. reinhardtii* [12,65,66]. The β-CAs were discovered during the 1990s, [67–69], and with the sequencing of the complete genome of *C. reinhardtii*, three novel γ-CAs were found in the latter part of the 2000s [57,70–72].

The alga *C. reinhardtii* has three α-CAs, nine β-CAs (including the recently discovered three homologs of the LCIB protein family), and three γ-CAs [10]. Among the CAs that are found in *C. reinhardtii*, the β-CAs are predominant, with the highest isozyme number in this organism. Details of all the CAs that have been discovered in *C. reinhardtii* to date are presented below (Table 2).

Table 2. Details of the 15 carbonic anhydrases found in *Chlamydomonas reinhardtii* belonging to the α, β, and γ gene families.

CA Protein	Chr	Gene Family	MW (kDa)	Location	Known/Predicted Physiological Roles of the CAs	References
CAH1 [a]	4	α	78	Periplasm/late secretory pathway	Supply of Ci in low CO_2	[66,73–79]
CAH2 [a]	4		84	Periplasm/late secretory pathway	Supply of Ci in high CO_2	[14,66,80,81]
CAH3 [a]	9		29.5	Chloroplasts	Growth in low CO_2	[14,82–88]
CAH4 [*,a]	5		21	Mitochondria	-	[14,89–91]
CAH5 [*,a]	5		21	Mitochondria	-	[14,40–42]
CAH6 [a]	12		31	Flagella	CCM	[14]
CAH7 [b]	13	β	35.79	Periplasm?	-	[92]
CAH8 [a]	9		35.79	Periplasm	-	[92]
CAH9 [a]	5		13.06	Cytosol	-	[14]
LCIB1 [b]			48 [c]	Chloroplasts	CO_2 uptake, CCM	[10]
LCIB2 [b]			48 [c]	Chloroplasts	CO_2 uptake, CCM	[10]
LCIB3 [b]			48 [c]	Chloroplasts	CO_2 uptake, CCM	[10]
CAG1 [b]	9	γ	24.29	Mitochondria	Transport of mitochondrial CO_2 to chloroplasts	[14,70,71,93]
CAG2 [b]	6		31.17	Mitochondria	Transport of mitochondrial CO_2 to chloroplasts	[14,71,72,93]
CAG3 [b]	12		32.69	Mitochondria	Transport of mitochondrial CO_2 to chloroplasts	[14,71,72,93]

* The amino acid sequences of these two β-CAs differ by a single amino acid. [a] CA activity is known; [b] CA activity is not known; [c] Predicted molecular weight. Chr = chromosome.

2.2.1. α-Carbonic Anhydrase 1

Among the CA genes, α-*Ca1* was the first gene that was identified in *C. reinhardtii* in the 1980s [65,66,94]; it was named *Ca1*, reflecting the order of discovery. Several groups have shown that CAH1 is localized in the periplasmic space of the alga [65,66,94]. *Cah1*, the gene encoding CAH1, has been cloned [78]. The cDNA encodes a polypeptide of 377 amino acid residues. It is composed of a signal peptide that is 20 amino acids long, a small subunit, a large subunit, and a spacer region between the subunits [76,78]. Fujiwara et al. [66] discovered that the gene sequence is 93.6% identical to the sequence of *Cah2*, which encodes CAH2. In addition, their intron insertion sites are identical. These findings indicate that *Cah1* and *Cah2* are paralogs or the products of gene duplication [66].

The expression of CAH1 can be induced in the presence of low amounts of CO_2 compared to high amounts of CO_2. However, Kucho et al. [75] showed that in addition to low amounts of CO_2, CAH1 requires the presence of light for its induction [75]. Additional studies have shown an accumulation of CAH1 when the CO_2 concentration is reduced in the presence of light [65,66]. Inhibition of the photosynthetic reaction using 3-(3,4-dichlorophenyl)-1,1-dimethylurea (DCMU) leads to a reduction in CAH1 mRNA, suggesting that the accumulation of CAH1 mRNA requires functioning photosynthesis [65]. CO_2 regulates the induction of CAH1 through various enhancer and silencer sites [75]. At least a 692-bp region from -651 to +41 relative to the transcription start site was detected to be adequate for the full induction of CAH1 in response to light and low CO_2 [75]. Kucho et al. [75] identified a crucial regulatory area (63 bp from −293 to −231 relative to the transcription start site) that contains two enhancer elements. In addition, they detected DNA-binding proteins that specifically interact with these enhancer elements in the presence of light and low CO_2 conditions [74]. Additionally, other silencers and enhancers have been found, but they are usually responsible for only small changes in the induction or downregulation of CAH1 [75].

The physiological role of CAH1 has already been extensively discussed in the earlier review [12]. CAH1 provides more C_i to the *C. reinhardtii* cell in a C_i-deficient environment [12]. Nonetheless, many studies have shown that CAH1 mutant cells are as viable as the wild type under the conditions assayed. In contrast, drug inhibition restricts the growth, which indicates that other CAs, such as CAH2 and CAH8, might maintain the necessary CA activity in CAH1-deficient cells [12].

2.2.2. α-Carbonic Anhydrase 2

CAH2 was discovered at same time as CAH1 by Fukuzawa et al. [65,80,81]. CAH2 is a periplasmic protein and is a heterotetramer, as is CAH1. CAH2 consists of two identical large subunits and two small subunits [66]. The molecular weight of the holoenzyme is approximately 84.5–87.9 kDa. The large subunit is 38 kDa in size, and the small one is 4.2 kDa, which are result of the cleavage of the subunits of the proprotein. Therefore, the subunits are comparatively larger than the corresponding units in CAH1 [80]. The genetic similarity has already been stated, but the similarity of the amino acid sequences is 91.8% [66]. Nevertheless, the catalytic activity of CAH2 is approximately 1.6 times that of CAH1, as that of CAH2 is 3300 units per mg protein compared to 2200 units per mg protein with CAH1 [80]. The subunits of CAH2 are bound to each other with disulfide bonds, as in CAH1. CAH2 also has similar glycosylation sites to those of CAH1 in the large subunit [80].

The expression of CAH2 is more abundant than that of CAH1, and the expression of CAH2 is greatly induced in low CO_2 conditions as opposed to CAH1, which is moderate in amount and is present in high CO_2 conditions [66]. Furthermore, Tachiki et al. [80] have suggested that CAH2 might be present in low as well as high CO_2 conditions, as *Cah2* mRNA is expressed in both conditions [80]. The function and role has been suggested to be the same as that of CAH1, and Rawat et al. [81] proposed that *Cah2* could represent a gene duplication, without a specific role of its own [81].

2.2.3. α-Carbonic Anhydrase 3

Among the α-CAs of *C. reinhardtii*, Karlsson et al. identified α-CAH3 in the late 1990s [82], and showed that it was localized in the thylakoid lumen [82,84,91]. CAH3 is a 29.5-kDa polypeptide that Karlsson et al. originally isolated in 1995 [84]. The longest cDNA clone obtained from the cDNA library consisted of 1383 bp and contained an open reading frame that encoded a polypeptide of 310 amino acids [82].

CAH3 functions in the thylakoid lumen, and has been suggested to be part of photosystem II (PSII) or CCM [13,88,95,96]. Hanson et al. [13] showed that cia3, which is a mutant line of *C. reinhardtii* lacking functioning CAH3, has a limiting effect on the function of Rubisco in vivo [13]. The physiological function of CAH3 is also related to the location within thylakoids; thus, in stromal thylakoids, CAH3 is probably associated with light reactions of photosynthesis, and in the intrapyrenoid thylakoids, CAH3 is presumably connected to the actions of Rubisco [88].

In addition, the actions of CAH3 are connected to the fatty acid composition of the thylakoid membranes [88]. In low CO_2 conditions, the activity of CAH3 is implicitly related to an increase in the relative amount of polyunsaturated fatty acids. The change in the fatty acid composition changes the fluidity of the membranes and, therefore, the ion transport across the thylakoid membrane. The desaturation of fatty acids also provides H^+ ions, and hence implies that there is a reaction where H^+ ions are needed [88].

The regulation of CAH3 in different CO_2 conditions differs from the regulation of CAH1 or CAH2. It has been found that the activity and localization of CAH3 changes according to the CO_2 conditions, unlike the case with CAH1 [85]. Blanco-Rivero et al. [85] discovered that the amount of mRNA or the actual protein did not increase significantly during acclimation to low CO_2 conditions [85]. However, the activity of CAH3 increased due to phosphorylation, as did the amount of CAH3 in intrapyrenoid thylakoids at the expense of stromal thylakoids [85].

Additionally, the optimal pH of CAH3 is more acidic [87] than that of other CAs of *C. reinhardtii*. Benlloch et al. [87] measured the activity of CAH3 at different pH values, and discovered that the optimum was approximately pH 6.5 compared with that of the other CAs, which function best around a neutral pH. The activity also persists at a higher level than the activity of the other CAs at lower pH values [87].

A recent study showed that CAH3a associates with TAT2 and TAT3 proteins of the twin arginine translocation (Tat) pathway, and delivers substrate proteins to the thylakoid lumen [14]. The study also showed that the interaction between CAH3 and STT7 phosphorylates CAH3, CAH3 increases its catalytic activity when CO_2 is low, and CAH3 converts HCO_3^- to CO_2 in thylakoid membranes that traverse the pyrenoid, supplying the pyrenoid with the high concentration of CO_2 that is essential for CCM [14,82].

2.2.4. β-Carbonic Anhydrase 4

Eriksson et al. reported the presence of a CA in *C. reinhardtii* that belongs to the β-CA family in 1995 [91]. The CAH4 is localized in the mitochondria of *C. reinhardtii*, and has a molecular mass of 20.7–22 kDa. The gene coding CAH4 is called *β-Ca1*, of which the whole nucleotide sequence has been examined, and was found to have 96% identity with another mitochondrial CA5 (CAH5) coding gene, *β-Ca2* [91]. *β-Ca1* is induced in low CO_2 conditions, but not in high CO_2 conditions; hence, it is present only when CO_2 levels are high [91].

There have been many theories about the physiological role of CAH4 and CAH5. On one hand, Eriksson et al. [91] suggested that they are used in buffering reactions in changing CO_2 conditions [91]. Glycine decarboxylation in photorespiration produces excessive amounts of CO_2 and NH_3 in low CO_2 conditions. H^+ is used because NH_3 forms NH_4^+ at the pH of the mitochondrial matrix. Due to the need for H^+, CAH4 catalyzes the hydration of CO_2 to be faster in order to maintain the pH in the matrix [91]. On the other hand, Raven hypothesized that there might be a HCO_3^- channel in the inner mitochondrial membrane; thus, both CAH4 and CAH5 have a role in preserving the CO_2 [89].

There is also a third hypothesis for the function of CAH4 and CAH5, suggesting that they might provide HCO_3^- for reactions catalyzed by phosphoenolpyruvate carboxylase where N is combined to C skeletons that can be later used in protein synthesis [89]. It has also been shown that because of this assumed function, the external NH_4^+ concentration is an essential regulator of the expression and function of CAH4 [89].

2.2.5. β-Carbonic Anhydrase 5

Eriksson at al. identified CAH4 and CAH5 simultaneously in *C. reinhardtii* [91]. The two clones that code for CAH4 and CAH5 differ only slightly in their nucleotide sequences. In the coding area, the difference is only seven nucleotides, leading to one amino acid change at position 53, where serine is replaced by alanine [91]. In addition, the upstream regulating sites of *β-Ca1* and *β-Ca2* are very similar. Due to the striking similarity of *β-Ca1* and *β-Ca2*, the genes are likely to be duplicates that were selected because they increased the quantity of mtCA [90]. CAH4 and CAH5 lack any known functional difference, which also supports the gene-duplication assumption [90].

2.2.6. β-Carbonic Anhydrase 6

Mirta et al. discovered CAH6 in 2004 [97], and showed it to be localized in the chloroplast stroma [97,98]. In contrast, localization studies performed by Mackinder et al. [14] recently showed that CAH6 is expressed in flagella and shows no detectable signal in chloroplasts [14]. To validate their findings, the authors analyzed the presence of CAH6 in proteomic datasets, and showed it in the flagellar proteome and in intraflagellar transport (IFT) cargo [14].

The cDNA of *Cah6* is 2886 bp long and encodes a 264-amino-acid-long polypeptide, CAH6. CAH6 has a calculated molecular mass of 26 kDa, but experimentally, it has a mass of 28.5 kDa in an SDS-polyacrylamide gel [97]. Some amount of CAH6 expression is induced in low CO_2 conditions, but its expression levels are high in high CO_2 conditions, similar to the case with many other CA isoenzymes in *C. reinhardtii*. CAH6 was believed to be involved in trapping CO_2 that is leaking out of pyrenoids by converting it to HCO_3^- and thus preventing C_i from leaving the chloroplast [97].

However, a recent study showing the localization of CAH6 to be in the flagella suggested that CAH6 is not required in the chloroplast, as its presence in the chloroplast may short circuit the CCM by converting CO_2 from HCO_3^- and its subsequent release away from Rubisco [14]. Indeed, this is the case at least in cyanobacteria, where the presence of CA disrupts the CCM [99]. *Chlamydomonas* are known to show chemotaxis toward HCO_3^-, and CAs have been implicated in C_i sensing, and hence may be directly involved in sensing C_i [14,100,101].

2.2.7. β-Carbonic Anhydrase 7

Ynalvez et al. identified CAH7 in 2008 [92] by examining the sequences of two genes that code for CAs, namely, CAH7 and CAH8 [92]. The identified gene sequence of *Cah7* contains 5077 bp. The protein product of the gene *Cah7* has 399 amino acids, including 23 amino acids that are well conserved in β-CAs, as well as two cysteines and one histidine, which coordinate Zn^{2+}. In addition, the researchers predicted that CAH7 has a transmembrane domain, and thus might be attached to a membrane [92].

The amount of CAH7 in the cell depends upon the levels of CO_2 in the surroundings. The CAH7 is more abundant in low than in high CO_2 conditions. Overall, CAH7 is expressed in lower amounts than most of the other CAs in *C. reinhardtii*. The location and physiological role of CAH7 in the cell is yet to be resolved [92].

2.2.8. β-Carbonic Anhydrase 8

Ynalvez et al. identified *C. reinhardtii* CAH8 [92] in 2008 with CAH7, and both sequences were found to be closely related to each other [92]. The cDNA coding for CAH8 contains 2649 bp corresponding to a 333-amino-acid-long polypeptide. Furthermore, CAH8 has the same β-CA characteristics as CAH7, except that CAH8 has 22 of the 23 well-conserved amino acid residues. The molecular mass of CAH8 is approximately 40 kDa. Additionally, CAH8 has the same transmembrane domain near the C-terminus, which is similar to the transmembrane domain of CAH7. Immunolocalization studies have shown that CAH8 is located in the periplasmic space along with CAH1 and CAH2, but that the localization of CAH8 appears closer to the cell membrane compared with that of CAH1 [92].

The expression of CAH8 is constant in the algal cell, but the amount of enzyme is present in higher amounts in the presence of abundant CO_2 than in the presence of lower CO_2 amounts [92]. The overall expression of CAH8 resembles that of CAH6, as it is moderate among the CAs in *C. reinhardtii*. There are some theories regarding the function of CAH8. First, it has been suggested that, because CAH8 is closely related to the cell membrane, it would ensure the presence of CO_2 near the membrane, despite the external pH conditions. Second, CAH8 has been proposed to be a part of the C_i delivery system as a carbon-binding protein. Third, an association with a pore or a channel has been proposed [92].

2.2.9. β-Carbonic Anhydrase 9

The presence of CAH9 in *C. reinhardtii* was first reported in 2005 by Cardol et al. [71] from the genome sequencing project [71]. The RNA-Seq data that are available suggest that CAH9 is expressed at low levels (http://genomes.mcdb.ucla.edu/Cre454/) under the growth conditions that were used in the experiment at that time [12]. Since then, no further studies have been done on CAH9 expression and its role in *C. reinhardtii*.

2.2.10. Limiting CO₂ Inducible-B Protein/β-Carbonic Anhydrase Family

Limiting CO_2 inducible-B protein (LCIB) is a key player in the eukaryotic algal CCM function in *Chlamydomonas reinhardtii* [100]. The LCIB gene encodes a novel chloroplast protein that consists of 448 amino acids with a predicted MW of 48 kDa, and forms a heteromultimeric complex with its close homolog LCIC; the complex may be tightly regulated or may require additional factors for proper functioning [14,98,100]. Interestingly, a recent study involving a double mutant analysis of LCIB/CAH3 showed that LCIB functions downstream of CAH3. It has been hypothesized that LCIB captures CO_2 that leaked from the pyrenoid, possibly by unidirectionally hydrating CO_2 back to HCO_3^- [101]. Recently, in order to study the function of LCIB, a phylogenetically diverse set of recombinant LCIB homologs were produced in *E. coli* and purified [10]. Structural characterization of the purified proteins showed that three of the six homologs were structurally similar to the β-CAs at the level of the overall fold, zinc binding motif, and active site architecture. However, none of the three proteins showed CA enzymatic activity, and the lack of CA activity could be due to the widening of the intersubunit cleft, which affects active site integrity by causing disordering of the important His162/161 and Arg194/193 residues in the protein [10].

Based on the results of the study, it is proposed that LCIB in association with LCIC acts as a noncatalytic structural barrier for the leaked CO_2 from the pyrenoid [10]. However, in order to elucidate the precise role of LCIB, further studies are needed involving the characterization of a LCIB–LCIC complex purified from a native source.

2.2.11. γ-Carbonic Anhydrases

The gene *Glp1* that encodes γ-CAH1 was discovered in 2005 using the γ-CA protein sequence of *M. thermophila* and expressed sequence tag (EST) databases [70]. Similarly, the presence of three γ-CAs (CAG1, CAG2, and CAG3) in *C. reinhardtii* was also shown by two other groups, and these were predicted to be localized in the mitochondrial matrix [71,72].

The *Glp1* gene that codes for γ-CAH has seven exons and six introns, and encodes a putative protein of 312 amino acids [70]. The localization studies using prediction programs showed that this enzyme is localized in the cytoplasm or is secreted outside the cell. γ-CAH1 has approximately 40% similarity with the γ-CAH of *M. thermophila*, and has three histidine residues coordinating zinc at the active site of the enzyme. The recombinant proteins expressed in *E. coli* show no CA activity in either crude cell extracts or purified fusion protein [70]. However, the γ-CAHs may be active in the parent organism.

There are two additional γ-CAHs that have been annotated as subunits of the mitochondrial NADH dehydrogenase complex [70]. The sequence analysis showed that these γ-CAHs do not contain three histidine residues that are required for the catalytic activity of the CAs [70]. Based on the available studies, the γ-CAHs of *C. reinhardtii* are localized in the mitochondrial matrix, and are part of mitochondrial complex I. Interestingly, complex I of the mitochondrial electron transport chain (mETC) in *Arabidopsis thaliana* also contains three different protein domains that are homologous to γ-CAs [102]. Double mutants of *Arabidopsis thaliana* lacking γ-CAH1 and γ-CAH2 were analyzed for their role in development and physiology. The analysis of mutant strains of *A. thaliana* showed a developmental delay and an upregulation of complex II and complex IV, with increased oxygen consumption in mitochondrial respiration [102]. The results of this study suggest that the three γ-CAHs in *C. reinhardtii* may perform similar functions. The few studies on γ-CAHs were conducted a decade ago; therefore, the information on the physiological roles of these CAs is incomplete. We need more studies using bioinformatic and molecular tools for the structural and functional analysis of these γ-CAHs in order to know their precise roles in *C. reinhardtii*. Based on the latest information we propose a hypothetical model showing the localizations and functions of CAs in *C. reinhardtii* (Figure 2).

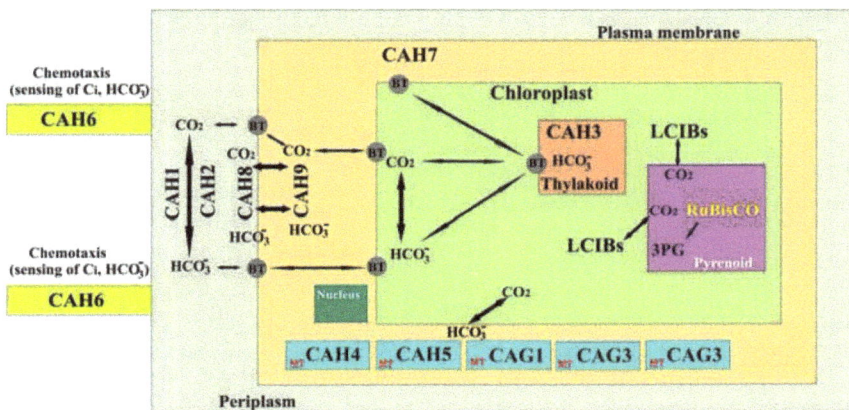

Figure 2. Schematic presentation of the *C. reinhardtii* model showing the roles of CAs in the cell and subcellular organelles. CAH1, CAH2, CAH3: α-Carbonic anhydrases; CAH4, CAH5, CAH6, CAH7, CAH8, and CAH9: β-Carbonic anhydrases; LCIB1, LICB2, and LCIB3: Low CO_2-inducible proteins (β-CAs); CAG1, CAG2, and CAG3: γ-Carbonic anhydrases; BT: Bicarbonate transporters; *RuBisCO*: Ribulose-1,5-bisphosphate carboxylase oxygenase; 3PG: 3-phosphoglycerate. MT: mitochondria.

3. Conclusions and Future Directions

The CA enzymes belonging to different classes of CA gene families are found in vertebrates, invertebrates, plants, unicellular marine and fresh water algae, bacteria, and archaea. The CAs are localized in almost all of the tissues of higher animals and subcellular organelles of eukaryotic cells, and perform a variety of metabolic and physiological roles. In plants, several classes of CAs are found that are localized in subcellular organelles, are involved in CCM for photosynthesis, and perform other metabolic functions. Plant biologists have used marine and fresh water unicellular photosynthetic model organisms in order to study the precise metabolic roles of CA enzymes. The freshwater alga *C. reinhardtii* is one such model organism that has helped us to understand the metabolic and physiological roles of CAs mainly on CCM. However, the precise metabolic roles of most of the CA enzymes in this alga remain to be studied.

There has been a continuous interest in CA research in unicellular photosynthetic organisms, especially as the genomes of these algae have become available. The availability of bioinformatic and molecular tools has helped to study the precise metabolic roles of CAs in photosynthetic model organisms. In *C. reinhardtii*, researchers have attempted to study the localization and metabolic roles of three α-CAs. Contradictory reports have emerged on the precise localizations of the CAs, and only limited information is available on the physiological roles of six β-CAs and the newly reported LCIB protein family that belongs to the β-CA group. No studies are available on γ-CAs except the presence of three forms of this enzyme and their predicated localization in the mitochondrial matrix. The challenge for future researchers will be to determine the precise localization and biochemical roles of all 12 CAs and the newly discovered three LCIB family proteins.

It is important to identify the precise physiological roles for all of the CAs found in *C. reinhardtii*, which is an important model organism for studying fundamental processes such as photosynthesis. *C. reinhardtii* is the most commonly studied species of *Chlamydomonas* and has a relatively simple genome, which has been sequenced in many different strains, including the nonmotile strains. More importantly, various strains of *C. reinhardtii* have been developed for specific research purposes. In photosynthetic organisms, including *C. reinhardtii*, the role of CAs in CCM have been studied extensively. In addition to the fundamental roles of CAs in the reversible hydration of CO_2, CAs have been shown to play important roles in defense mechanisms in plants and animals, protozoa, and bacteria. Therefore, it will be important to widen the perspective of CA studies in *C. reinhardtii* to cover not only pH regulation, but also other potential processes.

Acknowledgments: The work was supported by grants from the Sigrid Jusélius Foundation and the Academy of Finland to Seppo Parkkila and by a Finnish Cultural Foundation grant to Ashok Aspatwar.

Author Contributions: A.A. and S.H. are responsible for preparing the manuscript. A.A., S.H. and S.P. contributed to the writing of the article and have read and approved the final version.

Conflicts of Interest: The authors declare no conflict of interest.

References

1. Del Prete, S.; Vullo, D.; De Luca, V.; AlOthman, Z.; Osman, S.M.; Supuran, C.T.; Capasso, C. Biochemical characterization of recombinant beta-carbonic anhydrase (PgiCAb) identified in the genome of the oral pathogenic bacterium *Porphyromonas gingivalis*. *J. Enzyme Inhib. Med. Chem.* **2015**, *30*, 366–370. [CrossRef] [PubMed]
2. Supuran, C.T.; Capasso, C. The eta-class carbonic anhydrases as drug targets for antimalarial agents. *Expert Opin. Ther. Targets* **2015**, *19*, 551–563. [CrossRef] [PubMed]
3. Krishnamurthy, V.M.; Kaufman, G.K.; Urbach, A.R.; Gitlin, I.; Gudiksen, K.L.; Weibel, D.B.; Whitesides, G.M. Carbonic anhydrase as a model for biophysical and physical-organic studies of proteins and protein-ligand binding. *Chem. Rev.* **2008**, *108*, 946–1051. [CrossRef] [PubMed]
4. Kikutani, S.; Nakajima, K.; Nagasato, C.; Tsuji, Y.; Miyatake, A.; Matsuda, Y. Thylakoid luminal theta-carbonic anhydrase critical for growth and photosynthesis in the marine diatom *Phaeodactylum tricornutum*. *Proc. Natl. Acad. Sci. USA* **2016**, *113*, 9828–9833. [CrossRef] [PubMed]
5. Capasso, C.; Supuran, C.T. An overview of the alpha-, beta- and gamma-carbonic anhydrases from Bacteria: Can bacterial carbonic anhydrases shed new light on evolution of bacteria? *J. Enzyme Inhib. Med. Chem.* **2015**, *30*, 325–332. [CrossRef] [PubMed]
6. Floryszak-Wieczorek, J.; Arasimowicz-Jelonek, M. The multifunctional face of plant carbonic anhydrase. *Plant Physiol. Biochem.* **2017**, *11*, 362–368. [CrossRef] [PubMed]
7. Supuran, C.T.; Capasso, C. An overview of the bacterial carbonic anhydrases. *Metabolites* **2017**, *7*, 56. [CrossRef] [PubMed]
8. Supuran, C.T. Carbonic anhydrases: Novel therapeutic applications for inhibitors and activators. *Nat. Rev. Drug. Discov.* **2008**, *7*, 168–181. [CrossRef] [PubMed]
9. DiMario, R.J.; Clayton, H.; Mukherjee, A.; Ludwig, M.; Moroney, J.V. Plant carbonic anhydrases: Structures, locations, evolution, and physiological roles. *Mol. Plant* **2017**, *10*, 30–46. [CrossRef] [PubMed]
10. Jin, S.; Sun, J.; Wunder, T.; Tang, D.; Cousins, A.B.; Sze, S.K.; Mueller-Cajar, O.; Gao, Y.G. Structural insights into the LCIB protein family reveals a new group of beta-carbonic anhydrases. *Proc. Natl. Acad. Sci. USA* **2016**, *113*, 14716–14721. [CrossRef] [PubMed]
11. Lindskog, S. Structure and mechanism of carbonic anhydrase. *Pharmacol. Ther.* **1997**, *74*, 1–20. [CrossRef]
12. Moroney, J.V.; Ma, Y.; Frey, W.D.; Fusilier, K.A.; Pham, T.T.; Simms, T.A.; DiMario, R.J.; Yang, J.; Mukherjee, B. The carbonic anhydrase isoforms of *Chlamydomonas reinhardtii*: Intracellular location, expression, and physiological roles. *Photosynth. Res.* **2011**, *109*, 133–149. [CrossRef] [PubMed]
13. Hanson, D.T.; Franklin, L.A.; Samuelsson, G.; Badger, M.R. The *Chlamydomonas reinhardtii* cia3 mutant lacking a thylakoid lumen-localized carbonic anhydrase is limited by CO_2 supply to rubisco and not photosystem II function in vivo. *Plant Physiol.* **2003**, *132*, 2267–2275. [CrossRef] [PubMed]

14. Mackinder, L.C.M.; Chen, C.; Leib, R.D.; Patena, W.; Blum, S.R.; Rodman, M.; Ramundo, S.; Adams, C.M.; Jonikas, M.C. A spatial interactome reveals the protein organization of the algal CO_2-concentrating mechanism. *Cell* **2017**, *171*, 133–147.e14. [CrossRef] [PubMed]

15. Hoang, C.V.; Chapman, K.D. Biochemical and molecular inhibition of plastidial carbonic anhydrase reduces the incorporation of acetate into lipids in cotton embryos and tobacco cell suspensions and leaves. *Plant Physiol.* **2002**, *128*, 1417–1427. [CrossRef] [PubMed]

16. Henry, R.P. Multiple roles of carbonic anhydrase in cellular transport and metabolism. *Annu. Rev. Physiol.* **1996**, *58*, 523–538. [CrossRef] [PubMed]

17. Geers, C.; Gros, G. Carbon dioxide transport and carbonic anhydrase in blood and muscle. *Physiol. Rev.* **2000**, *80*, 681–715. [CrossRef] [PubMed]

18. Riccardi, D.; Yang, S.; Cui, Q. Proton transfer function of carbonic anhydrase: Insights from QM/MM simulations. *Biochim. Biophys. Acta* **2010**, *1804*, 342–351. [CrossRef] [PubMed]

19. Roy, A.; Taraphder, S. Role of protein motions on proton transfer pathways in human carbonic anhydrase II. *Biochim. Biophys. Acta* **2010**, *1804*, 352–361. [CrossRef] [PubMed]

20. Missner, A.; Kugler, P.; Saparov, S.M.; Sommer, K.; Mathai, J.C.; Zeidel, M.L.; Pohl, P. Carbon dioxide transport through membranes. *J. Biol. Chem.* **2008**, *283*, 25340–25347. [CrossRef] [PubMed]

21. Parkkila, S.; Parkkila, A.K.; Rajaniemi, H.; Shah, G.N.; Grubb, J.H.; Waheed, A.; Sly, W.S. Expression of membrane-associated carbonic anhydrase XIV on neurons and axons in mouse and human brain. *Proc. Natl. Acad. Sci. USA* **2001**, *98*, 1918–1923. [CrossRef] [PubMed]

22. Supuran, C.T. Structure-based drug discovery of carbonic anhydrase inhibitors. *J. Enzyme Inhib. Med. Chem.* **2012**, *27*, 759–772. [CrossRef] [PubMed]

23. Supuran, C.T. Carbonic anhydrases: Catalytic and inhibition mechanisms, distribution and physiological roles. In *Carbonic Anhydrase: Its Inhibitors and Activators*; Supuran, C.T., Scozzafava, A., Conway, J., Eds.; CRC Press: Boca Raton, FL, USA, 2004; pp. 1–23.

24. Chegwidden, W.R.; Dodgson, S.J.; Spencer, I.M. *The Roles of Carbonic Anhydrase in Metabolism, Cell Growth and Cancer in Animals*; Chegwidden, W.R., Carter, N.D., Edwards, Y.H., Eds.; Birkhäuser Verlag: Boston, MA, USA, 2000; Volume 90, pp. 343–363.

25. Woolley, P. Models for metal ion function in carbonic anhydrase. *Nature* **1975**, *258*, 677–682. [CrossRef] [PubMed]

26. Esbaugh, A.J.; Tufts, B.L. The structure and function of carbonic anhydrase isozymes in the respiratory system of vertebrates. *Respir. Physiol. Neurobiol.* **2006**, *154*, 185–198. [CrossRef] [PubMed]

27. Dodgson, S.J.; Forster, R.E., 2nd. Carbonic anhydrase: Inhibition results in decreased urea production by hepatocytes. *J. Appl. Physiol.* **1986**, *60*, 646–652. [CrossRef] [PubMed]

28. Gay, C.V.; Schraer, H.; Anderson, R.E.; Cao, H. Current studies on the location and function of carbonic anhydrase in osteoclasts. *Ann. N. Y. Acad. Sci.* **1984**, *429*, 473–478. [CrossRef] [PubMed]

29. Henry, R.P. The function of invertebrate carbonic anhydrase in ion transport. *Ann. N. Y. Acad. Sci.* **1984**, *429*, 544–546. [CrossRef] [PubMed]

30. Park, H.; Song, B.; Morel, F.M. Diversity of the cadmium-containing carbonic anhydrase in marine diatoms and natural waters. *Environ. Microbiol.* **2007**, *9*, 403–413. [CrossRef] [PubMed]

31. DiMario, R.J.; Machingura, M.C.; Waldrop, G.L.; Moroney, J.V. The many types of carbonic anhydrases in photosynthetic organisms. *Plant Sci.* **2018**, *268*, 11–17. [CrossRef] [PubMed]

32. Hewett-Emmett, D. *Evolution and Distribution of the Carbonic Anhydrase Gene Families*; Chegwidden, W.R., Carter, N.D., Edwards, Y.H., Eds.; Birkhäuser Verlag: Boston, MA, USA, 2000; Volume 90, pp. 29–76.

33. Xu, Y.; Feng, L.; Jeffrey, P.D.; Shi, Y.; Morel, F.M. Structure and metal exchange in the cadmium carbonic anhydrase of marine diatoms. *Nature* **2008**, *452*, 56–61. [CrossRef] [PubMed]

34. Smith, K.S.; Jakubzick, C.; Whittam, T.S.; Ferry, J.G. Carbonic anhydrase is an ancient enzyme widespread in prokaryotes. *Proc. Natl. Acad. Sci. USA* **1999**, *96*, 15184–15189. [CrossRef] [PubMed]

35. Roberts, S.B.; Lane, T.W.; Morel, F.M.M. Carbonic anhydrase in the marine diatom *Thalassiosira weissflogii* (Bacillariophyceae). *J. Phycol.* **1997**, *33*, 845–850. [CrossRef]

36. Soto, A.R.; Zheng, H.; Shoemaker, D.; Rodriguez, J.; Read, B.A.; Wahlund, T.M. Identification and preliminary characterization of two cDNAs encoding unique carbonic anhydrases from the marine alga *Emiliania huxleyi*. *Appl. Environ. Microbiol.* **2006**, *72*, 5500–5511. [CrossRef] [PubMed]

37. De Simone, G.; Di Fiore, A.; Capasso, C.; Supuran, C.T. The zinc coordination pattern in the eta-carbonic anhydrase from Plasmodium falciparum is different from all other carbonic anhydrase genetic families. *Bioorg. Med. Chem. Lett.* **2015**, *25*, 1385–1389. [CrossRef] [PubMed]

38. Mangani, S.; Hakansson, K. Crystallographic studies of the binding of protonated and unprotonated inhibitors to carbonic anhydrase using hydrogen sulphide and nitrate anions. *Eur. J. Biochem.* **1992**, *210*, 867–871. [CrossRef] [PubMed]

39. West, D.; Kim, C.U.; Tu, C.; Robbins, A.H.; Gruner, S.M.; Silverman, D.N.; McKenna, R. Structural and kinetic effects on changes in the CO_2 binding pocket of human carbonic anhydrase II. *Biochemistry* **2012**, *51*, 9156–9163. [CrossRef] [PubMed]

40. Cronk, J.D.; Rowlett, R.S.; Zhang, K.Y.; Tu, C.; Endrizzi, J.A.; Lee, J.; Gareiss, P.C.; Preiss, J.R. Identification of a novel noncatalytic bicarbonate binding site in eubacterial beta-carbonic anhydrase. *Biochemistry* **2006**, *45*, 4351–4361. [CrossRef] [PubMed]

41. Iverson, T.M.; Alber, B.E.; Kisker, C.; Ferry, J.G.; Rees, D.C. A closer look at the active site of gamma-class carbonic anhydrases: High-resolution crystallographic studies of the carbonic anhydrase from *Methanosarcina thermophila*. *Biochemistry* **2000**, *39*, 9222–9231. [CrossRef] [PubMed]

42. Kisker, C.; Schindelin, H.; Alber, B.E.; Ferry, J.G.; Rees, D.C. A left-hand beta-helix revealed by the crystal structure of a carbonic anhydrase from the archaeon *Methanosarcina thermophila*. *EMBO J.* **1996**, *15*, 2323–2330. [PubMed]

43. Brinkman, R.; Margaria, R.; Meldrum, N.; Roughton, F. The CO_2 catalyst present in blood. *J. Physiol.* **1932**, *75*, 3–4.

44. Meldrum, N.; Roughton, F. Some properties of carbonic anhydrase, the CO_2 enzyme present in blood. *J. Physiol.* **1932**, *75*, 15.

45. Aspatwar, A.; Tolvanen, M.E.; Parkkila, S. Phylogeny and expression of carbonic anhydrase-related proteins. *BMC Mol. Biol.* **2010**, *11*, 25. [CrossRef] [PubMed]

46. Neish, A.C. Studies on chloroplasts: Their chemical composition and the distribution of certain metabolites between the chloroplasts and the remainder of the leaf. *Biochem. J.* **1939**, *33*, 300–308. [CrossRef] [PubMed]

47. Supuran, C.T. Bacterial carbonic anhydrases as drug targets: Toward novel antibiotics? *Front. Pharmacol.* **2011**, *2*, 34. [CrossRef] [PubMed]

48. Cox, E.H.; McLendon, G.L.; Morel, F.M.; Lane, T.W.; Prince, R.C.; Pickering, I.J.; George, G.N. The active site structure of *Thalassiosira weissflogii* carbonic anhydrase 1. *Biochemistry* **2000**, *39*, 12128–12130. [CrossRef] [PubMed]

49. Suarez Covarrubias, A.; Larsson, A.M.; Hogbom, M.; Lindberg, J.; Bergfors, T.; Björkelid, C.; Mowbray, S.L.; Unge, T.; Jones, T.A. Structure and function of carbonic anhydrases from Mycobacterium tuberculosis. *J. Biol. Chem.* **2005**, *280*, 18782–18789. [CrossRef] [PubMed]

50. Kimber, M.S.; Pai, E.F. The active site architecture of *Pisum sativum* beta-carbonic anhydrase is a mirror image of that of alpha-carbonic anhydrases. *EMBO J.* **2000**, *19*, 1407–1418. [CrossRef] [PubMed]

51. Smith, K.S.; Cosper, N.J.; Stalhandske, C.; Scott, R.A.; Ferry, J.G. Structural and kinetic characterization of an archaeal beta-class carbonic anhydrase. *J. Bacteriol.* **2000**, *182*, 6605–6613. [CrossRef] [PubMed]

52. Strop, P.; Smith, K.S.; Iverson, T.M.; Ferry, J.G.; Rees, D.C. Crystal structure of the "cab"-type beta class carbonic anhydrase from the archaeon *Methanobacterium thermoautotrophicum*. *J. Biol. Chem.* **2001**, *276*, 10299–10305. [CrossRef] [PubMed]

53. Bury-Mone, S.; Mendz, G.L.; Ball, G.E.; Thibonnier, M.; Stingl, K.; Ecobichon, C.; Avé, P.; Huerre, M.; Labigne, A.; Thiberge, J.M.; et al. Roles of alpha and beta carbonic anhydrases of Helicobacter pylori in the urease-dependent response to acidity and in colonization of the murine gastric mucosa. *Infect. Immun.* **2008**, *76*, 497–509. [CrossRef] [PubMed]

54. Rose, S.J.; Bermudez, L.E. Identification of Bicarbonate as a Trigger and Genes Involved with Extracellular DNA Export in Mycobacterial Biofilms. *mBio* **2016**, *7*, 6. [CrossRef] [PubMed]

55. Johnson, B.K.; Colvin, C.J.; Needle, D.B.; Mba Medie, F.; Champion, P.A.; Abramovitch, R.B. The Carbonic Anhydrase Inhibitor Ethoxzolamide Inhibits the Mycobacterium tuberculosis PhoPR Regulon and Esx-1 Secretion and Attenuates Virulence. *Antimicrob. Agents Chemother.* **2015**, *59*, 4436–4445. [CrossRef] [PubMed]

56. Parisi, G.; Perales, M.; Fornasari, M.S.; Colaneri, A.; Gonzalez-Schain, N.; Gomez-Casati, D.; Zimmermann, S.; Brennicke, A.; Araya, A.; Ferry, J.G.; et al. Gamma carbonic anhydrases in plant mitochondria. *Plant Mol. Biol.* **2004**, *55*, 193–207. [CrossRef] [PubMed]

57. Alber, B.E.; Ferry, J.G. A carbonic anhydrase from the archaeon *Methanosarcina thermophila*. *Proc. Natl. Acad. Sci. USA* **1994**, *91*, 6909–6913. [CrossRef] [PubMed]

58. Macauley, S.R.; Zimmerman, S.A.; Apolinario, E.E.; Evilia, C.; Hou, Y.M.; Ferry, J.G.; Sowers, K.R. The archetype gamma-class carbonic anhydrase, Cam, contains iron when synthesized in vivo. *Biochemistry* **2009**, *48*, 817–819. [CrossRef] [PubMed]

59. Alber, B.E.; Colangelo, C.M.; Dong, J.; Stalhandske, C.M.; Baird, T.T.; Tu, C.; Fierke, C.A.; Silverman, D.N.; Scott, R.A.; Ferry, J.G. Kinetic and spectroscopic characterization of the gamma-carbonic anhydrase from the methanoarchaeon *Methanosarcina thermophila*. *Biochemistry* **1999**, *38*, 13119–13128. [CrossRef] [PubMed]

60. Zimmerman, S.A.; Ferry, J.G. Proposal for a hydrogen bond network in the active site of the prototypic gamma-class carbonic anhydrase. *Biochemistry* **2006**, *45*, 5149–5157. [CrossRef] [PubMed]

61. McGinn, P.J.; Morel, F.M. Expression and regulation of carbonic anhydrases in the marine diatom *Thalassiosira pseudonana* and in natural phytoplankton assemblages from Great Bay, New Jersey. *Physiol. Plant.* **2008**, *133*, 78–91. [CrossRef] [PubMed]

62. Alterio, V.; Langella, E.; Viparelli, F.; Vullo, D.; Ascione, G.; Dathan, N.A.; Morel, F.M.; Supuran, C.T.; De Simone, G.; Monti, S.M. Structural and inhibition insights into carbonic anhydrase CDCA1 from the marine diatom *Thalassiosira weissflogii*. *Biochimie* **2012**, *94*, 1232–1241. [CrossRef] [PubMed]

63. Pena, K.L.; Castel, S.E.; de Araujo, C.; Espie, G.S.; Kimber, M.S. Structural basis of the oxidative activation of the carboxysomal gamma-carbonic anhydrase, CcmM. *Proc. Natl. Acad. Sci. USA* **2010**, *107*, 2455–2460. [CrossRef] [PubMed]

64. Price, G.D.; Howitt, SM.; Harrison, K.; Badger, M.R. Analysis of a genomic DNA region from the cyanobacterium *Synechococcus* sp. strain PCC7942 involved in carboxysome assembly and function. *J. Bacteriol.* **1993**, *175*, 2871–2879. [CrossRef] [PubMed]

65. Fukuzawa, H.; Fujiwara, S.; Yamamoto, Y.; Dionisio-Sese, M.L.; Miyachi, S. cDNA cloning, sequence, and expression of carbonic anhydrase in *Chlamydomonas reinhardtii*: Regulation by environmental CO_2 concentration. *Proc. Natl. Acad. Sci. USA* **1990**, *87*, 4383–4387. [CrossRef] [PubMed]

66. Fujiwara, S.; Fukuzawa, H.; Tachiki, A.; Miyachi, S. Structure and differential expression of two genes encoding carbonic anhydrase in *Chlamydomonas reinhardtii*. *Proc. Natl. Acad. Sci. USA* **1990**, *87*, 9779–9783. [CrossRef] [PubMed]

67. Burnell, J.N.; Gibbs, M.J.; Mason, J.G. Spinach chloroplastic carbonic anhydrase: Nucleotide sequence analysis of cDNA. *Plant Physiol.* **1990**, *92*, 37–40. [CrossRef] [PubMed]

68. Fawcett, T.W.; Browse, J.A.; Volokita, M.; Bartlett, S.G. Spinach carbonic anhydrase primary structure deduced from the sequence of a cDNA clone. *J. Biol. Chem.* **1990**, *265*, 5414–5417. [PubMed]

69. Roeske, C.A.; Ogren, W.L. Nucleotide sequence of pea cDNA encoding chloroplast carbonic anhydrase. *Nucleic Acids Res.* **1990**, *18*, 3413. [CrossRef] [PubMed]

70. Mitra, M.; Mason, C.; Lato, S.M.; Ynalvez, R.A.; Xiao, Y.; Moroney, J.V. The carbonic anhydrase gene families of *Chlamydomonas reinhardtii*. *Can. J. Bot.* **2005**, *83*, 780–795. [CrossRef]

71. Cardol, P.; Gonzalez-Halphen, D.; Reyes-Prieto, A.; Baurain, D.; Matagne, R.F.; Remacle, C. The mitochondrial oxidative phosphorylation proteome of *Chlamydomonas reinhardtii* deduced from the Genome Sequencing Project. *Plant Physiol.* **2005**, *137*, 447–459. [CrossRef] [PubMed]

72. Price, G.D.; Badger, M.R.; Woodger, F.J.; Long, B.M. Advances in understanding the cyanobacterial CO_2-concentrating-mechanism, CCM, functional components, Ci transporters, diversity, genetic regulation and prospects for engineering into plants. *J. Exp. Bot.* **2008**, *59*, 1441–1461. [CrossRef] [PubMed]

73. Toguri, T.; Muto, S.; Miyachi, S. Biosynthesis and intracellular processing of carbonic anhydrase in *Chlamydomonas reinhardtii*. *Eur. J. Biochem.* **1986**, *158*, 443–450. [CrossRef] [PubMed]

74. Kucho, K.; Yoshioka, S.; Taniguchi, F.; Ohyama, K.; Fukuzawa, H. Cis-acting elements and DNA-binding proteins involved in CO_2-responsive transcriptional activation of Cah1 encoding a periplasmic carbonic anhydrase in *Chlamydomonas reinhardtii*. *Plant Physiol.* **2003**, *133*, 783–793. [CrossRef] [PubMed]

75. Kucho, K.; Ohyama, K.; Fukuzawa, H. CO_2-responsive transcriptional regulation of CAH1 encoding carbonic anhydrase is mediated by enhancer and silencer regions in Chlamydomonas reinhardtii. *Plant Physiol.* **1999**, *121*, 1329–1337. [CrossRef] [PubMed]

76. Juvale, P.S.; Wagner, R.L.; Spalding, M.H. Opportunistic proteolytic processing of carbonic anhydrase 1 from Chlamydomonas in Arabidopsis reveals a novel route for protein maturation. *J. Exp. Bot.* **2016**, *67*, 2339–2351. [CrossRef] [PubMed]

77. Ishida, S.; Muto, S.; Miyachi, S. Structural analysis of periplasmic carbonic anhydrase 1 of *Chlamydomonas reinhardtii*. *Eur. J. Biochem.* **1993**, *214*, 9–16. [CrossRef] [PubMed]

78. Kamo, T.; Shimogawara, K.; Fukuzawa, H.; Muto, S.; Miyachi, S. Subunit constitution of carbonic anhydrase from *Chlamydomonas reinhardtii*. *Eur. J. Biochem.* **1990**, *192*, 557–562. [CrossRef] [PubMed]

79. Yoshioka, S.; Taniguchi, F.; Miura, K.; Inoue, T.; Yamano, T.; Fukuzawa, H. The novel Myb transcription factor LCR1 regulates the CO_2-responsive gene Cah1, encoding a periplasmic carbonic anhydrase in *Chlamydomonas reinhardtii*. *Plant Cell* **2004**, *16*, 1466–1477. [CrossRef] [PubMed]

80. Tachiki, A.; Fukuzawa, H.; Miyachi, S. Characterization of carbonic anhydrase isozyme CA2, which is the CAH2 gene product, in *Chlamydomonas reinhardtii*. *Biosci. Biotechnol. Biochem.* **1992**, *56*, 794–798. [CrossRef] [PubMed]

81. Rawat, M.; Moroney, J.V. Partial, characterization of a new isoenzyme of carbonic anhydrase isolated from *Chlamydomonas reinhardtii*. *J. Biol. Chem.* **1991**, *266*, 9719–9723. [PubMed]

82. Karlsson, J.; Clarke, A.K.; Chen, Z.Y.; Hugghins, S.Y.; Park, Y.I.; Husic, H.D.; Moroney, J.V.; Samuelsson, G. A novel alpha-type carbonic anhydrase associated with the thylakoid membrane in Chlamydomonas reinhardtii is required for growth at ambient CO_2. *EMBO J.* **1998**, *17*, 1208–1216. [PubMed]

83. Funke, R.P.; Kovar, J.L.; Weeks, D.P. Intracellular carbonic anhydrase is essential to photosynthesis in *Chlamydomonas reinhardtii* at atmospheric levels of CO_2: Demonstration via genomic complementation of the high-CO_2-requiring mutant ca-1. *Plant Physiol.* **1997**, *114*, 237–244. [CrossRef] [PubMed]

84. Karlsson, J.; Hiltonen, T.; Husic, H.D.; Ramazanov, Z.; Samuelsson, G. Intracellular carbonic anhydrase of *Chlamydomonas reinhardtii*. *Plant Physiol.* **1995**, *109*, 533–539. [CrossRef] [PubMed]

85. Blanco-Rivero, A.; Shutova, T.; Roman, M.J.; Villarejo, A.; Martinez, F. Phosphorylation controls the localization and activation of the lumenal carbonic anhydrase in *Chlamydomonas reinhardtii*. *PLoS ONE* **2012**, *7*, e49063. [CrossRef] [PubMed]

86. Park, Y.I.; Karlsson, J.; Rojdestvenski, I.; Pronina, N.; Klimov, V.; Oquist, G.; Samuelsson, G. Role of a novel photosystem II-associated carbonic anhydrase in photosynthetic carbon assimilation in *Chlamydomonas reinhardtii*. *FEBS Lett.* **1999**, *444*, 102–105. [CrossRef]

87. Benlloch, R.; Shevela, D.; Hainzl, T.; Grundstrom, C.; Shutova, T.; Messinger, J.; Samuelsson, G.; Sauer-Eriksson, A.E. Crystal structure and functional characterization of photosystem II-associated carbonic anhydrase CAH3 in *Chlamydomonas reinhardtii*. *Plant Physiol.* **2015**, *167*, 950–962. [CrossRef] [PubMed]

88. Sinetova, M.A.; Kupriyanova, E.V.; Markelova, A.G.; Allakhverdiev, S.I.; Pronina, N.A. Identification and functional role of the carbonic anhydrase Cah3 in thylakoid membranes of pyrenoid of *Chlamydomonas reinhardtii*. *Biochim. Biophys. Acta* **2012**, *1817*, 1248–1255. [CrossRef] [PubMed]

89. Giordano, M.; Norici, A.; Forssen, M.; Eriksson, M.; Raven, J.A. An anaplerotic role for mitochondrial carbonic anhydrase in *Chlamydomonas reinhardtii*. *Plant Physiol.* **2003**, *132*, 2126–2134. [CrossRef] [PubMed]

90. Villand, P.; Eriksson, M.; Samuelsson, G. Carbon dioxide and light regulation of promoters controlling the expression of mitochondrial carbonic anhydrase in *Chlamydomonas reinhardtii*. *Biochem. J.* **1997**, *327*, 51–57. [CrossRef] [PubMed]

91. Eriksson, M.; Karlsson, J.; Ramazanov, Z.; Gardestrom, P.; Samuelsson, G. Discovery of an algal mitochondrial carbonic anhydrase: Molecular cloning and characterization of a low-CO_2-induced polypeptide in *Chlamydomonas reinhardtii*. *Proc. Natl. Acad. Sci. USA* **1996**, *93*, 12031–12034. [CrossRef] [PubMed]

92. Ynalvez, R.A.; Xiao, Y.; Ward, A.S.; Cunnusamy, K.; Moroney, J.V. Identification and characterization of two closely related beta-carbonic anhydrases from *Chlamydomonas reinhardtii*. *Physiol. Plant.* **2008**, *133*, 15–26. [CrossRef] [PubMed]

93. Cardol, P.; Vanrobaeys, F.; Devreese, B.; Van Beeumen, J.; Matagne, R.F.; Remacle, C. Higher plant-like subunit composition of mitochondrial complex I from *Chlamydomonas reinhardtii*: 31 conserved components among eukaryotes. *Biochim. Biophys. Acta* **2004**, *1658*, 212–224. [CrossRef] [PubMed]

94. Coleman, J.R.; Berry, J.A.; Togasaki, R.K.; Grossman, A.R. Identification of Extracellular Carbonic Anhydrase of *Chlamydomonas reinhardtii*. *Plant Physiol.* **1984**, *76*, 472–477. [CrossRef] [PubMed]

95. Villarejo, A.; Shutova, T.; Moskvin, O.; Forssen, M.; Klimov, V.V.; Samuelsson, G. A photosystem II-associated carbonic anhydrase regulates the efficiency of photosynthetic oxygen evolution. *EMBO J.* **2002**, *21*, 1930–1938. [CrossRef] [PubMed]

96. Shutova, T.; Kenneweg, H.; Buchta, J.; Nikitina, J.; Terentyev, V.; Chernyshov, S.; Andersson, B.; Allakhverdiev, S.I.; Klimov, V.V.; Dau, H.; et al. The photosystem II-associated Cah3 in Chlamydomonas enhances the O_2 evolution rate by proton removal. *EMBO J.* **2008**, *27*, 782–791. [CrossRef] [PubMed]

97. Mitra, M.; Lato, S.M.; Ynalvez, R.A.; Xiao, Y.; Moroney, J.V. Identification of a new chloroplast carbonic anhydrase in *Chlamydomonas reinhardtii*. *Plant Physiol.* **2004**, *135*, 173–182. [CrossRef] [PubMed]

98. Wang, Y.; Stessman, D.J.; Spalding, M.H. The CO_2 concentrating mechanism and photosynthetic carbon assimilation in limiting CO_2: How Chlamydomonas works against the gradient. *Plant J.* **2015**, *82*, 429–448. [CrossRef] [PubMed]

99. Price, G.D.; Badger, M.R. Expression of Human Carbonic Anhydrase in the *Cyanobacterium Synechococcus* PCC7942 Creates a High CO_2-Requiring Phenotype: Evidence for a Central Role for Carboxysomes in the CO_2 Concentrating Mechanism. *Plant Physiol.* **1989**, *91*, 505–513. [CrossRef] [PubMed]

100. Choi, H.I.; Kim, J.Y. Quantitative analysis of the chemotaxis of a green alga, *Chlamydomonas reinhardtii*, to bicarbonate using diffusion-based microfluidic device. *Biomicrofluidics* **2016**, *10*, 014121. [CrossRef] [PubMed]

101. Hu, H.; Boisson-Dernier, A.; Israelsson-Nordstrom, M.; Bohmer, M.; Xue, S.; Ries, A.; Godoski, J.; Kuhn, J.M.; Schroeder, J.I. Carbonic anhydrases are upstream regulators of CO_2-controlled stomatal movements in guard cells. *Nat. Cell Biol.* **2010**, *12*, 87–93. [CrossRef] [PubMed]

102. Fromm, S.; Braun, H.P.; Peterhansel, C. Mitochondrial gamma carbonic anhydrases are required for complex I assembly and plant reproductive development. *New Phytol.* **2016**, *211*, 194–207. [CrossRef] [PubMed]

Article

Activation Studies of the β-Carbonic Anhydrase from the Pathogenic Protozoan *Entamoeba histolytica* with Amino Acids and Amines

Silvia Bua [1,†] , **Susanna Haapanen** [2,†] , **Marianne Kuuslahti** [2] , **Seppo Parkkila** [2,3] and **Claudiu T. Supuran** [1,*]

[1]   Sezione di Scienze Farmaceutiche e Nutraceutiche, Dipartimento Neurofarba, Università degli Studi di Firenze, Via U. Schiff 6, 50019 Sesto Fiorentino, 50019 Florence, Italy; silvia.bua@unifi.it
[2]   Faculty of Medicine and Health Technology, Tampere University, 33100 Tampere, Finland; Haapanen.Susanna.E@student.uta.fi (S.H.); Marianne.Kuuslahti@staff.uta.fi (M.K.); seppo.parkkila@staff.uta.fi (S.P.)
[3]   Fimlab Ltd., Tampere University Hospital, 33100 Tampere, Finland
*   Correspondence: claudiu.supuran@unifi.it; Tel./Fax: +39-055-4573729
†   These authors equally contributed to the article.

Received: 24 December 2018; Accepted: 31 January 2019; Published: 1 February 2019

Abstract: The β-carbonic anhydrase (CA, EC 4.2.1.1) from the pathogenic protozoan *Entamoeba histolytica*, EhiCA, was investigated for its activation with a panel of natural and non-natural amino acids and amines. EhiCA was potently activated by D-His, D-Phe, D-DOPA, L- and D-Trp, L- and D-Tyr, 4-amino-L-Tyr, histamine and serotonin, with K_As ranging between 1.07 and 10.1 μM. The best activator was D-Tyr (K_A of 1.07 μM). L-Phe, L-DOPA, L-adrenaline, L-Asn, L-Asp, L-Glu and L-Gln showed medium potency activation, with K_As of 16.5–25.6 μM. Some heterocyclic- alkyl amines, such as 2-pyridyl-methyl/ethyl-amine and 4-(2-aminoethyl)-morpholine, were devoid of EhiCA activating properties with $K_As > 100$ μM. As CA activators have poorly been investigated for their interaction with protozoan CAs, our study may be relevant for an improved understanding of the role of this enzyme in the life cycle of *E. histolytica*.

Keywords: *Entamoeba histolytica*; carbonic anhydrase; metalloenzymes; protozoan; amine; amino acid; activator

---

1. Introduction

Recently, we have reported [1,2] the cloning, purification and characterization of a β-carbonic anhydrase (CA, EC 4.2.1.1) present in the genome of the pathogenic protozoan *Entamoeba histolytica*, the etiological agents provoking amebiasis, an endemic disease in developing countries and also affecting travelers returning from risk zones [3–5]. In addition, invasive forms of *E. histolytica* infection were reported to lead to liver cysts, associated frequently with complications such as pleural effusion due to the rupture of the cysts as well as dissemination to extra-intestinal organs, e.g., the brain or pericardium, which occasionally may have fatal consequences [3,6]. In the previous work [1,2] we also investigated the inhibition profile of the new enzyme (nominated EhiCA) with the main classes of CA inhibitors (CAIs) [7–10], the sulfonamides and the inorganic anions [11–14]. Our main scope was to identify agents that by interference with the activity of this enzyme, might lead to anti-infectives with a novel mechanism of action, considering the fact that many CAs are essential in the life cycle of microorganisms belonging to the bacteria, fungal or protozoan domains [15–17]. As β-CAs are not present in mammals [18,19], effective EhiCA inhibitors may represent an alternative therapeutic option for this protozoan infection. In fact, in the previous work we have shown that inhibition of other protozoan CAs, such as the β-class enzyme from *Leishmania donovani* [20,21] or the α-CA from *Trypanosoma cruzi* [20,22,23], has important antiparasitic effects in vitro and in vivo [21].

Indeed, various pathogenic organisms belonging to the bacteria, fungal or protozoan domains encode for CAs, which have been investigated in some detail ultimately, in the search of anti-infectives with a diverse mechanism of action [7–10,14–23]. CAs catalyze the reaction between CO_2 and water, with formation of bicarbonate (HCO_3^-) and protons (H^+), and are highly effective catalysts, among the most efficient known so far in nature [7–10]. CAs are involved in various biochemical and metabolic processes, among which are acid-base homeostasis, respiration, biosynthesis of various metabolites (urea, glucose, fatty acids, carbamoyl phosphate), electrolytes secretion, etc. [7–12]. Seven distinct CA families are known to date, the α, β, γ, δ, ζ, η and θ class CAs, which are widespread all over the phylogenetic tree, from simple organisms, such as bacteria and Archaea, to more complex ones, such as vertebrates [7–10,24–28]. These diverse CA genetic families do not share significant sequence homology or structural identity, being an interesting example of convergent evolution at the molecular level [7–10]. In humans, as in many other vertebrates, only α-CAs are present, and their inhibition has been exploited from the pharmacological viewpoint for decades, for drugs such as diuretics [29], anticonvulsants [29,30], antiobesity [30] and more recently, antitumor agents [31]. However, these enzymes may also be activated [32] but the CA activators (CAAs) have seen fewer applications up until now. However, recent studies [33] pointed out to the possible application of CAAs targeting human enzymes for the enhancement of cognition. The nonvertebrate CAs were on the other hand only in the last few years investigated in some detail [34–37]. Here we report the first activation study of the β-CA from *E. histolytica* with a panel of amines and amino acid derivatives. As CAAs have poorly been investigated for their interaction with protozoan CAs, our study may be relevant for an improved understanding of the role of this enzyme in the life cycle of *E. histolytica*.

2. Results and Discussion

The catalytic activity of the recombinant EhiCA (for the CO_2 hydration reaction), has been recently reported [1,2], being measured by using a stopped flow technique [38]. EhiCA showed a significant catalytic activity for the physiologic, CO_2 hydration reaction, with the following kinetic parameters: $k_{cat} = 6.7 \times 10^5$ s^{-1} and $k_{cat}/K_m = 8.9 \times 10^7$ M^{-1} × s^{-1}. Thus, EhiCA is 1.8 times more effective as a catalyst compared to the slow human (h) isoform hCA I (considering the k_{cat}/K_m values) or 3.35 times more effective than hCA I (considering only the kinetic constant k_{cat}) [1,2]. EhiCA activity was also inhibited by the standard, clinically used sulfonamide CA inhibitor acetazolamide (AZA, 5-acetamido-1,3,4-thiadiazole-2-sulfonamide), with a K_I of 509 nM (data not shown here) [1,2].

Similar to all β-CAs investigated to date, EhiCA has a catalytically crucial zinc ion and its conserved protein ligands, which for this enzyme are: Cys50, His103 and Cys106 [1,2] The fourth metal ion ligand is a water molecule/hydroxide ion, which acts as nucleophile in the catalytic cycle (Equation (1) below). A catalytic dyad constituted by the pair Asp52-Arg54 [1,2], conserved in all enzymes belonging to the β-class is also present in EhiCA, presumably with the role to enhance the nucleophilicity of the zinc-coordinated water molecule [18–20]. However, the rate-determining step for many CAs is the generation of the nucleophilic species of the enzyme, represented by Equation (2) below:

$$H_2O$$

$$EZn^{2+}-OH^- + CO_2 \Leftrightarrow EZn^{2+}-HCO_3^- \Leftrightarrow EZn^{2+}-OH_2 + HCO_3 \tag{1}$$

$$EZn^{2+}- \text{-}OH_2 \Leftrightarrow EZn^{2+}-OH^- + H^+ \tag{2}$$

In most CAs, this step (Equation (2)) is assisted by amino acid residues from the active site [32], becoming an intramolecular step (instead of an intermolecular one), which is favored thermodynamically. Furthermore, the activators (CAAs) may participate in this step, as outlined in Equation (3):

$$EZn^{2+}\!-\!-OH_2 + A \Leftrightarrow [EZn^{2+}\!-\!-OH_2 - A] \Leftrightarrow [EZn^{2+}\!-\!OH^- - AH^+] \Leftrightarrow EZn^{2+}\!-\!OH^- + AH^+ \quad (3)$$

enzyme - activator complexes

The enzyme forms with the activator complexes (E-A complexes, where E stands for enzyme and A for activator), in which the proton transfer step from the zinc-coordinated water to the environment is intramolecular and thus, more efficient than the corresponding intermolecular process shown schematically in Equation (2) [32]. In fact, X-ray crystal structures are available for many CAs to which activators are bound within the active site [32,39–41], but only for α-class enzymes these structures have been reported to date. The activator binding site for the α-CAs is situated at the entrance of the active site cavity not far away from His64, which acts as proton shuttle residue in the process described by Equation (2) [32,39–41].

We have performed detailed kinetic measurements of EhiCA activity in the presence of amine and amino acid activators (Figure 1), such as for example L-Trp (Table 1). Data of Table 1 show that the presence of L-Trp does not change the K_M, both for the α-class enzymes hCA I/II as well as the β-CA, EhiCA, investigated here. Interestingly, it has an effect on the k_{cat}, which at 10-μM concentration of activator leads to a 2.83 times enhancement of the kinetic constant for the protozoan enzyme, from 6.7 $\times 10^5$ s^{-1} to 1.9 $\times 10^6$ s^{-1} (Table 1).

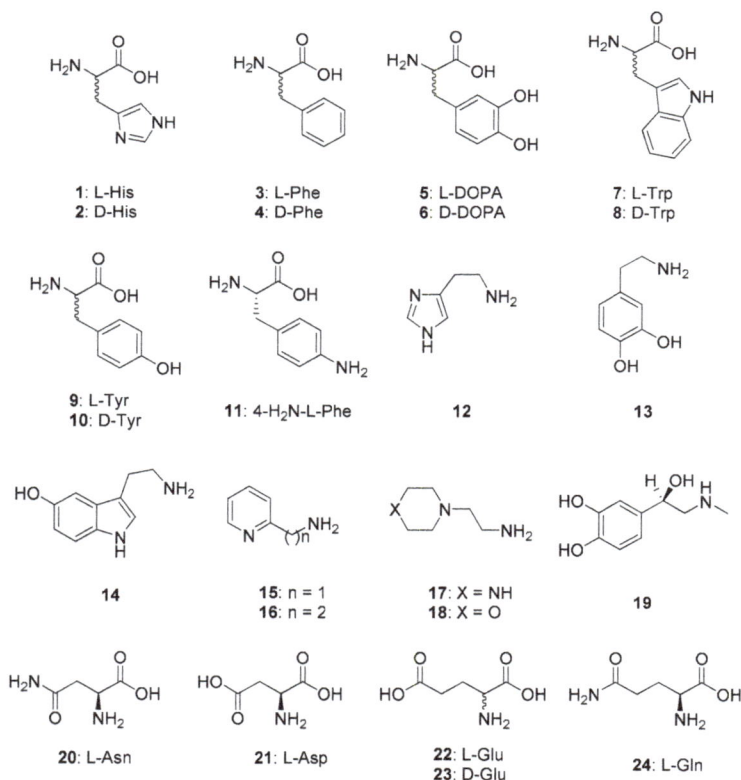

Figure 1. CAAs of type **1–24** used in the present study.

Table 1. Activation of human carbonic anhydrase (hCA) isozymes I, II, and EhiCA with L-Trp at 25 °C for the CO_2 hydration reaction [38].

Isozyme	k_{cat} *	K_M *	$(k_{cat})_{L-Trp}$ **	K_A *** (μM)
	(s^{-1})	(mM)	(s^{-1})	L-Trp
hCA I [a]	2.0×10^5	4.0	3.4×10^5	44.0
hCA II [a]	1.4×10^6	9.3	4.9×10^6	27.0
LdCA	9.35×10^5	15.8	1.9×10^6	4.02
EhiCA [b]	6.7×10^5	7.5	1.9×10^6	5.24

* Observed catalytic rate without activator. K_M values in the presence and the absence of activators were the same for the various CAs (data not shown).; ** Observed catalytic rate in the presence of 10 μM activator; *** The activation constant (K_A) for each enzyme was obtained by fitting the observed catalytic enhancements as a function of the activator concentration [41]. Mean from at least three determinations by a stopped-flow, CO_2 hydrase method [38]. Standard errors were in the range of 5–10% of the reported values (data not shown); [a] Human recombinant isozymes, from ref. [32]; [b] Protozoan recombinant enzyme, this work.

In order to obtain an activation profile of EhiCA with a wide range of amino acid and amine activators of types 11–24, we performed dose response curves of the activation of EhiCA in the presence of increasing concentrations of activators, in order to determine the activation constants K_A-s (see Materials and Methods for details). We included in our study the amino acids and amines which were investigated as activators of CAs belonging to various classes from diverse organisms [32–37,40–42]. These activation data are reported in Table 2, in which, for comparison reasons, the activation of the human isoforms hCA I and II and of the protozoan β-CA from *Leishmania donovani chagasi* are also presented.

Table 2. Activation constants of hCA I, hCA II and the protozoan enzymes LdcCA (*L. donovani chagasi*) and EhiCA (*E. histolytica*) with amino acids and amines **1–24**. Data for hCA I and II are from [32] and for LdcCA from [42].

No.	Compound	K_A (mM) *			
		hCA I [a]	hCA II [a]	LdcCA [b]	EhiCA [c]
1	L-His	0.03	10.9	8.21	78.7
2	D-His	0.09	43	4.13	9.83
3	L-Phe	0.07	0.013	9.16	16.5
4	D-Phe	86	0.035	3.95	10.1
5	L-DOPA	3.1	11.4	1.64	16.6
6	D-DOPA	4.9	7.8	5.47	4.05
7	L-Trp	44	27	4.02	5.24
8	D-Trp	41	12	6.18	4.95
9	L-Tyr	0.02	0.011	8.05	4.52
10	D-Tyr	0.04	0.013	1.27	1.07
11	4-H_2N-L-Phe	0.24	0.15	15.9	8.12
12	Histamine	2.1	125	0.74	7.38
13	Dopamine	13.5	9.2	0.81	30.8
14	Serotonin	45	50	0.62	4.94
15	2-Pyridyl-methylamine	26	34	0.23	>100
16	2-(2-Aminoethyl)pyridine	13	15	0.012	>100
17	1-(2-Aminoethyl)-piperazine	7.4	2.3	0.009	43.8
18	4-(2-Aminoethyl)-morpholine	0.14	0.19	0.94	>100
19	L-Adrenaline	0.09	96	4.89	25.6
20	L-Asn	11.3	>100	4.76	23.8
21	L-Asp	5.2	>100	0.3	23.9
22	L-Glu	6.43	>100	12.9	25.5
23	D-Glu	10.7	>100	0.082	30.3
24	L-Gln	>100	>50	2.51	20.1

* Mean from three determinations by a stopped-flow, CO_2 hydrase method [38]. Standard errors were in the range of 5–10% of the reported values (data not shown). [a] Human recombinant isozymes, from ref. [32]; [b] Protozoan recombinant enzyme, from ref. [42]; [c] This work.

The structure–activity relationship (SAR) for the activation of EhiCA with compounds **1–24** revealed the following observations:

(i)   Some heterocyclic-alkyl amines, such as 2-pyridyl-methyl/ethyl-amine **15**, **16** and 4-(2-aminoethyl)-morpholine, were devoid of EhiCA activating properties up to 100 µM concentration of activator in the assay system. All these compounds are structurally related, possessing a heterocyclic ring and aminomethyl/aminoethyl moieties in their molecules.

(ii) L-His, dopamine, 1-(2-aminoethyl)-piperazine and D-Glu were poor EhiCA activators, with activation constants ranging between 30.3 and 78.7 µM (Table 2). There is no strong structural correlation between these three compounds.

(iii) Many of the compounds investigated here showed medium potency efficacy as EhiCA activators, with K_As ranging between 16.5 and 25.6 µM. They include L-Phe, L-DOPA, L-adrenaline, L-Asn, L-Asp, L-Glu and L-Gln. It may be observed that there are no remarkable differences of activity between the pairs L-Asp/L-Asn and L-Glu/L-Gln, whereas D-Glu was more ineffective compared to L-Glu. This is in fact the exception, as for other L-/D-enantiomeric amino acids investigated here, the D-enantiomer was the most effective activator (see later in the text).

(iv) Effective EhiCA activating properties were detected for the following amino acids/amines: D-His, D-Phe, D-DOPA, L- and D-Trp, L- and D-Tyr, 4-amino-L-Tyr, histamine and serotonin, which showed K_As ranging between 1.07 and 10.1 µM. The best activator was D-Tyr (K_A of 1.07 µM). In fact for all aromatic amino acids investigated here, the D-enantiomer was more effective as EhiCA activator compared to the corresponding L-enantiomer. For the Phe-Tyr-DOPA subseries, the activity increased by hydroxylation of the Phe, achieving a maximum for Tyr and then slightly decreased with the introduction of an additional OH moiety in DOPA, but always the D-enantiomers were better activators compared to the L-ones. The loss of the carboxyl moiety, such as in histamine and serotonin, did not lead to important changes of activity compared to the corresponding D-amino acids, but in the case of dopamine, the activating efficacy was much lower compared to those of both L- and D-DOPA.

(v)   The activation profile of EhiCA with amino acid and amine derivatives is rather different from those of other CAs, among which the protozoan β-CA from *Leishmania donovani chagasi* (LdcCA) or the α-class human CAs, isoforms hCA I and II. For example **17** was a nanomolar activator for LdcCA whereas its affinity for EhiCA was of only 43.8 µM. For the moment, no EhiCA-selective activators were detected.

3. Materials and Methods

3.1. EhiCA Production and Purification

The protocol described in [1,2] has been used to obtain purified recombinant EhiCA. All activators were commercially available from Sigma-Aldrich (Milan, Italy) and were of the highest purity available.

3.2. CA activity and Activation Measurements

An Sx.18Mv-R Applied Photophysics (Oxford, UK) stopped-flow instrument has been used to assay the catalytic activity of various CA isozymes for CO_2 hydration reaction [38]. Phenol red (at a concentration of 0.2 mM) was used as indicator, working at the absorbance maximum of 557 nm, with 10 mM Hepes (pH 7.5, for α-CAs) or TRIS (pH 8.3, for β-CAs) as buffers, 0.1 M $NaClO_4$ (for maintaining constant ionic strength), following the CA-catalyzed CO_2 hydration reaction for a period of 10 s at 25 °C. The CO_2 concentrations ranged from 1.7 to 17 mM for the determination of the kinetic parameters and inhibition constants. For each activator at least six traces of the initial 5–10% of the reaction have been used for determining the initial velocity. The uncatalyzed rates were determined in the same manner and subtracted from the total observed rates. Stock solutions of activators (at 0.1 mM) were prepared in distilled-deionized water and dilutions up to 1 nM were made thereafter with the assay buffer. Enzyme and activator solutions were pre-incubated together for 15 min prior to

Metabolites **2019**, *9*, 26

assay, in order to allow for the formation of the enzyme–activator complexes. The activation constant (K_A), defined similarly with the inhibition constant K_I, can be obtained by considering the classical Michaelis–Menten equation (Equation (4)), which has been fitted by nonlinear least squares by using PRISM 3:

$$v = v_{max}/\{1 + (K_M/[S]) \, (1 + [A]_f/K_A)\} \tag{4}$$

where $[A]_f$ is the free concentration of activator.

Working at substrate concentrations considerably lower than K_M ($[S] << K_M$), and considering that $[A]_f$ can be represented in the form of the total concentration of the enzyme ($[E]_t$) and activator ($[A]_t$), the obtained competitive steady-state equation for determining the activation constant is given by Equation (5):

$$v = v_0 \cdot K_A/\{K_A + ([A]_t - 0.5\{([A]_t + [E]_t + K_A) - ([A]_t + [E]_t + K_A)^2 - 4[A]_t \cdot [E]_t)^{1/2}\}\} \tag{5}$$

where v_0 represents the initial velocity of the enzyme-catalyzed reaction in the absence of activator [32,41,42].

4. Conclusions

We report the first activation study of the β-CA from the protozoan parasite *Entamoeba histolytica*, EhiCA, with a panel of amino acids and amines, some of which are important autacoids. The enzyme was potently activated by D-His, D-Phe, D-DOPA, L- and D-Trp, L- and D-Tyr, 4-amino-L-Tyr, histamine and serotonin, with K_As ranging between 1.07 and 10.1 µM. The best activator was D-Tyr (K_A of 1.07 µM). L-Phe, L-DOPA, L-adrenaline, L-Asn, L-Asp, L-Glu and L-Gln showed medium potency activation, with K_As of 16.5–25.6 µM. Some heterocyclic-alkyl amines, such as 2-pyridyl-methyl/ethyl-amine and 4-(2-aminoethyl)-morpholine, were devoid of EhiCA activating properties with K_As > 100 µM. The X-ray crystal structure of this enzyme is not known for the moment, and in addition, no adducts of other parasite enzymes complexed with activators are available so far in order to rationalize our results. However, as CAAs have poorly been investigated for their interaction with protozoan CAs, our study may be relevant for an improved understanding of the role of this enzyme in the life cycle of *E. histolytica*.

Author Contributions: S.B., S.H. and M.K. performed the experiments. S.P. and C.T.S. designed the experiments, evaluated the data and wrote the manuscript. All authors participated to the writing of the work.

Funding: Academy of Finland, Sigrid Juselius Foundation and Jane and Aatos Erkko Foundation.

Acknowledgments: The authors acknowledge the Tampere Facility of Protein Services (PS) for their service. The work has been supported by grants from the Academy of Finland, Sigrid Juselius Foundation and Jane and Aatos Erkko Foundation.

Conflicts of Interest: The authors declare no conflict of interest.

References

1. Bua, S.; Haapanen, S.; Kuuslahti, M.; Parkkila, S.; Supuran, C.T. Sulfonamide Inhibition Studies of a New β-Carbonic Anhydrase from the Pathogenic Protozoan *Entamoeba histolytica*. *Int. J. Mol. Sci.* **2018**, *19*, E3946. [CrossRef] [PubMed]
2. Haapanen, S.; Bua, S.; Kuuslahti, M.; Parkkila, S.; Supuran, C.T. Cloning, Characterization and Anion Inhibition Studies of a β-Carbonic Anhydrase from the Pathogenic Protozoan *Entamoeba histolytica*. *Molecules* **2018**, *23*, E3112. [CrossRef] [PubMed]
3. Shirley, D.T.; Farr, L.; Watanabe, K.; Moonah, S. A Review of the Global Burden, New Diagnostics, and Current Therapeutics for Amebiasis. *Open Forum Infect. Dis.* **2018**, *5*, ofy161. [CrossRef]
4. Hashmey, N.; Genta, N.; White, N., Jr. Parasites and Diarrhea. I: Protozoans and Diarrhea. *J. Travel Med.* **1997**, *4*, 17–31. [CrossRef] [PubMed]
5. Loftus, B.; Anderson, I.; Davies, R.; Alsmark, U.C.; Samuelson, J.; Amedeo, P.; Roncaglia, P.; Berriman, M.; Hirt, R.P.; Mann, B.J.; et al. The genome of the protist parasite *Entamoeba histolytica*. *Nature* **2005**, *433*, 865–868. [CrossRef] [PubMed]

6. Andrade, R.M.; Reed, S.L. New drug target in protozoan parasites: The role of thioredoxin reductase. *Front. Microbiol.* **2015**, *6*, 975. [CrossRef]
7. Supuran, C.T. Structure and function of carbonic anhydrases. *Biochem. J.* **2016**, *473*, 2023–2032. [CrossRef]
8. Supuran, C.T. Carbonic Anhydrases and Metabolism. *Metabolites* **2018**, *8*, E25. [CrossRef]
9. Capasso, C.; Supuran, C.T. An overview of the alpha-, beta- and gamma-carbonic anhydrases from Bacteria: Can bacterial carbonic anhydrases shed new light on evolution of bacteria? *J. Enzyme Inhib. Med. Chem.* **2015**, *30*, 325–332. [CrossRef]
10. Supuran, C.T.; Capasso, C. An Overview of the Bacterial Carbonic Anhydrases. *Metabolites* **2017**, *7*, E56. [CrossRef]
11. Supuran, C.T. Carbonic anhydrases: Novel therapeutic applications for inhibitors and activators. *Nat. Rev. Drug Discov.* **2008**, *7*, 168–181. [CrossRef] [PubMed]
12. Neri, D.; Supuran, C.T. Interfering with pH regulation in tumours as a therapeutic strategy. *Nat. Rev. Drug Discov.* **2011**, *10*, 767–777. [CrossRef] [PubMed]
13. Supuran, C.T. Carbonic Anhydrase Inhibition and the Management of Hypoxic Tumors. *Metabolites* **2017**, *7*, E48. [CrossRef] [PubMed]
14. Supuran, C.T. Advances in structure-based drug discovery of carbonic anhydrase inhibitors. *Expert Opin. Drug Discov.* **2017**, *12*, 61–88. [CrossRef] [PubMed]
15. Nishimori, I.; Onishi, S.; Takeuchi, H.; Supuran, C.T. The α and β-Classes Carbonic Anhydrases from Helicobacter pylori as Novel Drug Targets. *Curr. Pharm. Des.* **2008**, *14*, 622–630. [PubMed]
16. Supuran, C.T.; Capasso, C. New light on bacterial carbonic anhydrases phylogeny based on the analysis of signal peptide sequences. *J. Enzyme Inhib. Med. Chem.* **2016**, *31*, 1254–1260. [CrossRef] [PubMed]
17. Supuran, C.T.; Capasso, C. Biomedical applications of prokaryotic carbonic anhydrases. *Expert Opin. Ther. Pat.* **2018**, *28*, 745–754. [CrossRef] [PubMed]
18. Zolfaghari Emameh, R.; Barker, H.; Hytönen, V.P.; Tolvanen, M.E.E.; Parkkila, S. Beta carbonic anhydrases: Novel targets for pesticides and anti-parasitic agents in agriculture and livestock husbandry. *Parasites Vect.* **2014**, *7*, 403. [CrossRef]
19. Syrjänen, L.; Parkkila, S.; Scozzafava, A.; Supuran, C.T. Sulfonamide inhibition studies of the β carbonic anhydrase from Drosophila melanogaster. *Bioorg. Med. Chem. Lett.* **2014**, *24*, 2797–2801. [CrossRef]
20. Vermelho, A.B.; Capaci, G.R.; Rodrigues, I.A.; Cardoso, V.S.; Mazotto, A.M.; Supuran, C.T. Carbonic anhydrases from Trypanosoma and Leishmania as anti-protozoan drug targets. *Bioorg. Med. Chem.* **2017**, *25*, 1543–1555. [CrossRef]
21. Da Silva Cardoso, V.; Vermelho, A.B.; Ricci Junior, E.; Almeida Rodrigues, I.; Mazotto, A.M.; Supuran, C.T. Antileishmanial activity of sulphonamide nanoemulsions targeting the β-carbonic anhydrase from Leishmania species. *J. Enzyme Inhib. Med. Chem.* **2018**, *33*, 850–857. [CrossRef] [PubMed]
22. Vermelho, A.B.; da Silva Cardoso, V.; Ricci Junior, E.; Dos Santos, E.P.; Supuran, C.T. Nanoemulsions of sulfonamide carbonic anhydrase inhibitors strongly inhibit the growth of Trypanosoma cruzi. *J. Enzyme Inhib. Med. Chem.* **2018**, *33*, 139–146. [CrossRef] [PubMed]
23. De Menezes Dda, R.; Calvet, C.M.; Rodrigues, G.C.; de Souza Pereira, M.C.; Almeida, I.R.; de Aguiar, A.P.; Supuran, C.T.; Vermelho, A.B. Hydroxamic acid derivatives: A promising scaffold for rational compound optimization in Chagas disease. *J. Enzyme Inhib. Med. Chem.* **2016**, *31*, 964–973. [CrossRef] [PubMed]
24. Rowlett, R.S. Structure and catalytic mechanism of the β-carbonic anhydrases. *Biochim. Biophys. Acta Prot. Proteom.* **2010**, *1804*, 362–373. [CrossRef] [PubMed]
25. Covarrubias, A.S.; Bergfors, T.; Jones, T.A.; Högbom, M. Structural mechanics of the pH-dependent activity of beta-carbonic anhydrase from Mycobacterium tuberculosis. *J. Biol. Chem.* **2006**, *281*, 4993–4999. [CrossRef] [PubMed]
26. Murray, A.B.; Aggarwal, M.; Pinard, M.; Vullo, D.; Patrauchan, M.; Supuran, C.T.; McKenna, R. Structural Mapping of Anion Inhibitors to β-Carbonic Anhydrase psCA3 from Pseudomonas aeruginosa. *Chem. Med. Chem.* **2018**, *13*, 2024–2029. [CrossRef] [PubMed]
27. Zimmerman, S.A.; Ferry, J.G.; Supuran, C.T. Inhibition of the archaeal beta-class (Cab) and gamma-class (Cam) carbonic anhydrases. *Curr. Top. Med. Chem.* **2007**, *7*, 901–908. [CrossRef]
28. De Simone, G.; Supuran, C.T. (In) organic anions as carbonic anhydrase inhibitors. *J. Inorg. Biochem.* **2012**, *111*, 117–129. [CrossRef]

29. Supuran, C.T. Applications of carbonic anhydrases inhibitors in renal and central nervous system diseases. *Expert Opin. Ther. Pat.* **2018**, *28*, 713–721. [CrossRef]

30. Supuran, C.T. Carbonic anhydrase inhibitors and their potential in a range of therapeutic areas. *Expert Opin. Ther. Pat.* **2018**, *28*, 709–712. [CrossRef]

31. Nocentini, A.; Supuran, C.T. Carbonic anhydrase inhibitors as antitumor/antimetastatic agents: A patent review (2008–2018). *Expert Opin Ther Pat.* **2018**, *28*, 729–740. [CrossRef] [PubMed]

32. Supuran, C.T. Carbonic anhydrase activators. *Future Med. Chem.* **2018**, *10*, 561–573. [CrossRef] [PubMed]

33. Canto de Souza, L.; Provensi, G.; Vullo, D.; Carta, F.; Scozzafava, A.; Costa, A.; Schmidt, S.D.; Passani, M.B.; Supuran, C.T.; Blandina, P. Carbonic anhydrase activation enhances object recognition memory in mice through phosphorylation of the extracellular signal-regulated kinase in the cortex and the hippocampus. *Neuropharmacology* **2017**, *118*, 148–156. [CrossRef] [PubMed]

34. Angeli, A.; Kuuslahti, M.; Parkkila, S.; Supuran, C.T. Activation studies with amines and amino acids of the α-carbonic anhydrase from the pathogenic protozoan *Trypanosoma cruzi*. *Bioorg. Med. Chem.* **2018**, *26*, 4187–4190. [CrossRef] [PubMed]

35. Angeli, A.; Del Prete, S.; Alasmary, F.A.S.; Alqahtani, L.S.; AlOthman, Z.; Donald, W.A.; Capasso, C.; Supuran, C.T. The first activation studies of the η-carbonic anhydrase from the malaria parasite *Plasmodium falciparum* with amines and amino acids. *Bioorg. Chem.* **2018**, *80*, 94–98. [CrossRef] [PubMed]

36. Stefanucci, A.; Angeli, A.; Dimmito, M.P.; Luisi, G.; Del Prete, S.; Capasso, C.; Donald, W.A.; Mollica, A.; Supuran, C.T. Activation of β- and γ-carbonic anhydrases from pathogenic bacteria with tripeptides. *J. Enzyme Inhib. Med. Chem.* **2018**, *33*, 945–950. [CrossRef] [PubMed]

37. Angeli, A.; Alasmary, F.A.S.; Del Prete, S.; Osman, S.M.; AlOthman, Z.; Donald, W.A.; Capasso, C.; Supuran, C.T. The first activation study of a δ-carbonic anhydrase: TweCAδ from the diatom *Thalassiosira weissflogii* is effectively activated by amines and amino acids. *J. Enzyme Inhib. Med. Chem.* **2018**, *33*, 680–685. [CrossRef] [PubMed]

38. Khalifah, R.G. The carbon dioxide hydration activity of carbonic anhydrase. I. Stop-flow kinetic studies on the native human isoenzymes B and C. *J. Biol. Chem.* **1971**, *246*, 2561–2573. [PubMed]

39. Briganti, F.; Mangani, S.; Orioli, P.; Scozzafava, A.; Vernaglione, G.; Supuran, C.T. Carbonic anhydrase activators: X-ray crystallographic and spectroscopic investigations for the interaction of isozymes I and II with histamine. *Biochemistry* **1997**, *36*, 10384–10392. [CrossRef] [PubMed]

40. Clare, B.W.; Supuran, C.T. Carbonic anhydrase activators. 3: Structure-activity correlations for a series of isozyme II activators. *J. Pharm. Sci.* **1994**, *83*, 768–773. [CrossRef] [PubMed]

41. Temperini, C.; Scozzafava, A.; Vullo, D.; Supuran, C.T. Carbonic anhydrase activators. Activation of isoforms I, II, IV, VA, VII, and XIV with L- and D-phenylalanine and crystallographic analysis of their adducts with isozyme II: Stereospecific recognition within the active site of an enzyme and its consequences for the drug design. *J. Med. Chem.* **2006**, *49*, 3019–3027. [PubMed]

42. Angeli, A.; Donald, W.A.; Parkkila, S.; Supuran, C.T. Activation studies with amines and amino acids of the β-carbonic anhydrase from the pathogenic protozoan. *Leishmania donovani chagasi*. *Bioorg. Chem.* **2018**, *78*, 406–410. [CrossRef] [PubMed]

metabolites

MDPI

Review
An Overview of the Bacterial Carbonic Anhydrases

Claudiu T. Supuran [1,*] (ID) and Clemente Capasso [2,*] (ID)

[1]   Dipartimento Neurofarba, Sezione di Scienze Farmaceutiche, Laboratorio di Chimica Bioinorganica,
       Università degli Studi di Firenze, Polo Scientifico, Via U. Schiff 6, Sesto Fiorentino, 50019 Florence, Italy
[2]   Istituto di Bioscienze e Biorisorse (IBBR), Consiglio Nazionale delle Ricerche (CNR),
       via Pietro Castellino 111, 80131 Napoli, Italy
*   Correspondence: claudiu.supuran@unifi.it (C.T.S.); clemente.capasso@ibbr.cnr.it (C.C.);
       Tel.: +39-55-457-3729 (C.T.S.); +39-81-613-2559 (C.C.)

Received: 25 October 2017; Accepted: 8 November 2017; Published: 11 November 2017

Abstract: Bacteria encode carbonic anhydrases (CAs, EC 4.2.1.1) belonging to three different genetic families, the α-, β-, and γ-classes. By equilibrating CO_2 and bicarbonate, these metalloenzymes interfere with pH regulation and other crucial physiological processes of these organisms. The detailed investigations of many such enzymes from pathogenic and non-pathogenic bacteria afford the opportunity to design both novel therapeutic agents, as well as biomimetic processes, for example, for CO_2 capture. Investigation of bacterial CA inhibitors and activators may be relevant for finding antibiotics with a new mechanism of action.

Keywords: bacterial carbonic anhydrases; inhibitors; antibiotic; CO_2 capture; engineered bacteria

1. Introduction

In the time of emerging antibiotic resistance, the improvement of pharmacological arsenal against bacterial pathogens is of pivotal importance [1,2]. Among the strategies adopted for fighting antibiotic resistance, the effectiveness is a structural upgrade of the current clinical drugs for generating novel antibiotics [1,2]. The limit of this strategy is that the newly generated drugs could have a limited lifespan due to the possible resistance that they will develop sooner or later. Fortunately, in the last years, the DNA sequencing approach applied to the bacterial genome allowed the discovery of numerous genes encoding for enzymes which catalyze metabolic pathways essential for the life cycle and/or the virulence of these microbes [3]. Thus, scientists possess in vitro essential bacterial targets for finding and designing new antiinfectives able to disarm pathogens through their inhibition, as well as to bypass their resistance to conventional antimicrobials. In fact, the inhibition of the new bacterial targets takes place through mechanisms different from those usually represented by the block of DNA gyrase, the inhibition of the ribosomal function, and the shut down of the cell-wall biosynthesis, as most clinically used antibiotics act [3]. Moreover, this strategy will result in the development of new antiinfectives, which can replace those already used in clinics with increasing bacterial resistance. In this context, the superfamily of carbonic anhydrases (CAs, EC 4.2.1.1) represents a valuable member of such new macromolecules affecting the growth of microorganisms or making them vulnerable to the host defense mechanisms [4–8]. These metalloenzymes catalyze the simple but physiologically crucial reaction of carbon dioxide hydration to bicarbonate and protons: $CO_2 + H_2O \rightleftharpoons HCO_3^- + H+$ [4,8–14], and they are involved in the transport and supply of CO_2 or HCO_3^- in pH homeostasis, the secretion of electrolytes, biosynthetic processes, and photosynthesis [15,16]. Moreover, CAs are target molecules of some antibacterial drugs, such as sulfanilamide.

Since CAs are very effective catalysts for the conversion of CO_2 to bicarbonate, the CA superfamily might be involved in the capture/sequestration of CO_2 from combustion gases with the goal of alleviating the global warming effects through a reduction of CO_2 emissions in the atmosphere [17]. The production of CO_2 is linked to the industrial development that must necessarily reduce its production. To decrease the amount of CO_2 in the atmosphere, a number of CO_2 sequestration methods have been proposed [17]. Most of them require that the CO_2 captured from the flue gases is compressed, transported to the sequestration site, and injected into specific areas for long-term storage [17]. All these procedures lead to an increase in the costs of the capture and storage processes [17]. For this reason, the biomimetic approach represents a valid strategy for CO_2 capture. It allows the CO_2 conversion to water-soluble ions and offers many advantages over other methods, such as its eco-compatibility and the possibility to use the reaction products for multiple applications, with no added costs. Furthermore, thermophilic CAs are still active at high temperatures compared to their mesophilic counterparts, and their use is preferred in environments characterized by hard conditions, such as those of the carbon capture process (high temperature, high salinity, extreme pH) [18–21].

2. Classification and Structure

2.1. CA-Classes

The CAs make up a widely distributed class of metalloenzymes with the catalytically active species represented by a metal hydroxide derivative [4–8]. CAs are grouped into seven genetically distinct families, named α-, β-, γ-, δ-, ζ-, η-, and θ-CAs, with different folds and structures but common CO_2 hydratase activity, coupled to low sequence similarity. Bacteria encode for enzymes belonging to α-, β-, and γ-CA classes [8,22–27]. Bacteria have a very intricate CA distribution pattern because some of them encode CAs belonging to only one family, whilst others encode those from two or even three different genetic families. The α- and β-CAs are metalloenzymes, which use the Zn(II) ion as a catalytic metal; γ-CAs are Fe(II) enzymes, but they are also active with bound Zn(II) or Co(II) ions [28–35]. The metal ion from the CA active site is coordinated by three His residues in the α- and γ-classes (Figures 1 and 2), and by one His and two Cys residues in the β-class (Figure 3). The fourth ligand is a water molecule/hydroxide ion acting as a nucleophile in the catalytic cycle of the enzyme [8,24,25,36–39]. The rate-determining step of the entire catalytic process is the formation of the metal hydroxide species of the enzyme by the transfer of a proton from the metal-coordinated water molecule to the surrounding solvent, possibly via proton-shuttling residues [5,8,22,24,25].

2.2. α-CA Structure

Bacterial α-CAs have only been poorly characterized with respect to the mammalian α-CAs. In fact, the CAs from *Neisseria gonorrhoeae*, *Sulfurihydrogenibium yellowstonense*, *Sulfurihydrogenibium azorense*, and *Thermovibrio ammonificans* are the only bacterial α-CAs with a known three-dimensional structure [30,33,40,41]. An example of the typical structural organization of a bacterial α-CA is offered by the X-ray crystal structure of the CA identified in the thermophilic bacterium *Sulfurihydrogenibium yellowstonense* YO3AOP1 (Figure 4) [30,33]. This three-dimensional structure generally resembles those of human α-CAs and it was obtained in the presence of the classical inhibitor of CAs, the sulfonamide acetazolamide (AAZ). In particular, it shows a homodimeric arrangement stabilized by a large number of hydrogen bonds and several hydrophobic interactions. The crystallized α-CAs are active as monomers and dimers (Figure 4). The active site is located in a deep cavity, which extends from the protein surface to the center of the molecule, and is characterized by hydrophilic and hydrophobic regions. The hydrophilic part assists in the transfer of the proton from the Zn-bound water to the solvent, while the hydrophobic district is involved in CO_2 binding and ligand recognition. The catalytic zinc ion is located at the bottom of this cavity and is tetrahedrally coordinated by three histidine residues and by the N atom of the sulfonamide moiety of the inhibitor (or probably by the water molecule in the uninhibited enzyme). Intriguingly, the bacterial α-CAs show a more compact structure with respect to the mammalian counterpart, which is characterized by the presence of three insertions (Figure 1) [30,33]. Due to the absence of these inserts, an active site larger than that of human enzymes characterizes the bacterial CAs. Moreover, the structure of the thermostable CAs, such as

SspCA (from *Sulfurihydrogenibium yellowstonense*) and SazCA (from *Sulfurihydrogenibium azorense*) identified in thermophilic bacteria, are characterized by a higher content of secondary-structural elements and an increased number of charged residues, which are all elements responsible for protein thermostability [30,33]. It is interesting to note that the crystal structure of TaCA from *Thermovibrio ammonificans* is tetrameric, with a central core stabilized by two intersubunit disulfides and a single lysine residue from each monomer, which is involved in intersubunit ionic interactions [40].

```
                       signal peptide

SspCA    MRKILISAVLVLSSISIS---------FAEHEWSYEG-EKGPEHWAQLKPEFFWCK-LKN
SazCA    -MKKFILSILSLSIVSIAGEHAILQKNAEVHHWSYEG-ENGPENWAKLNPEYFWCN-LKN
NgonCA   MPRFPRTLPRLTAVLLLACTAFSAAAHGNHTHWGYTG-HDSPESWGNLSEEFRLCSTGKN
VchCA    -------MKKTTWVLAMAASMSFGVQAS---EWGYEG-EHAPEHWG---KVAPLCAEGKN
HpylCA   ----------MKKTFLIALALTASLIGAENTKWDYKNKENGPHRWDKLHKDFEVCKSGKS
HumCAII  ----------------------------MSHHWGYGK-HNGPEHWHKDFPIAKGER----
HumCAI   ----------------------------MASPDWGYDD-KNGPEQWSKLYPIANGNN----
                                   .*.*   ...*. *
                                                         64

SspCA    QSPINIDKKY-KVKANLPKLNLYYKTAKESEVVNNGHTIQINIK-EDNTLNYLG----EK
SazCA    QSPVDISDNY-KVHAKLEKLHINYNKAVNPEIVNNGHTIQVNVL-EDFKLNIKG----KE
NgonCA   QSPVNITET---VSGKLPAIKVNYKPSMVD-VENNGHTIQVNYPEGGNTLTVNG----RT
VchCA    QSPIDVSQS---VEADLQPFTLNYQGQVVG-LLNNGHTLQAIVS-GNNPLQIDG----KT
HpylCA   QSPINIEHYY-HTQDKAD-LQFKYAASKPKAVFFPTHHTLKASFE-PTNHINYRG----HD
HumCAII  QSPVDIDTHTAKYDPSLKPLSVSYDQATSLRILNNGHAFNVEFDDSQDKAVLKGGPLDGT
HumCAI   QSPVDIKTSETKHDTSLKPISVSYNPATAKEIINVGHSFHVNFEDNDNRSVLKGGPFSDS
         ***:::    .   :  .*      :        *::::              *
          94 96         106        119

SspCA    YQLKQFHFHTP------SEHTIEKKSYPLEIHFVHKT-----------EDGKILVVGVM
SazCA    YHLKQFHFHAP------SEHTVNGKYYPLEMHLVHKD-----------KDGNIAVIGVF
NgonCA   YTLKQFHFHVP------SENQIKGRTFPMEARFVHLD-----------ENKQPLVLAVL
VchCA    FQLKQFHFHTP------SENLLKGKQFPLEARFVHAD-----------EGGNLAVVAVM
HpylCA   YVLDNVHFHAP------MEFLINNKTRPLSAHFVHKD-----------AKGRLLVLAIG
HumCAII  YRLIQFHFHWGSLDGQGSEHTVDKKKYAAELHLVHWN-TKYGDFGKAVQQPDGLAVLGIF
HumCAI   YRLFQFHFHWGSTNEHGSEHTVDGVKYSAELHVAHWNSAKYSSLAEAASKADGLAVIGVL
         : * :.***     *  :.   .  . *..*        *:.:
                                                      119

SspCA    AKLGKTNKELDKILNVAP--AEEGEKILDKNLNLNNLIPKDKRYMTYSGSLTPPCTEGV
SazCA    FKEGKANPELDKVFKNAL--KEEGSKVFDGSININALLPPVKNYYTYSGSLTPPCTEGV
NgonCA   YEAGKTNGRLSSIWNVMP--MTAGKVKLNQPFDASTLLPKRLKYYRFAGSLTPPCTEGV
VchCA    YQVGSENPLLKALTADMP--TKGNSTQLTQGIPLADWIPESKHYYRFNGSLTPPCSEGV
HpylCA   FEEGKENPNLDPILEGIQ--KKQN----LKEVALDAFLPKSINYYHFNGSLTAPPCTEGV
HumCAII  LKVGSAKPGLQKVVDVLDSIKTKGKSADFTNFAARGLLPESLDYWTYPGSLTTPPLLECV
HumCAI   MKVGEANPKLQKVLDALQAIKTKGKRAPFTNFDPSTLLPSSLDFWTYPGSLTHPPLYESV
         : *. :  *. :        .      :*  :   : **** ** * *

SspCA    RWIVLKKPISISKQQLEKLKSVMVN---------PNNRPVQEINSRWIIEGF----
SazCA    LWIVLKQPITASKQQIELFKSIMKH---------NNNRPTQPINSRYILESN----
NgonCA   SWLVLKTYDHIDQAQAEKFTRAVGS---------ENNRPVQPLNARVVIE------
VchCA    RWIVLKEPAHVSNQQEQQLSAVMG---------HNNRPVQPHNARLVLQAD----
HpylCA   AWFVIEEPLEVSAKQLAEIKKRMKNS---------PNQRPVQPDYNTVIIKSSAETR
HumCAII  TWIVLKEPISVSSEQVLKFRKLNFNGEGEPEELMVDNWRPAQPLKNRQIKASFK---
HumCAI   TWIICKESISVSSEQLAQFRSLLSNVEGDNAVPMQHNNRPTQPLKGRTVRASF----
         *:: :      .   *  :       * **.*    :
```

Figure 1. Multi-alignment of the amino acid sequences of two human α-CAs (hCAI and hCAII) and of five bacterial α-CAs (SspCA, SazCA, NgoCA, VchCA, and HypyCA) was performed with the ClustalW program, version 2.1. The hCA I numbering system was used. Black bold indicates the amino acid residues of the catalytic triad; blue bold represents the "gate-keeper" residues; and red bold shows the "proton shuttle residue". Box indicates the signal peptide. The asterisk (*) indicates identity at a position; the symbol (:) designates conserved substitutions, while (.) indicates semi-conserved substitutions. Multi-alignment was performed with the program Clustal W, version 2.1. Legend: hCAI, α-CA isoform I from *Homo sapiens*; hCAII, α-CA isoform II from *Homo sapiens*; SspCA, α-CA from *Sulfurihydrogenibium yellowstonense*; SazCA, α-CA from *Sulfurihydrogenibium azorense*; NgonCA, α-CA from *Neisseria gonorrhea*; VchCA, α-CA from *Vibrio cholerae*; HpyCA, α-CA from *Helicobacter pylori*.

```
                              signal peptide
EcoCA    -------------------|----------------------------------|----MKDIDTLIS
VchCA    -------------------|----------------------------------|----MPEIKQLFE
bSuCA    -------------------|----------------------------|----MKNDHSPDQRTLSELFE
HpylCA   -------------------|------------------------------------------MKAFL
PgiCA    |MKKIVLFSAAMAMLIACG|NQTTQTKSDTPTAAVEGRISEVLTQDIQQGLTPEAVLVGLQE

                                    160
EcoCA    NNALWSKMLVEEDPGFFEKLA--QAQKPRFLWIGCSDSRVPAERLTGLEPGELFVHRNVA
VchCA    NNSKWSASIKAETPEYFAKLA--KGQNPDFLWIGCADSRVPAERLTGLYSGELFVHRNVA
bSuCA    HNRQWAAEKQEKDPEYFSRLS--SSQRPEFLWIGCSDSRVPANVVTGLQPGEVFVHRNVA
HpylCA   GALEFQENEYEELKELYESLK--AKQKPHTLFISCVDSRVVPNLITGTKPGELYVIRNMG
PgiCA    GNARYVANKQLPRDLNAQAVAGLEGQFPEAIILSCIDSRVPVEYIFDKGIGDLFVGRVAG
                 :           :       * *  ::.* ****  ::.    *:::* * .

                                 220 223
EcoCA    NLV-----IHTDLNCLSVVQYAVDVLEVEHIIICGHYGCGGVQAAVEN------PELGLI
VchCA    NQV-----IHTDLNCLSVVQYAVDVLQVKHIIVCGHYGCGGVTAAIDN------PQLGLI
bSuCA    NLV-----HRADLNLLSVLEFAVGVLEIKHIIVCGHYGCGGVRAAMDG------YGHGII
HpylCA   NIIPPKASYKESLSTIASIEYAIMHVGVKNLIICGHSDCGACGSVHLINDETTKAKTPYI
PgiCA    NVV--------DDHMLGSLEYACEVSGSKVLLVLGHEDCGAIKSAIKG------VEMGNI
         * :    .  :. :::*  : ::: ** .**.  :.               *

EcoCA    NNWLLHIRDIWFKHSSLLGEMP-QERRLDTLCELNVMEQVYNLGHSPIMQSAWKRGQKVT
VchCA    NNWLLHIRDYYLKHREYLDKMP-AEDRSDKLAEINVAEQVYNLANSTVLQNAWERGQAVE
bSuCA    DNWLQPIRDIAQANQAELDTIENTQDRLDRLCELSVSSQVESLSRTPVLQSAWKDGKDII
HpylCA   ANWIQFLEPIKEELKNHPQFSNHFAKRSWLTERLNARLQLNNLLSYDFIQKRVVDN-ELK
PgiCA    TSLMEEIKPSVEATQYMGERTYANKEFADAVVKENVIQTMDEIRRDSPILKKLEEEGKIK
         . :  :.             . ..  . .:  . .          :

EcoCA    IHGWAYGIHDGLLRDLDVTATNRETLEQRYRHGISNLKLKHANHK----
VchCA    VHGPVYGIEDGRLEYLGVRCASRSAVEDNYHKALEKILNPNHRLLCR
bSuCA    VHGWMYNLKDGLLRDIGCDCTRNALQFACQPAE------------
HpylCA   IFGWRYIIETGRIYNYNFESHFFEPIEETIKQRKSHENF--------
PgiCA    ICGAIYEMSTGKVHFL-----------------------------
         :*  *  *:  *   :
```

Figure 2. Alignment of the amino acid sequences of bacterial β-CAs from different species. Zinc ligands are indicated in black bold; amino acids involved in the enzyme catalytic cycle are indicated in blue bold. Box indicates the signal peptide. The asterisk (*) indicates identity at a position; the symbol (:) designates conserved substitutions, while (.) indicates semi-conserved substitutions. Multi-alignment was performed with the program Clustal W, version 2.1. *Pisum sativum* numbering system was used. Legend: EcoCA, β-CA from *Escherichia coli*; VchCA, β-CA from *Vibrio cholerae*; bSuCA, β-CA from *Brucella suis*; HpyCA, β-CA from *Helicobacter pylori*; PgiCA, β-CA from *Porphyromonas gingivalis*.

```
                              signal peptide
VchCA    -------------------|----------------------------|----MMSSIRSYKG--IVPK
SspCA    -------------------|----------------------------|----MAIIKPYKG--IHPK
PgiCA    -------------------|-------------------------|----MAQRENSDYLTTKMALIQSVRG--FTPI
CAM      |MFNKQIFTILILSLSLALAG|SGCISEGAEDNVAQEITVDEFSNIRENPVTPWNPEPSAPV
                                                            :  .    .    *

                                          73 75     81
VchCA    LGEGVYVDSSAVLVGDIELGDDASIWPLVAARGDVNB-IRIGKRTHIQDGSVLHVTHKNA
SspCA    IDQTVFVAENAVIIGDVEIGKDSSIWYNVVIRGDVNY-IRIGERTNIQDGTIIHVDHK--
PgiCA    IGEDTFLAENATIVGDVVMGKGCSVWFNAVLRGDVNS-IRIGDNVNIQDGSILNTLYQ--
CAM      IDPTAYIDPQASVIGEVTIGANVHVSPMASIRSDEGMPIFVGDRSHVQDGVVLHALETIN
         1:. .:1  .* ::*::1 1*  .  1    *.*  .1*.. *:***1 1*.

                                117   122
VchCA    E------------NPNGYPLCIGDDVTIGHKVMLHG-CTIHDRVLVGMGSIVLDGAVIEN
SspCA    ------------RYPTIIGNNVTVGHKVMLNA-CTIEDYCLIGMSATVMDGVIVGK
PgiCA    ------------KSTIEIGDNVSVGHNVVING-AKICDYALIGMGAVVLDHVVVGE
CAM      EEGEPIEDNIVEVDGKEYAVYIGNNVSLAHQSQVHGPAAVGDDTFIGMQAFVFK-SKVGN
                 . **:1*11.*1  1*. . 1 *  1**  1 *1. 1 1

                                              202
VchCA    DVMIGAGSLVPPGKRLESGFLYMGSPVKQARPLS-DKERAFLVKSSSN--YVQSKNDYLN
SspCA    YSIVAAGALVTPGKVIEPYSLWAGVPAKFVRKLT-EEEIAWLEKSAEN--YVKYKNSYLE
PgiCA    GAIVAAGSVVLTGTQIEPNSIYAGAPARFIKKVDPEQSREMNFRIAHN--YRMYASWFKD
CAM      NCVLEPRSAAIGVTIPDGRYIPAGMVVTSQAEADKLPEVTDDYAYSHTNEAVVYVNVHLA
         11 .1.  .    1 1 * .      .       1 .      . .

VchCA    DVKTVRE-
SspCA    EGLG----
PgiCA    ESSEIDNP
CAM      EGYKETS-
         :
```

Figure 3. Amino acid sequence alignment of the γ-CAs from different bacterial sources, such as *Vibrio cholerae, Sulfurihydrogenibium yellowstonense, Porphyromonas gingivalis*, and *Methanosarcina thermophila*. The metal ion ligands (His81, His117, and His122) are indicated in black bold; the catalytically relevant residues of CAM, such as Asn73, Gln75, and Asn202, which participate in a network of hydrogen bonds with the catalytic water molecule, are indicated in red bold; the acidic loop residues containing the proton shuttle residues (Glu89) are colored in blue bold, but are missing in PgiCA. The CAM numbering system was used. Box indicates the signal peptide. Legend: VchCA (γ-CA from *Vibrio cholerae*), SspCA (γ-CA from *Sulfurihydrogenibium yellowstonense*), PgiCA (γ-CA from *Porphyromonas gingivalis*), and CAM (γ-CA from *Methanosarcina thermophila*). The asterisk (*) indicates identity at all aligned positions; the symbol (:) relates to conserved substitutions, while (.) means that semi-conserved substitutions are observed. The multi-alignment was performed with the program Clustal W.

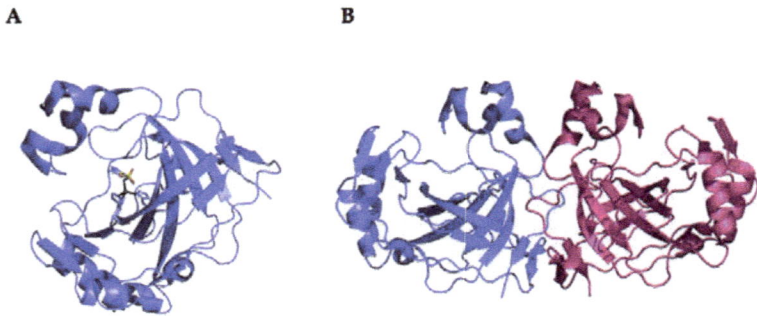

Figure 4. Ribbon representation of the overall fold of α-CA (SspCA) from *Sulfurihydrogenibium yellowstonense*. (**A**): SspCA active monomer with the inhibitor acetazolamide (**AAZ**) showed; (**B**): SspCA active dimer.

2.3. β-CA Structure

X-ray crystal structures are available for several of β-CAs, such as those from *Escherichia coli, Haemophilus influenzae, Mycobacterium tuberculosis, Salmonella enterica,* and *Vibrio cholerae* [29,42–45]. The 3-D folds of these enzymes are rather conserved, although some of them are dimers whereas others are tetramers. All bacterial β-CAs crystallized so far are active as dimers or tetramers, with two or four identical active sites. Their shape is that of a rather long channel at the bottom of which the catalytic zinc ion is found, tetrahedrally coordinated by two cysteines, one-histidine and one-aspartic amino acid residue (the so called "closed active site"). Interesting, the enzyme structure from *Vibrio cholerae* (VchCAβ) was determined in its closed active site form at pH values <8.3 (Figure 5) [29]. The "closed active site" is named in this way as these enzymes are not catalytically active (at pH values <8.3). Interesting, in its inactive form, the bicarbonate is bound in a pocket close to the zinc ion [29]. However, at pH values >8.3, the "closed active site" is converted to the "open active site" (with gain of catalytic activity), which is associated with a movement of the Asp residue from the catalytic Zn(II) ion, with the concomitant coordination of an incoming water molecule approaching the metal ion [29]. This water molecule (as hydroxide ion) is, in fact, responsible for the catalytic activity, as shown above for the α-CAs.

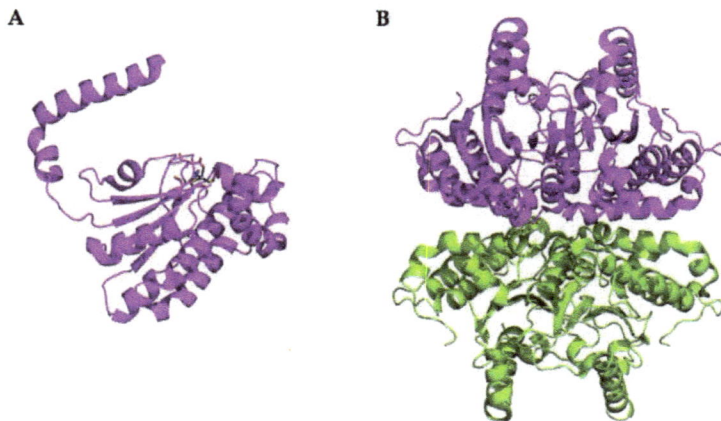

Figure 5. Ribbon representation of the catalytically inactive monomer (**A**) and active tetramer (**B**) of β-CA (VchCA) from *Vibrio cholerae*.

2.4. γ-CA Structure

CAM (Carbonic Anhydrase Methanosarcina) from *Methanosarcina thermophila* is the prototype of the γ-class carbonic anhydrase and the only enzyme from this class that has been crystallized so far (Figure 6) [46]. This enzyme adopts a left-handed parallel β-helix fold and crystallizes as a trimer with three zinc-containing active sites, each located at the interface between two monomers. The metalloenzyme is only active as a trimer (Figure 6) [46]. Interestingly, in this class of enzyme, instead of a histidine (as in α-CAs), there is a glutamic acid residue acting as a proton shuttle residue (Figure 3). In fact, the high-resolution crystal of CAM showed that Glu89 has two orientations, similar to those of His64 in α-CAs (Figure 3) [46].

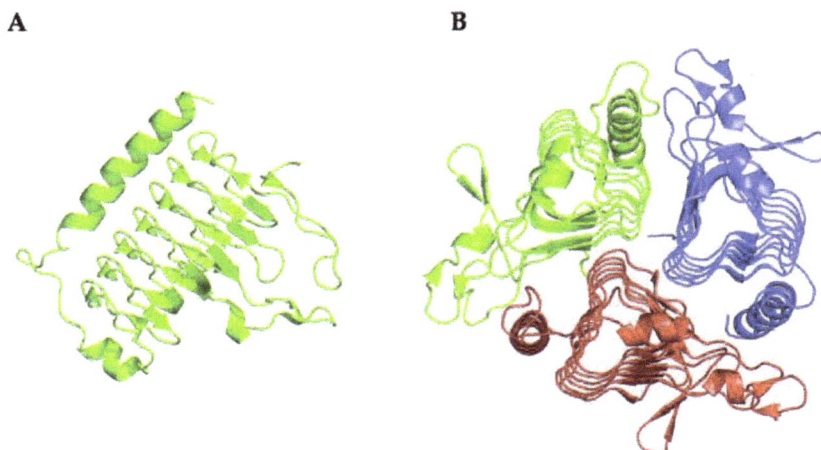

Figure 6. Structural representation of the catalytically inactive monomer (**A**) and active trimer (**B**) of the CAM (γ-CA) enzyme from *Methanosarcina thermophila*.

3. Catalytic Activity

The spontaneous reversible CO_2 hydration reaction in the absence of the catalyst has an effective first-order rate constant of $0.15\ s^{-1}$, while the reverse reaction shows a rate constant of $50\ s^{-1}$ [36,47]. In the living organisms, the CO_2 hydration and the HCO_3^- dehydration are connected to very fast processes, such as those related to transport/secretory processes. The main metabolic role of CAs is to catalyze the carbon dioxide hydration at a very high rate, with a pseudo first order kinetic constant (k_{cat}) ranging from 10^4 to $10^6\ s^{-1}$ [36,47]. Thus, the CA superfamily significantly accelerates the hydration reaction to support the metabolic processes involving dissolved inorganic carbon. Until 2012, the most active CA was the human isoform hCA II ($k_{cat} = 1.40 \times 10^6\ s^{-1}$), belonging to the α-class and abundantly present in the human erythrocytes [36,48]. The hCA II, at the level of the peripheral tissues, converts the CO_2 into carbonic acid, while when the blood reaches the lungs, dehydrates the HCO_3^- to CO_2 for it exhalation. In 2012, a new α-CA was identified, and was shown to be a highly and catalytically effective catalyst for the CO_2 hydration reaction (Figure 7) [49]. To our surprise, this CA (SazCA) was identified in the genome of the thermophilic bacterium *Sulfurihydrogenibium azorense* and showed a $k_{cat} = 4.40 \times 10^6\ s^{-1}$, thus being 2.33 times more active than the human isoform hCA II (Figure 7) [30,49]. In general, the bacterial CAs belonging to the three known classes (α, β, and γ) are efficient catalysts for the CO_2 hydration reaction. Analyzing the three-dimensional structures of the bacterial CAs, it has been observed that the catalytic pocket is rather small for the γ-CAs, gets bigger for β-CAs, and becomes quite large in the α-CAs (Figures 4–6) [5,8,25,50]. As a consequence, the catalytic constant of the γ-CAs is usually low compared to the β-CAs, which is lower when compared to many bacterial α-CAs (Figure 7). Sometimes, there are γ-CAs with a catalytic turnover number that is higher with respect to that shown by the β-class, such as the γ-CAs from *Porphyromonas gingivalis* and *Vibrio cholerae* (Figure 7).

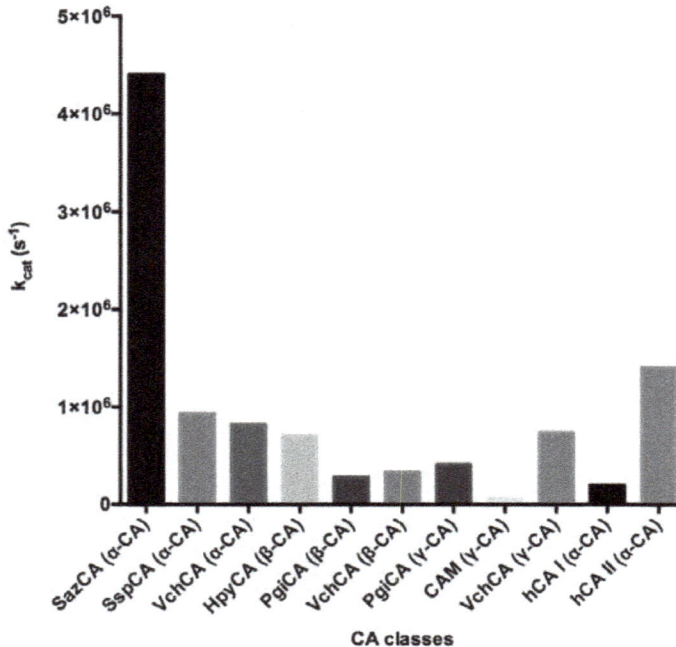

Figure 7. Kinetic parameters for the CO_2 hydration reaction catalyzed by the human cytosolic isozymes hCA I and II (α-class CAs) and bacterial α-, β-, and γ-CAs, such as SazCA (α-CAs from *Sulfurihydrogenibium azorense*), SspCA (α-CAs from *Sulfurihydrogenibium yellowstonense*), HpyCA (α- and β-CAs from *Helicobacter pylori*), VchCA (α-, β-, and γ-CAs from *Vibrio cholerae*), PgiCA (β- and γ-CAs from *Porphyromonas gingivalis*), and CAM (γ-CA from *Methanosarcina thermophila*). All the measurements were done at 20 °C, pH 7.5 (α-class enzymes), and pH 8.3 (β- and γ-CAs) by a stopped flow CO_2 hydratase assay method.

4. CA Inhibitors

Different types of CA inhibitors (CAIs) exist [47,48] and they can be grouped into: (1) the metal ion binders (anion, sulfonamides, and their bioisosteres, dithiocarbamates, xanthates, etc.); (2) compounds which anchor to the zinc-coordinated water molecule/hydroxide ion (phenols, polyamines, thioxocoumarins, sulfocumarins); (3) compounds occluding the active site entrance, such as coumarins and their isosteres; and (4) compounds binding out of the active site [47]. This subdivision has been made considering the way that the inhibitors bind the catalytic metal ion, the metal coordinated-water molecule, and the occlusion of the active site [47]. The most investigated CAIs are anions and sulfonamides [36,47,51,52]. Sulfonamides were discovered by Domagk in 1935 [53], and were the first antimicrobial drugs. The first sulfonamide showing effective antibacterial activity was Prontosil, a sulfanilamide prodrug isosteric/isostructural with p-aminobenzoic acid (PABA) [54]. In the following years, a range of analogs constituting the so-called sulfa drug class of anti-bacterials entered into clinical use, and many of these compounds are still widely used. A library of 40 compounds, 39 sulfonamides, and one sulfamate was used to provide CAIs (Figure 8) [6,10,13,14,55–66].

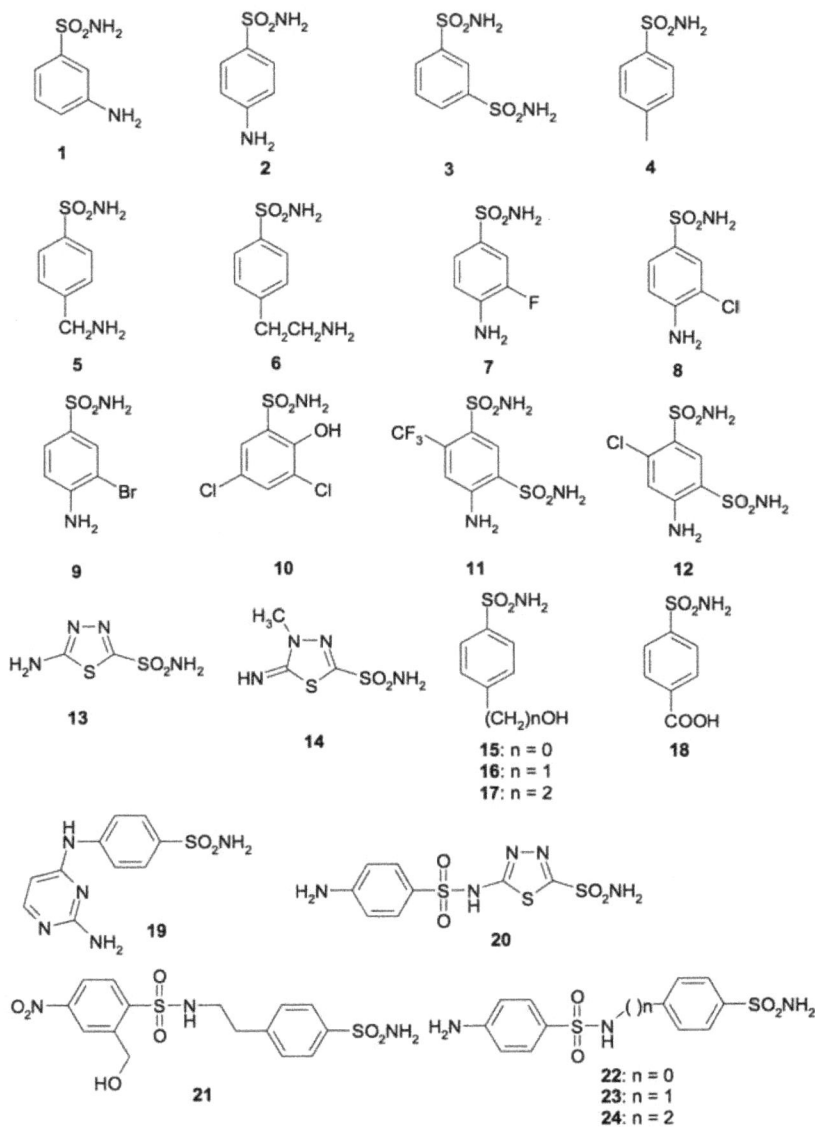

Figure 8. *Cont.*

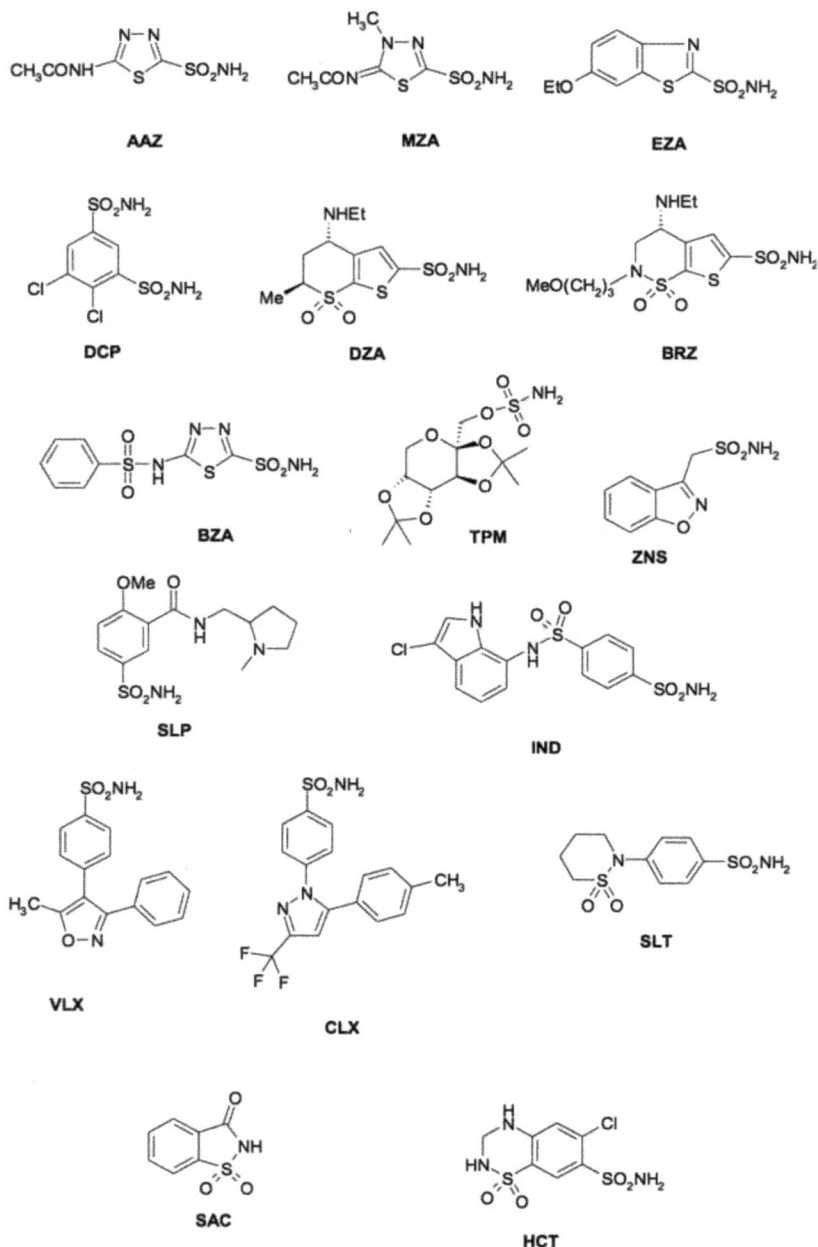

Figure 8. Sulfonamides, sulfamates, and some of their derivatives investigated as bacterial CA inhibitors.

Derivatives **1–24** and **AAZ-HCT** are either simple aromatic/heterocyclic sulfonamides widely used as building blocks for obtaining new families of such pharmacological agents, or they are clinically used agents, among which acetazolamide (**AAZ**), methazolamide (**MZA**), ethoxzolamide (**EZA**), and dichlorophenamide (**DCP**) are the classical, systemically acting antiglaucoma CAIs. Dorzolamide (**DZA**) and brinzolamide (**BRZ**) are topically acting antiglaucoma agents; benzolamide (**BZA**) is an orphan drug belonging to this class of pharmacological agents. Moreover, the zonisamide (**ZNS**), sulthiame (**SLT**), and the sulfamic acid ester topiramate (**TPM**) are widely used antiepileptic drugs. Sulpiride (**SLP**) and indisulam (**IND**) were also shown by our group to belong to this class of pharmacological agents, together with the COX2 selective inhibitors celecoxib (**CLX**) and valdecoxib (**VLX**). Saccharin (**SAC**) and the diuretic hydrochlorothiazide (**HCT**) are also known to act as CAIs. Sulfonamides, such as the clinically used derivatives acetazolamide, methazolamide, ethoxzolamide, dichlorophenamide, dorzolamide, and brinzolamide, bind in a tetrahedral geometry to the Zn(II) ion in the deprotonated state, with the nitrogen atom of the sulfonamide moiety coordinated to Zn(II) and an extended network of hydrogen bonds, involving amino acid residues of the enzyme, also participating in the anchoring of the inhibitor molecule to the metal ion [36,47,48,67]. The aromatic/heterocyclic part of the inhibitor interacts with the hydrophilic and hydrophobic residues of the catalytic cavity [36,47,51,52].

Anions, such as inorganic metal-complexing anions or more complicated species such as carboxylates, are also known to bind to CAs [47,48]. These anions may bind either the tetrahedral geometry of the metal ion or as trigonal–bipyramidal adducts. Anion inhibitors are important both for understanding the inhibition/catalytic mechanisms of these enzymes fundamental for many physiologic processes, and for designing novel types of inhibitors which may have clinical applications for the management of a variety of disorders in which CAs are involved [47,48].

In the last ten years, numerous results concerning the inhibition profile of the three bacterial CA classes (α, β, and γ) have been reported using anions and sulfonamides. Most of these studies were carried out on bacterial CAs from pathogenic bacteria, such as *Francisella tularensis*, *Burkholderia pseudomallei*, *Vibrio cholerae*, *Streptococcus mutans*, *Porphyromonas gingivalis*, *Legionella pneumophila*, *Clostridium perfringens*, and *Mycobacterium turberculosis*, etc. [6,14,68–70]. The results indicated that certain CAIs were able to highly inhibit most of the CAs identified in the genome of the aforementioned bacteria (for details see associate bibliography) [4,62,71–74]. Moreover, certain CAIs, such as acetazolamide and methazolamide, were shown to effectively inhibit bacterial growth in cell cultures [75]. The inhibition profile with simple and complex anions, as well as small molecules inhibiting other CAs, showed that the most efficient inhibitors detected so far are sulfamide, sulfamate, phenylboronic acid, and phenylarsonic acid [24,62,76]. Generally, halides, cyanide, bicarbonate, nitrite, selenate, diphosphate, divanadate, tetraborate, peroxodisulfate, hexafluorophosphate, and triflate exhibit weak inhibitory activity against the bacterial CAs [22,24,25,76,77].

5. Activators

An interesting feature of the CA superfamily is that they can bind within the middle-exit part of the active site molecules known as "activators" (CAA). They are biogenic amines (histamine, serotonin, and catecholamines), amino acids, oligopeptides, or small proteins (Figure 9 shows the small molecule CAAs mostly investigated) [78–81]. By means of electronic spectroscopy, X-ray crystallography, and kinetic measurements, it has been demonstrated that CAAs do not influence the binding of CO_2 to the CA active site but mediate the rate-determining step of the catalysis hurrying the transfer of protons from the active site to the environment. The final result is an overall increase of the catalytic turnover. Thus, the CA activators enhance the k_{cat} of the enzyme, with no effect on KM [78,80,81]. Numerous studies concerning the activation of the mammalian enzymes with amines and amino acids are reported in the literature [78,80,81]. In fact, CAAs may have pharmacologic applications in therapy memory, neurodegenerative diseases (Alzheimer's disease), or the treatment of genetic CA deficiency syndromes [78,80,81]. On the other hand, the activation of CA classes different from those belonging to mammals has been poorly investigated. Considering the limited data available at this moment on the activation of other classes of CAs and using a series of structurally related amino acids and amines of types **25–43** (Figure 9), Supuran and coworkers have investigated the activation profiles of some bacterial CAs [82,83]. More precisely, the activation profile of the γ-CA (BpsCA) identified in the genome of the pathogenic bacteria *Burkholderia pseudomallei* has been investigated for understanding the role of the CAs in the lifecycle and virulence of these bacteria [82]. Moreover, the activation profile of the thermophilic α-CAs (SspCA, from *Sulfurihydrogenibium yellowstonense* and SazCA, from *Sulfurihydrogenibium azorense*) has also been explored [83]. From Figure 10, it is readily apparent that the activators L-Tyr for BpsCA and L-Phe for SspCA enhanced the values of the k_{cat} by one order of magnitude compared to those without activators.

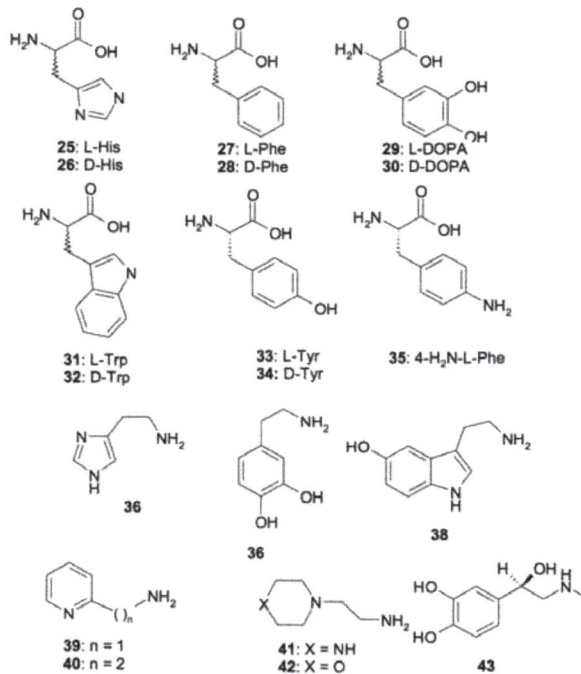

Figure 9. Amino acid and amine CA activators **25–43** investigated for their interaction with bacterial CAs.

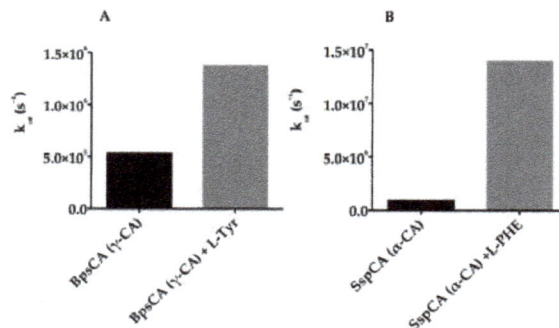

Figure 10. (**A**): Activation of the bacterial BpsCA (γ-CA) with L-Tyr; (**B**): Activation of the bacterial SspCA (α-CA) L-Phe. All measurements were carried out at 25 °C and pH 7.5, for the CO_2 hydration reaction.

6. Phylogenetic Analysis

The complex distribution of the various CA classes in Gram-positive and -negative bacteria allowed us to find a correlation between the evolutionary history of the bacteria and the three CA classes (α, β, and γ) identified in their genome. Prokaryotes appeared on the Earth 3.5–3.8 billion years ago, while eukaryotes were dated to 1.8 billion years ago [84]. During the first 2.0–2.5 billion years, the Earth's atmosphere did not contain oxygen, and the first organisms were thus anaerobic. Eukaryotic organisms' almost aerobes developed on the Earth when the atmosphere was characterized by a stable and relatively high oxygen content [84]. The oldest part of the evolutionary history of the planet and more than 90% of the phylogenetic diversity of life can be attributed to the microbial world. Moreover, the fact that the Archaea are distinct from other prokaryotes is demonstrated by the existence of protein sequences that are present in Archaea, but not in eubacteria [85]. Many phylogenetic methods support a close correlation of Archaea with Gram-positive bacteria, while Gram-negative bacteria form a separate clade, indicating their phylogenetic distinction. Gupta et al. believe that the Gram-positive bacteria occupy an intermediate position between Archaea and Gram-negative bacteria, and that they evolved precisely from Archaea [25,77]. Phylogenetic analysis of carbonic anhydrases identified bacteria Gram-positive and negatively showed that the ancestral CA is represented by the γ-class. In fact, the γ-CA is the only CA class, which has been identified in Archaea [86–89]. This is consistent with the theory that maintains a close relationship between the Archaea and the Gram-positive bacteria, considering that Gram-negative arised from the latter. Furthermore, phylogenetic analysis of bacterial CAs showed that the α-CAs, exclusively present in Gram-negative bacteria, were the most recent CAs. These results have been corroborated by the enzymatic promiscuity theory, which is the ability of an enzyme to catalyze a side reaction in addition to the main reaction [90,91]. In fact, as reported in the literature, the α-CAs can catalyze a secondary reaction, such as the hydrolysis of p-NpA or a thioester, in addition to the primary reaction consisting of CO_2 hydration.

7. Localization and Physiological Role

A common feature of all bacterial α-CAs known to date is the presence of an N-terminal signal peptide, which suggests a periplasmic or extracellular location (Figure 1). From these findings, we have speculated that in Gram-negative bacteria, the α-CA are able to convert the CO_2 to bicarbonate diffused in the periplasmic space ensuring the survival and/or satisfying the metabolic needs of the microorganism [25,77]. In fact, several essential metabolic pathways require either CO_2 or bicarbonate as a substrate, and probably, the spontaneous diffusion of CO_2 to the outer membrane and the conversion to bicarbonate inside the cell are not sufficient for the metabolic needs of the microorganism. On the contrary, β- or γ-classes have a cytoplasmic localization and are responsible for CO_2 supply for carboxylase enzymes, pH homeostasis, and other intracellular functions [25,77]. Not all the Gram-negative bacteria, however, have α-CAs. Probably, the α-CAs are not required when the Gram-negative bacteria colonize habitats defined as not metabolic limiting or adverse to their survival [77]. Recently, we analyzed the amino acid sequence of the β-CAs encoded by the genome of Gram-negative bacteria with SignalP version 4.1, which is a program designed to discriminate between signal peptides and transmembrane regions of proteins. The program is available as a web tool at http://www.cbs.dtu.dk/services/SignalP/ [92]. We noted that the primary structure of some β-CAs identified in the genome of some pathogenic Gram-negative bacteria, such as such as HpyCA (from *Helicobacter pylori*), VchCA (from *Vibrio cholerae*), NgonCA (from *Neisseria gonorrhoeae*), and SsalCA (from *Streptococcus salivarius*), present a pre-sequence of 18 or more amino acid residues at the N-terminal part, which resulted in a secretory signal peptide [25,77]. Intriguingly, during the writing of this review, we saw that the CAM enzyme also contained a short putative signal peptide at its N-terminus (Figure 3). Since the signal peptide is essential for the translocation across the cytoplasmic membrane in prokaryotes, it has been suggested that the β- and/or γ-CAs found in Gram-negative bacteria and characterized by the presence of a signal peptide might exhibit a periplasmic localization and a role similar to that described previously for the α-CAs [25,77].

In the past ten years, the understanding of the function of the bacterial CAs has increased significantly [25,77]. We suggested that the activity of CAs is connected with the survival of the microbes because the metabolic reaction catalyzed by CA is essential for supporting numerous physiological functions involving dissolved inorganic carbon. For example, in non-pathogenic bacteria such as *Ralstonia eutropha* (Gram-negative bacterium found in soil and water) and *Escherichia coli* (a facultative Gram-negative bacterium), it has been demonstrated in vivo that the bacterial growth at an ambient CO_2 concentration is dependent on CA activity [93,94]. In fact, the CO_2 and bicarbonate are both produced and consumed by bacterial metabolism. Since CO_2 is rapidly lost from the bacterial cells by passive diffusion, their rate is maintained individually in balance by the CA activity. In fact, the reversible spontaneous CO_2 hydratase reaction is insufficient to restore the amount of dissolved inorganic carbon. More interesting is the in vivo evidence concerning the involvement of CAs for the growth of pathogenic bacteria. For example, CAs encoded by the genome of *Helicobacter pylori*, a Gram-negative, microaerophilic bacterium colonizing the human stomach, are essential for the acid acclimatization of the pathogen within the stomach and thus, for bacterial survival in the host [15,16,95,96]. In the case of the pathogenic bacterium *Vibrio cholera* (Gram-negative bacterium responsible of cholera), its CAs are involved in the production of sodium bicarbonate, which induces cholera toxin expression [15,24,61,95–100]. Probably, *V. cholera* uses the CAs as a system to colonize the host [6,12,14,101]. Again, the causative agent of brucellosis *Brucella suis*, a non-motile Gram-negative coccobacillus, and the *Mycobacterium tuberculosis*, an obligate pathogenic bacterium responsible for tuberculosis, are needed for the growth of functional CAs [66,102–104].

8. Engineered Bacteria with a Thermostable CA for CO_2 Capture

Recently, the heterologous expression of the recombinant thermostable SspCA by the high-density fermentation of *Escherichia coli* cultures, in order to produce a usable biocatalyst for CO_2 capture, has been described [20]. The enzyme was covalently immobilized onto the surface of magnetic Fe_3O_4 nanoparticles (MNP) by using the carbodiimide activation reaction [20]. This approach offered two main advantages: 1) the magnetic nanoparticles-immobilized SspCA via carbodiimide increased the stability and the long-term storage of the biocatalyst; and 2) the immobilized biocatalyst can be recovered and reused from the reaction mixture by simply applying a magnet or an electromagnet field because of the strong ferromagnetic properties of Fe_3O_4 [20]. The main issues of this method are the costs connected to biocatalyst purification and the support used for enzyme immobilization. Often, all these aspects may discourage the utilization of enzymes in industrial applications. In 2017, a system able to overexpress and immobilize the protein directly on the outer membrane of *Escherichia coli* for lowering the costs of the purification of the biocatalyst and immobilization has been proposed [105]. To accomplish this, the *Escherichia coli* cells have been engineered using the well-described INP (Ice Nucleation Protein) technique [105]. Briefly, an expression vector composed of a chimeric gene resulting from the fusion of a signal peptide, the *Pseudomonas syringae* INP domain (INPN), and the SspCA gene encoding for the thermostable α-CA, SspCA, has been prepared. During protein overexpression, the signal peptide makes possible the translocation of the neo-synthetized protein through the cytoplasmic membrane, while the INPN domain is necessary for guiding and anchoring the protein to the bacterial outer membrane. The results demonstrated that the anchored SspCA was efficiently overexpressed and active on the bacterial surface of *E. coli* [105]. Moreover, the anchored SspCA was stable and active for 15 h at 70 °C and for days at 25 °C [105]. This approach with respect to the covalent immobilization of the enzyme onto the surface of magnetic Fe_3O_4 nanoparticles (MNP) clearly has important advantages. It is a one-step procedure for overexpressing and immobilizing the enzyme simultaneously on the outer membrane, and it drastically reduces the costs needed for enzyme purification, enzyme immobilization, and the support necessary for biocatalyst immobilization [105]. In addition, the biocatalyst could be recovered by a simple centrifugation step from the reaction mixture. The strategy of the INPN-SspCA obtained by engineering *E. coli* could be considered as a good method for approaching the biomimetic capture of CO_2 and other biotechnological applications in which a highly effective, thermostable catalyst is needed.

9. Conclusions

Bacterial CAs were rather poorly investigated until recently. However, in the last years the cloning, purification, and characterization of many representatives, belonging to all three genetic families present in Bacteria, has led to crucial advances in the field. The role of CAs in many pathogenic as well as non-pathogenic bacteria is thus beginning to be better understood. Apart from pH regulation and adaptation to various niches in which bacteria live (e.g., the highly acidic environment in the stomach, in the case of *Helicobacter pylori*, the alkaline one in the gut for *Vibrio cholerae*, etc.), CAs probably participate in biosynthetic processes in which bicarbonate or CO_2 are substrates, as in the case of other organisms for which these roles are demonstrated. The inhibition and activation of bacterial CAs may be exploited either from pharmacological or environmental viewpoints. On the one hand, the inhibitors of such enzymes may lead to antibiotics with a new mechanism of action, devoid of the drug resistance problems encountered with the various classes of clinically used agents. Moreover, catalytically highly efficient, thermally stable bacterial CAs may have interesting applications for biomimetic CO_2 capture in the context of global warming due to the accumulation of this gas in the atmosphere as a consequence of anthropic activities. Furthermore, Ca activators of such enzymes may represent an even more attractive option for mitigating global warming.

Acknowledgments: This work was supported in part by an FP7 European Union Project (Gums & Joints, Grant agreement Number HEALTH-F2-2010-261460).

Author Contributions: Clemente Capasso and Claudiu T. Supuran wrote, edited, and supervised the manuscript.

Conflicts of Interest: The authors declare no conflict of interest.

References

1. Turk, V.E.; Simic, I.; Likic, R.; Makar-Ausperger, K.; Radacic-Aumiler, M.; Cegec, I.; Juricic Nahal, D.; Kraljickovic, I. New drugs for bad bugs: What's new and what's in the pipeline? *Clin. Ther.* **2016**, *38*, e9. [CrossRef] [PubMed]
2. Decker, B.; Masur, H. Bad bugs, no drugs: Are we part of the problem, or leaders in developing solutions? *Crit. Care Med.* **2015**, *43*, 1153–1155. [CrossRef] [PubMed]
3. Payne, D.J.; Gwynn, M.N.; Holmes, D.J.; Pompliano, D.L. Drugs for bad bugs: Confronting the challenges of antibacterial discovery. *Nat. Rev. Drug. Dis.* **2007**, *6*, 29–40. [CrossRef] [PubMed]
4. Annunziato, G.; Angeli, A.; D'Alba, F.; Bruno, A.; Pieroni, M.; Vullo, D.; De Luca, V.; Capasso, C.; Supuran, C.T.; Costantino, G. Discovery of new potential anti-infective compounds based on carbonic anhydrase inhibitors by rational target-focused repurposing approaches. *Chem. Med. Chem.* **2016**, *11*, 1904–1914. [CrossRef] [PubMed]
5. Ozensoy Guler, O.; Capasso, C.; Supuran, C.T. A magnificent enzyme superfamily: Carbonic anhydrases, their purification and characterization. *J. Enzyme Inhibi. Med. Chem.* **2016**, *31*, 689–694. [CrossRef] [PubMed]
6. Del Prete, S.; Vullo, D.; De Luca, V.; Carginale, V.; Ferraroni, M.; Osman, S.M.; AlOthman, Z.; Supuran, C.T.; Capasso, C. Sulfonamide inhibition studies of the beta-carbonic anhydrase from the Pathogenic bacterium *Vibrio cholerae*. *Bioorg. Med. Chem.* **2016**, *24*, 1115–1120. [CrossRef] [PubMed]
7. Del Prete, S.; De Luca, V.; De Simone, G.; Supuran, C.T.; Capasso, C. Cloning, expression and purification of the complete domain of the eta-carbonic anhydrase from *Plasmodium falciparum*. *J. Enzyme Inhib. Med. Chem.* **2016**, *31*, 54–59. [CrossRef] [PubMed]
8. Capasso, C.; Supuran, C.T. An overview of the carbonic anhydrases from two pathogens of the oral cavity: *Streptococcus mutans* and *Porphyromonas gingivalis*. *Curr. Top. Med. Chem.* **2016**, *16*, 2359–2368. [CrossRef] [PubMed]
9. Supuran, C.T. Carbonic anhydrase inhibitors. *Bioorg. Med. Chem. Lett.* **2010**, *20*, 3467–3474. [CrossRef] [PubMed]
10. Del Prete, S.; Vullo, D.; De Luca, V.; Carginale, V.; Osman, S.M.; AlOthman, Z.; Supuran, C.T.; Capasso, C. Cloning, expression, purification and sulfonamide inhibition profile of the complete domain of the eta-carbonic anhydrase from *Plasmodium falciparum*. *Bioorg. Med. Chem. Lett.* **2016**, *26*, 4184–4190. [CrossRef] [PubMed]

11. Del Prete, S.; Vullo, D.; De Luca, V.; Carginale, V.; Di Fonzo, P.; Osman, S.M.; AlOthman, Z.; Supuran, C.T.; Capasso, C. Anion inhibition profiles of the complete domain of the η-carbonic anhydrase from *Plasmodium falciparum*. *Bioorg. Med. Chem.* **2016**, *24*, 4410–4414. [CrossRef] [PubMed]

12. Del Prete, S.; Vullo, D.; De Luca, V.; Carginale, V.; di Fonzo, P.; Osman, S.M.; AlOthman, Z.; Supuran, C.T.; Capasso, C. Anion inhibition profiles of α-, β- and γ-carbonic anhydrases from the pathogenic bacterium *Vibrio cholerae*. *Bioorg. Med. Chem.* **2016**, *24*, 3413–3417. [CrossRef] [PubMed]

13. Abdel Gawad, N.M.; Amin, N.H.; Elsaadi, M.T.; Mohamed, F.M.; Angeli, A.; De Luca, V.; Capasso, C.; Supuran, C.T. Synthesis of 4-(thiazol-2-ylamino)-benzenesulfonamides with carbonic anhydrase I, II and IX inhibitory activity and cytotoxic effects against breast cancer cell lines. *Bioorg. Med. Chem.* **2016**, *24*, 3043–3051. [CrossRef] [PubMed]

14. Del Prete, S.; Vullo, D.; De Luca, V.; Carginale, V.; Osman, S.M.; AlOthman, Z.; Supuran, C.T.; Capasso, C. Comparison of the sulfonamide inhibition profiles of the α-, β- and γ -carbonic anhydrases from the pathogenic bacterium *Vibrio cholerae*. *Bioorg. Med Chem. Lett.* **2016**, *26*, 1941–1946. [CrossRef] [PubMed]

15. Nishimori, I.; Onishi, S.; Takeuchi, H.; Supuran, C.T. The α and β classes carbonic anhydrases from *Helicobacter pylori* as novel drug targets. *Curr. Pharm. Des.* **2008**, *14*, 622–630. [PubMed]

16. Morishita, S.; Nishimori, I.; Minakuchi, T.; Onishi, S.; Takeuchi, H.; Sugiura, T.; Vullo, D.; Scozzafava, A.; Supuran, C.T. Cloning, polymorphism, and inhibition of β-carbonic anhydrase of *Helicobacter pylori*. *J. Gastroenterol.* **2008**, *43*, 849–857. [CrossRef] [PubMed]

17. Cheah, W.Y.; Ling, T.C.; Juan, J.C.; Lee, D.J.; Chang, J.S.; Show, P.L. Biorefineries of carbon dioxide: From carbon capture and storage (CCS) to bioenergies production. *Bioresour. Technol.* **2016**, *215*, 346–356. [CrossRef] [PubMed]

18. Migliardini, F.; De Luca, V.; Carginale, V.; Rossi, M.; Corbo, P.; Supuran, C.T.; Capasso, C. Biomimetic CO2 capture using a highly thermostable bacterial α-carbonic anhydrase immobilized on a polyurethane foam. *J. Enzyme Inhib. Med. Chem.* **2014**, *29*, 146–150. [CrossRef] [PubMed]

19. Perfetto, R.; Del Prete, S.; Vullo, D.; Sansone, G.; Barone, C.; Rossi, M.; Supuran, C.T.; Capasso, C. Biochemical characterization of the native α-carbonic anhydrase purified from the mantle of the Mediterranean mussel, *Mytilus galloprovincialis*. *J. Enzyme Inhib. Med. Chem.* **2017**, *32*, 632–639. [CrossRef] [PubMed]

20. Perfetto, R.; Del Prete, S.; Vullo, D.; Sansone, G.; Barone, C.M.A.; Rossi, M.; Supuran, C.T.; Capasso, C. Production and covalent immobilisation of the recombinant bacterial carbonic anhydrase (SspCA) onto magnetic nanoparticles. *J. Enzyme Inhib. Med. Chem.* **2017**, *32*, 759–766. [CrossRef] [PubMed]

21. Vullo, D.; De Luca, V.; Scozzafava, A.; Carginale, V.; Rossi, M.; Supuran, C.T.; Capasso, C. The first activation study of a bacterial carbonic anhydrase (CA). The thermostable α-CA from *Sulfurihydrogenibium yellowstonense* YO3AOP1 is highly activated by amino acids and amines. *Bioorg. Med. Chem. Lett.* **2012**, *22*, 6324–6327. [CrossRef] [PubMed]

22. Capasso, C.; Supuran, C.T. Bacterial, fungal and protozoan carbonic anhydrases as drug targets. *Expert Opin. Ther. Targets* **2015**, *19*, 1689–1704. [CrossRef] [PubMed]

23. Supuran, C.T.; Capasso, C. The η-class carbonic anhydrases as drug targets for antimalarial agents. *Exp. Opin. Ther. Targets* **2015**, *19*, 551–563. [CrossRef] [PubMed]

24. Capasso, C.; Supuran, C.T. An Overview of the selectivity and efficiency of the bacterial carbonic anhydrase inhibitors. *Curr. Med. Chem.* **2015**, *22*, 2130–2139. [CrossRef] [PubMed]

25. Capasso, C.; Supuran, C.T. An overview of the α-, β- and γ-carbonic anhydrases from Bacteria: Can bacterial carbonic anhydrases shed new light on evolution of bacteria? *J. Enzyme Inhib. Med. Chem.* **2015**, *30*, 325–332. [CrossRef] [PubMed]

26. Capasso, C.; Supuran, C.T. Sulfa and trimethoprim-like drugs-antimetabolites acting as carbonic anhydrase, dihydropteroate synthase and dihydrofolate reductase inhibitors. *J. Enzyme Inhib. Med. Chem.* **2014**, *29*, 379–387. [CrossRef] [PubMed]

27. Capasso, C.; Supuran, C.T. Anti-infective carbonic anhydrase inhibitors: A patent and literature review. *Exp. Opin. Ther. Pat.* **2013**, *23*, 693–704. [CrossRef] [PubMed]

28. Pinard, M.A.; Lotlikar, S.R.; Boone, C.D.; Vullo, D.; Supuran, C.T.; Patrauchan, M.A.; McKenna, R. Structure and inhibition studies of a type II beta-carbonic anhydrase psCA3 from *Pseudomonas aeruginosa*. *Bioorg. Med. Chem.* **2015**, *23*, 4831–4838. [CrossRef] [PubMed]

29. Ferraroni, M.; Del Prete, S.; Vullo, D.; Capasso, C.; Supuran, C.T. Crystal structure and kinetic studies of a tetrameric type II beta-carbonic anhydrase from the pathogenic bacterium *Vibrio cholerae*. *Acta Crystallogr. Sect. D Biolog. Crystallogr.* **2015**, *71*, 2449–2456. [CrossRef] [PubMed]

30. De Simone, G.; Monti, S.M.; Alterio, V.; Buonanno, M.; De Luca, V.; Rossi, M.; Carginale, V.; Supuran, C.T.; Capasso, C.; Di Fiore, A. Crystal structure of the most catalytically effective carbonic anhydrase enzyme known, SazCA from the thermophilic bacterium *Sulfurihydrogenibium azorense*. *Bioorg. Med. Chem. Lett.* **2015**, *25*, 2002–2006. [CrossRef] [PubMed]

31. Zolnowska, B.; Slawinski, J.; Pogorzelska, A.; Chojnacki, J.; Vullo, D.; Supuran, C.T. Carbonic anhydrase inhibitors. synthesis, and molecular structure of novel series N-substituted N'-(2-arylmethylthio-4-chloro-5-methylbenzenesulfonyl)guanidines and their inhibition of human cytosolic isozymes I and II and the transmembrane tumor-associated isozymes IX and XII. *Euro. J. Med. Chem.* **2014**, *71*, 135–147.

32. De Luca, L.; Ferro, S.; Damiano, F.M.; Supuran, C.T.; Vullo, D.; Chimirri, A.; Gitto, R. Structure-based screening for the discovery of new carbonic anhydrase VII inhibitors. *Euro. J. Med. Chem.* **2014**, *71*, 105–111. [CrossRef] [PubMed]

33. Di Fiore, A.; Capasso, C.; De Luca, V.; Monti, S.M.; Carginale, V.; Supuran, C.T.; Scozzafava, A.; Pedone, C.; Rossi, M.; De Simone, G. X-ray structure of the first 'extremo-alpha-carbonic anhydrase', a dimeric enzyme from the thermophilic bacterium *Sulfurihydrogenibium yellowstonense* YO3AOP1. *Acta Crystallogr. Sect. D Biol. Crystallogr.* **2013**, *69*, 1150–1159. [CrossRef] [PubMed]

34. Supuran, C.T. Structure-based drug discovery of carbonic anhydrase inhibitors. *J. Enzyme Inhib. Med. Chem.* **2012**, *27*, 759–772. [CrossRef] [PubMed]

35. Supuran, C.T. Carbonic anhydrases—An overview. *Curr. Pharm. Des.* **2008**, *14*, 603–614. [CrossRef]

36. Supuran, C.T. Structure and function of carbonic anhydrases. *Biochem. J.* **2016**, *473*, 2023–2032. [CrossRef] [PubMed]

37. Buzas, G.M.; Supuran, C.T. The history and rationale of using carbonic anhydrase inhibitors in the treatment of peptic ulcers. In memoriam Ioan Puscas (1932–2015). *J. Enzyme Inhib. Med. Chem.* **2016**, *31*, 527–533. [CrossRef] [PubMed]

38. Carta, F.; Supuran, C.T.; Scozzafava, A. Sulfonamides and their isosters as carbonic anhydrase inhibitors. *Future Med. Chem.* **2014**, *6*, 1149–1165. [CrossRef] [PubMed]

39. Supuran, C.T. Carbonic anhydrases: Novel therapeutic applications for inhibitors and activators. *Nat. Rev. Drug Dis.* **2008**, *7*, 168–181. [CrossRef] [PubMed]

40. James, P.; Isupov, M.N.; Sayer, C.; Saneei, V.; Berg, S.; Lioliou, M.; Kotlar, H.K.; Littlechild, J.A. The structure of a tetrameric alpha-carbonic anhydrase from *Thermovibrio ammonificans* reveals a core formed around intermolecular disulfides that contribute to its thermostability. *Acta Crystallogr. Sect. D Biol. Crystallogr.* **2014**, *70*, 2607–2618. [CrossRef] [PubMed]

41. Huang, S.; Xue, Y.; Sauer-Eriksson, E.; Chirica, L.; Lindskog, S.; Jonsson, B.H. Crystal structure of carbonic anhydrase from *Neisseria gonorrhoeae* and its complex with the inhibitor acetazolamide. *J. Mol. Biol.* **1998**, *283*, 301–310. [CrossRef] [PubMed]

42. Cronk, J.D.; O'Neill, J.W.; Cronk, M.R.; Endrizzi, J.A.; Zhang, K.Y. Cloning, crystallization and preliminary characterization of a β-carbonic anhydrase from *Escherichia coli*. *Acta Crystallogr. Sec. D Biol. Crystallogr.* **2000**, *56*, 1176–1179. [CrossRef]

43. Rowlett, R.S.; Tu, C.; Lee, J.; Herman, A.G.; Chapnick, D.A.; Shah, S.H.; Gareiss, P.C. Allosteric site variants of *Haemophilus influenzae* β-carbonic anhydrase. *Biochemistry* **2009**, *48*, 6146–6156. [CrossRef] [PubMed]

44. Covarrubias, A.S.; Bergfors, T.; Jones, T.A.; Hogbom, M. Structural mechanics of the pH-dependent activity of β-carbonic anhydrase from *Mycobacterium tuberculosis*. *J. Biol. Chem.* **2006**, *281*, 4993–4999. [CrossRef] [PubMed]

45. Nishimori, I.; Minakuchi, T.; Vullo, D.; Scozzafava, A.; Supuran, C.T. Inhibition studies of the β-carbonic anhydrases from the bacterial pathogen *Salmonella enterica* serovar *Typhimurium* with sulfonamides and sulfamates. *Bioorg. Med. Chem.* **2011**, *19*, 5023–5030. [CrossRef] [PubMed]

46. Kisker, C.; Schindelin, H.; Alber, B.E.; Ferry, J.G.; Rees, D.C. A left-hand β-helix revealed by the crystal structure of a carbonic anhydrase from the archaeon *Methanosarcina thermophila*. *EMBO J.* **1996**, *15*, 2323–2330. [PubMed]

47. Supuran, C.T. Advances in structure-based drug discovery of carbonic anhydrase inhibitors. *Exp. Opin. Drug Dis.* **2017**, *12*, 61–88. [CrossRef] [PubMed]

48. Supuran, C.T. How many carbonic anhydrase inhibition mechanisms exist? *J. Enzyme Inhib. Med. Chem.* **2016**, *31*, 345–360. [CrossRef] [PubMed]

49. Vullo, D.; De Luca, V.; Scozzafava, A.; Carginale, V.; Rossi, M.; Supuran, C.T.; Capasso, C. Anion inhibition studies of the fastest carbonic anhydrase (CA) known, the extremo-CA from the bacterium *Sulfurihydrogenibium azorense. Bioorg. Med. Chem. Lett.* **2012**, *22*, 7142–7145. [CrossRef] [PubMed]

50. Capasso, C.; Supuran, C.T. Inhibition of bacterial carbonic anhydrases as a novel approach to escape drug resistance. *Curr. Top. Med. Chem.* **2017**, *17*, 1237–1248. [CrossRef] [PubMed]

51. Supuran, C.T. Carbonic anhydrase inhibition and the management of neuropathic pain. *Exp. Rev. Neurother* **2016**, *16*, 961–968. [CrossRef] [PubMed]

52. Supuran, C.T. Drug interaction considerations in the therapeutic use of carbonic anhydrase inhibitors. *Exp. Opin. Drug Metab. Toxicol.* **2016**, *12*, 423–431. [CrossRef] [PubMed]

53. Otten, H. Domagk and the development of the sulphonamides. *J. Antimicrob. Chem.* **1986**, *17*, 689–696. [CrossRef]

54. Achari, A.; Somers, D.O.; Champness, J.N.; Bryant, P.K.; Rosemond, J.; Stammers, D.K. Crystal structure of the anti-bacterial sulfonamide drug target dihydropteroate synthase. *Nat. Struct. Biol.* **1997**, *4*, 490–497. [CrossRef] [PubMed]

55. Vullo, D.; Del Prete, S.; Fisher, G.M.; Andrews, K.T.; Poulsen, S.A.; Capasso, C.; Supuran, C.T. Sulfonamide inhibition studies of the η-class carbonic anhydrase from the malaria pathogen *Plasmodium falciparum. Bioorg. Med. Chem.* **2015**, *23*, 526–531. [CrossRef] [PubMed]

56. Vullo, D.; De Luca, V.; Del Prete, S.; Carginale, V.; Scozzafava, A.; Capasso, C.; Supuran, C.T. Sulfonamide inhibition studies of the γ-carbonic anhydrase from the Antarctic bacterium *Pseudoalteromonas haloplanktis. Bioorg. Med. Chem. Lett.* **2015**, *25*, 3550–3555. [CrossRef] [PubMed]

57. Vullo, D.; De Luca, V.; Del Prete, S.; Carginale, V.; Scozzafava, A.; Capasso, C.; Supuran, C.T. Sulfonamide inhibition studies of the γ-carbonic anhydrase from the Antarctic cyanobacterium *Nostoc commune. Bioorg. Med. Chem.* **2015**, *23*, 1728–1734. [CrossRef] [PubMed]

58. Dedeoglu, N.; DeLuca, V.; Isik, S.; Yildirim, H.; Kockar, F.; Capasso, C.; Supuran, C.T. Sulfonamide inhibition study of the β-class carbonic anhydrase from the caries producing pathogen *Streptococcus mutans. Bioorg. Med. Chem. Lett.* **2015**, *25*, 2291–2297. [CrossRef] [PubMed]

59. Alafeefy, A.M.; Ceruso, M.; Al-Tamimi, A.M.; Del Prete, S.; Supuran, C.T.; Capasso, C. Inhibition studies of quinazoline-sulfonamide derivatives against the γ-CA (PgiCA) from the pathogenic bacterium, *Porphyromonas gingivalis. J. Enzyme Inhib. Med. Chem.* **2015**, *30*, 592–596. [CrossRef] [PubMed]

60. Alafeefy, A.M.; Abdel-Aziz, H.A.; Vullo, D.; Al-Tamimi, A.M.; Awaad, A.S.; Mohamed, M.A.; Capasso, C.; Supuran, C.T. Inhibition of human carbonic anhydrase isozymes I, II, IX and XII with a new series of sulfonamides incorporating aroylhydrazone-, [1,2,4]triazolo[3,4-b][1,3,4]thiadiazinyl- or 2-(cyanophenylmethylene)-1,3,4-thiadiazol-3(2H)-yl moieties. *J. Enzyme Inhib. Med. Chem.* **2015**, *30*, 52–56. [CrossRef] [PubMed]

61. Diaz, J.R.; Fernandez Baldo, M.; Echeverria, G.; Baldoni, H.; Vullo, D.; Soria, D.B.; Supuran, C.T.; Cami, G.E. A substituted sulfonamide and its Co (II), Cu (II), and Zn (II) complexes as potential antifungal agents. *J. Enzyme Inhib. Med. Chem.* **2016**, *31*, 51–62. [CrossRef] [PubMed]

62. Supuran, C.T. *Legionella pneumophila* carbonic anhydrases: Underexplored antibacterial drug targets. *Pathogens* **2016**, *5*. [CrossRef] [PubMed]

63. Nishimori, I.; Vullo, D.; Minakuchi, T.; Scozzafava, A.; Capasso, C.; Supuran, C.T. Sulfonamide inhibition studies of two β-carbonic anhydrases from the bacterial pathogen *Legionella pneumophila. Bioorg. Med. Chem.* **2014**, *22*, 2939–2946. [CrossRef] [PubMed]

64. Vullo, D.; Sai Kumar, R.S.; Scozzafava, A.; Capasso, C.; Ferry, J.G.; Supuran, C.T. Anion inhibition studies of a β-carbonic anhydrase from *Clostridium perfringens. Bioorg. Med. Chem. Lett.* **2013**, *23*, 6706–6710. [CrossRef] [PubMed]

65. Nishimori, I.; Minakuchi, T.; Maresca, A.; Carta, F.; Scozzafava, A.; Supuran, C.T. The β-carbonic anhydrases from *Mycobacterium tuberculosis* as drug targets. *Curr. Pharm. Des.* **2010**, *16*, 3300–3309. [CrossRef]

66. Carta, F.; Maresca, A.; Covarrubias, A.S.; Mowbray, S.L.; Jones, T.A.; Supuran, C.T. Carbonic anhydrase inhibitors. Characterization and inhibition studies of the most active β-carbonic anhydrase from *Mycobacterium tuberculosis*, Rv3588c. *Bioorg. Med. Chem. Lett.* **2009**, *19*, 6649–6654. [CrossRef] [PubMed]

67. Supuran, C.T. Acetazolamide for the treatment of idiopathic intracranial hypertension. *Exp. Rev. Neurother* **2015**, *15*, 851–856. [CrossRef] [PubMed]

68. Del Prete, S.; Vullo, D.; Osman, S.M.; AlOthman, Z.; Supuran, C.T.; Capasso, C. Sulfonamide inhibition profiles of the β-carbonic anhydrase from the pathogenic bacterium *Francisella tularensis* responsible of the febrile illness tularemia. *Bioorg. Med. Chem.* **2017**, *25*, 3555–3561. [CrossRef] [PubMed]

69. Vullo, D.; Del Prete, S.; Di Fonzo, P.; Carginale, V.; Donald, W.A.; Supuran, C.T.; Capasso, C. Comparison of the sulfonamide inhibition profiles of the β- and γ-carbonic anhydrases from the pathogenic bacterium *Burkholderia pseudomallei*. *Molecules* **2017**, *22*. [CrossRef] [PubMed]

70. Dedeoglu, N.; DeLuca, V.; Isik, S.; Yildirim, H.; Kockar, F.; Capasso, C.; Supuran, C.T. Cloning, characterization and anion inhibition study of a β-class carbonic anhydrase from the caries producing pathogen *Streptococcus mutans*. *Bioorg. Med. Chem.* **2015**, *23*, 2995–3001. [CrossRef] [PubMed]

71. Cau, Y.; Mori, M.; Supuran, C.T.; Botta, M. Mycobacterial carbonic anhydrase inhibition with phenolic acids and esters: Kinetic and computational investigations. *Org. Biomol. Chem.* **2016**, *14*, 8322–8330. [CrossRef] [PubMed]

72. Modak, J.K.; Liu, Y.C.; Supuran, C.T.; Roujeinikova, A. Structure-activity relationship for sulfonamide inhibition of *Helicobacter pylori* α-carbonic anhydrase. *J. Med. Chem.* **2016**, *59*, 11098–11109. [CrossRef] [PubMed]

73. Supuran, C.T. Bortezomib inhibits bacterial and fungal β-carbonic anhydrases. *Bioorg. Med. Chem.* **2016**, *24*, 4406–4409. [CrossRef] [PubMed]

74. Vullo, D.; Kumar, R.S.S.; Scozzafava, A.; Ferry, J.G.; Supuran, C.T. Sulphonamide inhibition studies of the β-carbonic anhydrase from the bacterial pathogen *Clostridium perfringens*. *J. Enzyme Inhib. Med. Chem.* **2018**, *33*, 31–36. [CrossRef] [PubMed]

75. Shahidzadeh, R.; Opekun, A.; Shiotani, A.; Graham, D.Y. Effect of the carbonic anhydrase inhibitor, acetazolamide, on *Helicobacter pylori* infection in vivo: A pilot study. *Helicobacter* **2005**, *10*, 136–138. [CrossRef] [PubMed]

76. Capasso, C.; Supuran, C.T. Carbonic anhydrase from *Porphyromonas Gingivalis* as a drug target. *Pathogens* **2017**, *6*, 30–42.

77. Supuran, C.T.; Capasso, C. New light on bacterial carbonic anhydrases phylogeny based on the analysis of signal peptide sequences. *J. Enzyme Inhib. Med. Chem.* **2016**, *31*, 1254–1260. [CrossRef] [PubMed]

78. Licsandru, E.; Tanc, M.; Kocsis, I.; Barboiu, M.; Supuran, C.T. A class of carbonic anhydrase I—Selective activators. *J. Enzyme Inhib. Med. Chem.* **2017**, *32*, 37–46. [CrossRef] [PubMed]

79. Supuran, C.T. Carbonic anhydrase inhibitors and activators for novel therapeutic applications. *Future Med. Chem.* **2011**, *3*, 1165–1180. [CrossRef] [PubMed]

80. Supuran, C.T. Carbonic anhydrases: From biomedical applications of the inhibitors and activators to biotechnological use for $CO_{(2)}$ capture. *J. Enzyme Inhib. Med. Chem.* **2013**, *28*, 229–230. [CrossRef] [PubMed]

81. Vullo, D.; Del Prete, S.; Capasso, C.; Supuran, C.T. Carbonic anhydrase activators: Activation of the β-carbonic anhydrase from *Malassezia* globosa with amines and amino acids. *Bioorg. Med. Chem. Lett.* **2016**, *26*, 1381–1385. [CrossRef] [PubMed]

82. Vullo, D.; Del Prete, S.; Osman, S.M.; AlOthman, Z.; Capasso, C.; Donald, W.A.; Supuran, C.T. *Burkholderia pseudomallei* γ-carbonic anhydrase is strongly activated by amino acids and amines. *Bioorg. Med. Chem. Lett.* **2017**, *27*, 77–80. [CrossRef] [PubMed]

83. Akdemir, A.; Vullo, D.; De Luca, V.; Scozzafava, A.; Carginale, V.; Rossi, M.; Supuran, C.T.; Capasso, C. The extremo-alpha-carbonic anhydrase (CA) from *Sulfurihydrogenibium azorense*, the fastest CA known, is highly activated by amino acids and amines. *Bioorg. Med. Chem. Lett.* **2013**, *23*, 1087–1090. [CrossRef] [PubMed]

84. Forterre, P. Looking for the most "primitive" organism(s) on Earth today: The state of the art. *Planet Space Sci.* **1995**, *43*, 167–177. [CrossRef]

85. Gupta, R.S. Protein phylogenies and signature sequences: A reappraisal of evolutionary relationships among archaebacteria, eubacteria, and eukaryotes. *Microbiol. Mol. Biol. Rev. MMBR* **1998**, *62*, 1435–1491. [PubMed]

86. Zimmerman, S.; Innocenti, A.; Casini, A.; Ferry, J.G.; Scozzafava, A.; Supuran, C.T. Carbonic anhydrase inhibitors. Inhibition of the prokariotic β and γ-class enzymes from *Archaea* with sulfonamides. *Bioorg. Med. Chem. Lett.* **2004**, *14*, 6001–6006. [CrossRef] [PubMed]

87. Tripp, B.C.; Bell, C.B., 3rd; Cruz, F.; Krebs, C.; Ferry, J.G. A role for iron in an ancient carbonic anhydrase. *J. Biol. Chem.* **2004**, *279*, 6683–6687. [CrossRef] [PubMed]

88. Tripp, B.C.; Smith, K.; Ferry, J.G. Carbonic anhydrase: New insights for an ancient enzyme. *J. Biol. Chem.* **2001**, *276*, 48615–48618. [CrossRef] [PubMed]
89. Smith, K.S.; Jakubzick, C.; Whittam, T.S.; Ferry, J.G. Carbonic anhydrase is an ancient enzyme widespread in prokaryotes. *Proc. Natl. Acad. Sci. USA* **1999**, *96*, 15184–15189. [CrossRef]
90. Torrance, J.W.; Bartlett, G.J.; Porter, C.T.; Thornton, J.M. Using a library of structural templates to recognise catalytic sites and explore their evolution in homologous families. *J. Mol. Biol.* **2005**, *347*, 565–581. [CrossRef] [PubMed]
91. Torrance, J.W.; Holliday, G.L.; Mitchell, J.B.; Thornton, J.M. The geometry of interactions between catalytic residues and their substrates. *J. Mol. Biol.* **2007**, *369*, 1140–1152. [CrossRef] [PubMed]
92. Petersen, T.N.; Brunak, S.; von Heijne, G.; Nielsen, H. SignalP 4.0: Discriminating signal peptides from transmembrane regions. *Nat. Methods* **2011**, *8*, 785–786. [CrossRef] [PubMed]
93. Kusian, B.; Sultemeyer, D.; Bowien, B. Carbonic anhydrase is essential for growth of *Ralstonia eutropha* at ambient $CO_{(2)}$ concentrations. *J. Bact.* **2002**, *184*, 5018–5026. [CrossRef] [PubMed]
94. Merlin, C.; Masters, M.; McAteer, S.; Coulson, A. Why is carbonic anhydrase essential to *Escherichia coli*? *J. Bact.* **2003**, *185*, 6415–6424. [CrossRef] [PubMed]
95. Cobaxin, M.; Martinez, H.; Ayala, G.; Holmgren, J.; Sjoling, A.; Sanchez, J. Cholera toxin expression by El Tor *Vibrio cholerae* in shallow culture growth conditions. *Microb. Pathog.* **2014**, *66*, 5–13. [CrossRef] [PubMed]
96. Abuaita, B.H.; Withey, J.H. Bicarbonate induces *Vibrio cholerae* virulence gene expression by enhancing ToxT activity. *Infect. Immun.* **2009**, *77*, 4111–4120. [CrossRef] [PubMed]
97. Joseph, P.; Ouahrani-Bettache, S.; Montero, J.L.; Nishimori, I.; Minakuchi, T.; Vullo, D.; Scozzafava, A.; Winum, J.Y.; Kohler, S.; Supuran, C.T. A new β-carbonic anhydrase from *Brucella suis*, its cloning, characterization, and inhibition with sulfonamides and sulfamates, leading to impaired pathogen growth. *Bioorg. Med. Chem.* **2011**, *19*, 1172–1178. [CrossRef] [PubMed]
98. Modak, J.K.; Liu, Y.C.; Machuca, M.A.; Supuran, C.T.; Roujeinikova, A. Structural basis for the inhibition of *Helicobacter pylori* α-carbonic anhydrase by sulfonamides. *PLoS ONE* **2015**, *10*, e0127149. [CrossRef] [PubMed]
99. Nishimori, I.; Vullo, D.; Minakuchi, T.; Morimoto, K.; Onishi, S.; Scozzafava, A.; Supuran, C.T. Carbonic anhydrase inhibitors: Cloning and sulfonamide inhibition studies of a carboxyterminal truncated α-carbonic anhydrase from *Helicobacter pylori*. *Bioorg. Med. Chem. Lett.* **2006**, *16*, 2182–2188. [CrossRef] [PubMed]
100. Del Prete, S.; Isik, S.; Vullo, D.; De Luca, V.; Carginale, V.; Scozzafava, A.; Supuran, C.T.; Capasso, C. DNA cloning, characterization, and inhibition studies of an α-carbonic anhydrase from the pathogenic bacterium *Vibrio cholerae*. *J. Med. Chem.* **2012**, *55*, 10742–10748. [CrossRef] [PubMed]
101. Vullo, D.; Del Prete, S.; De Luca, V.; Carginale, V.; Ferraroni, M.; Dedeoglu, N.; Osman, S.M.; AlOthman, Z.; Capasso, C.; Supuran, C.T. Anion inhibition studies of the β-carbonic anhydrase from the pathogenic bacterium *Vibrio cholerae*. *Bioorg. Med. Chem. Lett.* **2016**, *26*, 1406–1410. [CrossRef] [PubMed]
102. Kohler, S.; Ouahrani-Bettache, S.; Winum, J.Y. *Brucella suis* carbonic anhydrases and their inhibitors: Towards alternative antibiotics? *J. Enzyme Inhibit Med. Chem.* **2017**, *32*, 683–687. [CrossRef] [PubMed]
103. Singh, S.; Supuran, C.T. 3D-QSAR CoMFA studies on sulfonamide inhibitors of the Rv3588c β-carbonic anhydrase from *Mycobacterium tuberculosis* and design of not yet synthesized new molecules. *J. Enzyme Inhib. Med. Chem.* **2014**, *29*, 449–455. [CrossRef] [PubMed]
104. Ceruso, M.; Vullo, D.; Scozzafava, A.; Supuran, C.T. Sulfonamides incorporating fluorine and 1,3,5-triazine moieties are effective inhibitors of three beta-class carbonic anhydrases from *Mycobacterium tuberculosis*. *J. Enzyme Inhib. Med. Chem.* **2014**, *29*, 686–689. [CrossRef] [PubMed]
105. Del Prete, S.; Perfetto, R.; Rossi, M.; Alasmary, F.A.S.; Osman, S.M.; AlOthman, Z.; Supuran, C.T.; Capasso, C. A one-step procedure for immobilising the thermostable carbonic anhydrase (SspCA) on the surface membrane of *Escherichia coli*. *J. Enzyme Inhib. Med. Chem.* **2017**, *32*, 1120–1128. [CrossRef] [PubMed]

MDPI

St. Alban-Anlage 66

4052 Basel

Switzerland

Tel. +41 61 683 77 34

Fax +41 61 302 89 18

www.mdpi.com

Metabolites Editorial Office

E-mail: metabolites@mdpi.com

www.mdpi.com/journal/metabolites

www.ingramcontent.com/pod-product-compliance
Lightning Source LLC
Chambersburg PA
CBHW051856210326
41597CB00033B/5917